MECHANISMS OF ANESTHETIC ACTION IN SKELETAL, CARDIAC, AND SMOOTH MUSCLE

ADVANCES IN EXPERIMENTAL MEDICINE AND BIOLOGY

Recent Volumes in this Series

MECHANISMS OF ANESTHETIC ACTION IN SKELETAL, CARDIAC, AND SMOOTH MUSCLE

Edited by

Thomas J. J. Blanck

and

David M. Wheeler

Johns Hopkins University
Baltimore, Maryland

PLENUM PRESS • NEW YORK AND LONDON

Library of Congress Cataloging-in-Publication Data

Mechanisms of anesthetic action in skeletal, cardiac, and smooth
  muscle / edited by Thomas J.J. Blanck and David M. Wheeler.
      p.   cm. -- (Advances in experimental medicine and biology ; v.
  301)
    Based on the proceedings of a conference held Sept. 13-16, 1990 in
  Baltimore, Md.
    Includes bibliographical references and index.
    ISBN 0-306-44011-3
    1. Anesthetics--Mechanism of action--Congresses. 2. Muscles-
  -Effect of drugs on--Congresses. 3. Heart--Muscle--Effect of drugs
  on--Congresses. 4. Smooth muscle--Effect of drugs on--Congresses.
  I. Blanck, Thomas J. J. II. Wheeler, David Martyn, 1953-
  III. Series.
    [DNLM: 1. Anesthetics--pharmacology--congresses. 2. Muscles--drug
  effects--congresses. 3. Muscles--physiology--congresses. W1 AD559
  v. 301 / WE 500 M4865]
  RD85.5.M43 1991
  615'.781--dc20
  DNLM/DLC
  for Library of Congress   91-24013
                                                                  CIP

Proceedings of the Mechanisms of Anesthetic Action in Muscle '90 symposium,
held September 13-16, 1990, in Baltimore, Maryland

ISBN-13: 978-1-4684-5981-4   e-ISBN-13: 978-1-4684-5979-1
DOI: 10.1007/978-1-4684-5979-1

© 1991 Plenum Press, New York
Softcover reprint of the hardcover 1st edition 1991
A Division of Plenum Publishing Corporation
233 Spring Street, New York, N.Y. 10013

Preface

The volatile anesthetics continue to be one of the most mysterious yet commonly used class of drugs in medical practice today. A prominent and troublesome side effect of volatile anesthetics is their ability to alter hemodynamics. This arises from two diverse but interrelated phenomena, depression of cardiac contractility and dilation of the vasculature. These effects of volatile anesthetics on cardiac and smooth muscle plus the action of volatile anesthetics on skeletal muscle in the malignant hyperthermic syndrome have led to concern about the interaction of volatile anesthetics (and other anesthetic agents) with calcium metabolism in the muscle cell. Many of the phenomena caused by anesthetics appear to have common mechanisms in all of the muscle types; however, the differences among skeletal, cardiac and smooth muscle also lead to distinct effects of the anesthetics in each. Given the diverse research disciplines which have been brought to bear on the mechanism of anesthetic alteration of contractility, the symposium from which this book originates was convened for the purpose of gathering those with common interests in anesthetic agents and their cellular and subcellular actions in muscle.

The recent symposium had its origins in a small but exciting meeting that took place at the University of Texas at Houston in 1984. At that time, Robert Merin and Jacques Chelly convened a group of people who had interest in cardiac muscle and calcium antagonists. From this initial small meeting, a larger meeting developed with a specific focus on the action of volatile anesthetics on muscle. This second meeting took place at the Medical College of Georgia in Augusta in 1988. At that time it was felt that a more formal mechanism of renewal should be in place, and the group at Johns Hopkins was commissioned to organize the next symposium.

This recent symposium, entitled "Mechanisms of Anesthetic Action in Muscle, '90," involved those people who are most active in the area of anesthetic action on the three types of muscle. One result of the symposium is the present volume, which is divided into three parts. The first part deals with anesthetic action on skeletal muscle; this is introduced by Thomas D. Nelson, Ph.D., who has a long-standing interest in malignant hyperthermia and the effect of volatile anesthetics on skeletal muscle. He summarizes in his first chapter the presentations by the other investigators involved in skeletal muscle research. The second section of the book deals with effect of anesthetics on cardiac muscle. This is the largest section of the book and is introduced by Zeljko J. Boznjak, Ph.D. Most contributions relate to the interference of volatile anesthetics with calcium delivery and recognition in cardiac muscle. The final section of the book is devoted to smooth muscle and is introduced by Sheila Muldoon, M.D., who summarizes anesthetic effects on vascular smooth muscle and on the endothelium of the vascular system. While this three-way division is useful organizationally, the overall purpose of this symposium was to bring together investigators who make use of diverse methodology and who have interests in muscle, whether it be cardiac, skeletal or smooth muscle. It is hoped that, as was true during the symposium, this

v

book also will generate active discussion among various research disciplines and serve to solidify our understanding of anesthetics and muscle action.

The editors would like to thank Dr. Mark Rogers and members of the Department of Anesthesiology and Critical Care Medicine for their kind and generous support which enabled the symposium to take place. We also thank Anaquest for their generous support and acknowledge the support of Organon and Janssen pharmaceuticals for helping underwrite some of the expenses of the symposium. The editors would like to thank Ms. Peggy Riley for her assistance and leadership in organizing the nuts and bolts of the symposium. We acknowledge the excellent secretarial support of Margot Emmett, Angela Clinton and Angela Liggins for their superb care in preparation of the manuscripts. The symposium was a successful venture into the science of the anesthetics and promises to continue in 1992 at the Medical College of Wisconsin in Milwaukee.

Thomas J. J. Blanck
David M. Wheeler

Baltimore

Contents

Part III — Smooth Muscle

Part I — Skeletal Muscle

1

Skeletal Muscle Targets for the Action of Anesthetic Agents

Thomas E. Nelson

Prior to its meeting in Augusta, Georgia in 1988, this group decided to focus on anesthetic action on skeletal muscle at sites distal to the myoneural junction. Therefore, in this summary chapter, targets for anesthetic action will be described beginning at the sarcolemma (excluding the motor endplate) and ending with the intracellular contractile elements. This section is not intended as a review, but rather as an overview and preview to some of the interesting investigations on skeletal muscle presented at this conference.

In order to gain an understanding of how anesthetics alter muscle function, considerable time and effort must be devoted toward obtaining an appropriately designed experiment. The clinician might argue that studies should be designed to explain anesthetic effects in the intact organism such that the new knowledge could directly benefit those who administer these agents and their patients. The basic scientist could argue that any new knowledge, whether directly applicable or not, has value for knowledge's sake alone. Thus, the basic approach may study biological effects of anesthetics at concentrations far in excess of those used for clinical anesthesia and many interesting and explainable results may be obtained. However, such observations probably have little, if anything, to do with the "anesthetic" actions of these compounds. The first premise for experimental design is, therefore, to study these agents within the range of concentrations that are used in clinical anesthesia. The next consideration would be whether to study anesthetic action in the intact organism, at the molecular level, or somewhere in between. In this series of investigations on anesthetic action on skeletal muscle, the investigators have utilized preparations that range from intact muscle tissue removed from the body to single protein molecules that have become "sandwiched" within a planar lipid bilayer. A fundamental question to be considered is: do the observed effects of anesthetics at these varying levels of biological organization have any relevance to anesthetic effects on skeletal muscle in the intact organism? This is an important, yet difficult question to answer, and the biological complexity involved is appropriately addressed in a study by Waud and Waud[1] (see especially figure 9 in this ref.). While ultimately the components must be studied as an intact, integrated system, there is good reason to investigate a system at its most simple level of organization because in the intact organism or even in the intact muscle cell, there are many processes or anesthetic

THOMAS E. NELSON, Department of Anesthesiology, University of Texas Health Science Center, Houston, Texas 77030.

Mechanisms of Anesthetic Action in Skeletal, Cardiac, and Smooth Muscle
Edited by T.J.J. Blanck and D.M. Wheeler, Plenum Press, New York, 1991

targets which could be contributing to the observed effect. In such complex systems even when a single endpoint such as muscle twitch or contracture tension is measured, it is often very difficult to discern which basic processes the anesthetic may be affecting.

For intact skeletal muscle, the gross system would begin with the sarcolemmal functions of the maintenance of the resting potential and the generation of action potentials, and perhaps end with resequestration of Ca^{2+} by the sarcoplasmic reticulum and the dissociation of the actin and myosin units to a resting state. The number of structural units involved in the function of this excitation-contraction/relaxation cycle provide many targets on which anesthetic agents may exert their effects. Our challenge is to explore each of these targets at all levels of functional organization and to provide essential pieces for the very complex puzzle of anesthetic action.

In this preview an attempt is made to develop a conceptual model of a functional pathway along the excitation-contraction (E-C) coupling system and various steps in this pathway on which anesthetics may have an action. These concepts will develop around skeletal muscle but could be applicable to cardiac and smooth muscle as well. For reference, an extremely simple diagram illustrating some of the structural components of skeletal muscle E-C coupling has been constructed (figure 1). We emphasize that the diagram is incomplete, and many structures (i.e., Cl channel, receptors) and intracellular signal transduction systems have not been included to maintain clarity and simplicity. Starting at the sarcolemma, anesthetic action on two functions of this structure will be considered: 1) maintenance of the resting membrane potential and 2) formation and propagation of the action potential. Measuring resting membrane potentials of gastrocnemius in the intact rat, Kendig and Bunker[2] reported that halothane caused a −6.2 mV depolarization of the sarcolemma whereas cyclopropane repolarized the sarcolemma by +4.7 mV. These investigators interpreted the anesthetic effects as possibly being indirectly mediated by anesthetic depression (halothane) and stimulation (cyclopropane) of β-adrenergic function and the consequent circulating catecholamine effects on the sodium pump.[2] Utilizing biopsied muscle, enflurane had no effect on resting membrane potentials in frog M. tibialis anterior[3] and similarly, halothane was without effect on the resting membrane potential of pig digital extensor muscle.[4] Although the number of studies is small, it appears that in intact animals the administration of volatile anesthetics can produce effects on a non-muscle system that in turn alters the resting membrane potential of skeletal muscle, but when the muscle is removed from the body no anesthetic effects on the resting membrane potential are observed. Comparison of these studies readily illustrates the problems of studying anesthetic action in the intact animal as opposed to studies on isolated tissue and the different results that can be obtained.

The next step, the action potential of excitable cells, is a thoroughly studied system, and the elaborate methodology for studying the different components that produce the action potential provides a means of investigating anesthetic action on this phenomenon. The effect of enflurane on the compound action potential of bundles of dissected frog M. tibialis anterior has been measured.[3] At a concentration of 5.9 mM (ca. 4 MAC), enflurane decreased amplitude and delayed inactivation of the action potential. At a concentration of 1 MAC, enflurane had no measurable effect on the action potential. Since the compound action potential is the net effect of ion fluxes through their specific channels (primarily the Na^+ and K^+ channels), anesthetics could alter one or both to produce the observed effect. The effects of volatile anesthetics on isolated components of the action potential have been extensively investigated in nerve tissue[5] but very few studies exist for skeletal muscle. Ruppersberg and Rudel[6] measured the effects of halothane on sodium channels in myoballs cultured from biopsied human skeletal muscle. The effect of halothane on

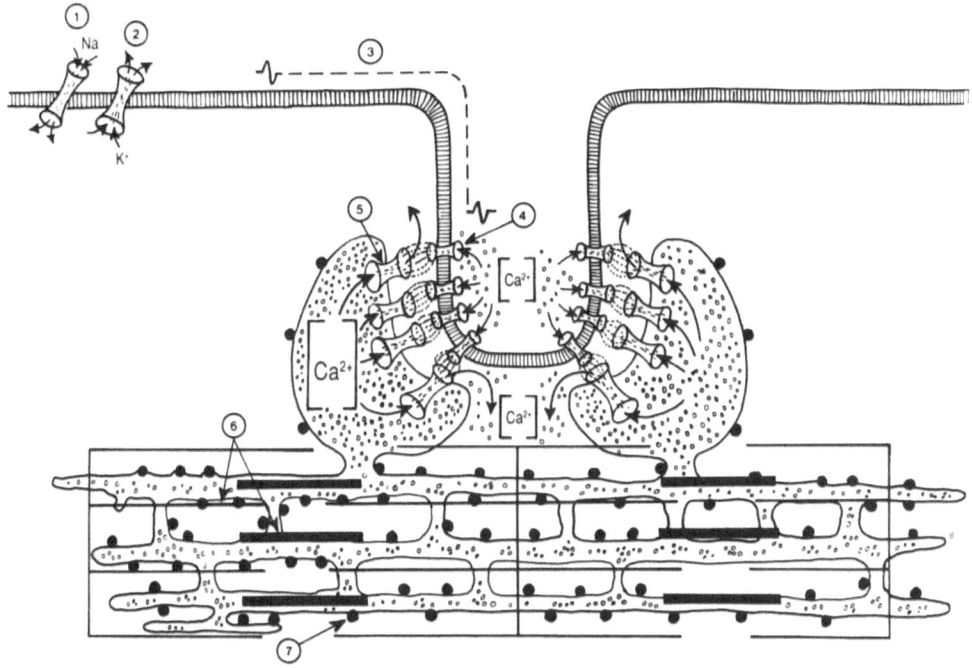

FIGURE 1. A schematic model for sites for anesthetic action on the excitation-contraction coupling process in skeletal muscle. Sites (1) and (2) depict sodium and potassium channels and their function in producing the action potential (3); site 4 is the dihydropyridine-sensitive Ca^{2+} channel residing in the transverse tubule membrane; (5) is the ryanodine binding Ca^{2+}-release channel that is thought to be the primary Ca^{2+} release site for activating (6), the interaction of thick (myosin) and thin (actin) filaments to produce contracture; (7) represents the Ca^{2+} pump protein that uses ATP to actively pump Ca^{2+} back to the sarcoplasmic reticulum membrane. The small open circles depict Ca^{2+} ions stored in highest concentration in the large terminal sac (terminal cisternae) that is bridged to the T-tubule by the ryanodine binding protein. This figure is a modification of figure 1 from T. E. Nelson and T. Sweo, *Anesthesiology* 69:571-577 (1988). Used with permission.

the skeletal muscle sodium channel is similar to that described for nerve sodium channels; *i.e.*, halothane decreases the amplitude of the sodium current. In the study by Ruppersberg and Rudel, the threshold concentration of halothane that produced this effect was between 1.3 and 3.4 mM (about 2.6 to 6.8 MAC), with 3.4 mM halothane representing the C_{50} for the sodium current depressant effect. The investigation of halothane effects on sodium channels in nerve tissue concluded that the observed effects were likely explained by an effect of the anesthetic to alter the gating properties of the sodium channel.[5] In contrast, Ruppersberg and Rudel interpreted their data as an effect of halothane to completely block the sodium channel and thereby reduce the number of conducting channels.

In the study presented at this symposium by Wieland and associates (Chapter 2), patch clamp methodology is used to study human skeletal muscle sodium channels. This type of experiment allows the researcher to "pinch off" a piece of sarcolemma from a cell and experimentally isolate specific ion channels, and, like an enzymologist performs enzyme kinetic studies on a protein molecule, the electrophysiologist can study the kinetics of a single protein molecule isolated in the patch. There are many experiments waiting to be done regarding the effects of anesthetics on sarcolemmal function in the excitation-contraction coupling of muscle, and these electrophysiologic techniques provide an elegant approach to single channel investigation. Dr. Weiland and coworkers isolated sodium channels in cultured muscle biopsied from normal and MH-susceptible patients and studied the effect of lipid soluble agents on the properties of this channel. Interestingly, the effects of halothane on the voltage-activated sodium current (decreased amplitude and increased rate of inactivation) were similar to the effect of enflurane on the compound action potential of intact muscle.[3] Dr. Wieland also investigated the effect of various fatty acids on the channel properties, suggesting that lipid-soluble agents can have modulatory effects on channels. In addition, these investigators compared the regulation of the sodium channel between muscle from normal and MH-susceptible human subjects. Based on their findings, it is suggested that mutations in the regulatory mechanisms may be linked to malignant hyperthermia.

The next conceptual step along the E-C coupling pathway takes us to a specialized area of the sarcolemma membrane. As the action potential is propagated along the sarcolemma it reaches the point at which the membrane invaginates to the cell's interior forming a structure known as the transverse-tubule. This specialized membrane contains other ion channels involved in E-C coupling and represents the linkage for transferring signal from the sarcolemma to the intracellular compartment. A thoroughly recognized and studied E-C coupling component of the T-tubule is the dihydropyridine-sensitive (DHP) calcium channel. The extent of this Ca^{2+} channel's role in E-C coupling in skeletal muscle is yet to be determined, but the methodology is available for detailed investigation of anesthetic action on its single channel and ligand binding properties. The DHP-sensitive Ca^{2+} channel may interact with another Ca^{2+} channel that forms a structural bridge between the T-tubule and the terminal cisternae of the internal sarcoplasmic reticulum membrane system. This bridging protein was purified through the use of its avid binding of the alkaloid ryanodine,[4] and consequently is often referred to as the ryanodine receptor protein. Present theory states that this Ca^{2+} channel is the primary pathway for Ca^{2+} release from the terminal cisternae storage site to the myoplasm for initiation of contraction. Single channel and ligand binding studies show a defect in this Ca^{2+} channel from skeletal muscle of malignant hyperthermia susceptible humans and pigs.[8,9] The effects of halothane on certain properties of the ryanodine receptor Ca^{2+} channel protein from normal and human malignant hyperthermia skeletal muscle are described in planar lipid bilayer experiments done in my laboratory (Chapter 3). One of the properties of this Ca^{2+} channel is that it can have different levels of conductivity, often referred to as substates. We discovered that the distribution of these apparent substates is different between normal and MH human Ca^{2+} release channels and that halothane (4 μM) alters this distribution only in channels from MH muscle. Multiple regulatory sites on the ryanodine receptor protein represent an example of how anesthetic action on a single protein molecule can have one or more targets. Thus, for the ryanodine receptor protein there is evidence for two different Ca^{2+} binding sites, one intra- and one extra-luminal to the SR membrane; and in addition, binding sites for ATP, calmodulin, and annexin VI have been reported. Each of these ligands has a regulatory effect on the properties of the ryanodine receptor protein as a calcium channel and each of these binding domains on this enormously large protein molecule represent a target for anesthetic action. Therefore it is evident, even at the single protein molecule level, that action of anesthetics may be complex because of the potential number of targets.

Ca^{2+} is released from the terminal cisternae of the sarcoplasmic reticulum and its concentration in the myoplasm begins to increase from the resting state pCa^{2+} level of 7.0 to a pCa^{2+} of 6.0 at which the contractile apparatus becomes activated. Simply described, Ca^{2+} binds to troponin C, causing this molecule to undergo a conformational change that releases its inhibition of the myosin ATPase site. This then allows actin and myosin to interact and generate shortening of the crossbridging structures. Regulation of contraction and relaxation of the skeletal muscle contractile apparatus is a complex series of events which have been greatly oversimplified for this overview, but represents a number of complex molecular targets on which anesthetics may act. The full cycle of E-C coupling is completed when Ca^{2+} is removed from the myoplasm and a resting level of $pCa^{2+} = 7.0$ is maintained. The removal of Ca^{2+} is accomplished by another sarcoplasmic reticulum membrane protein often referred to as the Ca^{2+} pump protein. This 100,000 dalton protein is a Ca^{2+}-Mg^{2+}-activated ATPase that transports 2 moles of Ca^{2+} from the myoplasm to the lumen of the sarcoplasmic reticulum membrane for each mole of ATP consumed. In order for the efficiency ratio of $2Ca^{2+}/ATP$ to exist for the pump, it is mandatory that the Ca^{2+} channel be in a closed state. If a concommitant closure of the Ca^{2+} channel does not occur, then as Ca^{2+} is pumped back into the SR it is readily "leaked" back to the myoplasm, creating a futile ATPase cycle. Consequently, the net amount of Ca^{2+} pumped into and retained by the SR becomes some value less than 2 Ca^{2+} per mole ATP consumed. The pumping of Ca^{2+} back into the SR storage sites and return of the contractile elements completes the oversimplified cycle of E-C coupling in the conceptual model.

The remaining studies on skeletal muscle presented at this symposium represent more than one element of E-C coupling and previews of these chapters will follow. Salviati and coworkers (Chapter 4) have used chemically skinned single psoas rabbit fibers to study the effects of halothane on calcium regulation in the muscle cell. In this preparation, the sarcolemma is not functional and basically the only intact mechanisms are active calcium uptake and calcium release by the sarcoplasmic reticulum membrane and the contractile apparatus. These investigators also used isolated sarcoplasmic reticulum membranes to determine halothane effects on Ca^{2+}-ATPase activity and protein composition of the SR membrane. Utilizing various pharmacologic agents, Salviati *et al.* differentiate halothane effects on Ca^{2+} uptake from Ca^{2+} release and conclude that halothane acts primarily on the Ca^{2+}-induced-Ca^{2+}-release process of the sarcoplasmic reticulum membrane. A most interesting aspect of this study is results showing that calmodulin, a modulator of the Ca^{2+}-release channel, may be involved in the halothane effect on the Ca^{2+}-release process. They show, biochemically, that halothane can displace calmodulin from its sarcoplasmic reticulum binding site.

In a review of several previously published studies, Sudo and Suarez-Kurtz (Chapter 5) demonstrate the efficacy of rapid-cooling-induced contractures as a model for studying anesthetic action. The effects of halothane on rapid cooling contractures were investigated in innervated *vs.* denervated fast and slow types of muscle, chemically skinned fibers, and in biopsied human muscle. At room temperature, equilibration with halothane, 0.3 - 3% vol/vol, produced little or no contracture in the muscles studied. However, following equilibration with halothane, rapidly cooling the muscle to 2°C produces transient contractures in muscle that normally has little or no change in tension when rapidly cooled in the absence of halothane. The amplitude of these halothane-induced cooling contractures (HCCs) is dependent on halothane concentration, final cooling temperature, and on the muscle fiber type. The studies show that normal human skeletal muscle produces HCCs in a reproducible manner and suggest that this mechanism may have value in discriminating between normal and malignant hyperthermia susceptible muscle. Experiments are described which apparently exclude the contractile elements and the calcium uptake system as processes contributing to HCCs. These investigators' general consensus is that HCCs result from a synergistic interaction among halothane, cooling and the Ca^{2+}-induced-Ca^{2+}-release process.

In Chapter 6, by Fletcher, Rosenberg and Beech, anesthetic action *per se* is not addressed but the investigators present their data and theories about the effect of altered lipid metabolism in skeletal muscle and how it might change the response of muscle to anesthetic agents. The implications of these experiments are directed primarily at malignant hyperthermia in humans, pigs and horses and at how environmental (anesthetic) and mutational (MH) factors can act together to produce malfunction of the muscle cell.

The number of targets in skeletal muscle is enormous and in this overview we have simplistically glanced at just one aspect: that being a model for excitation-contraction coupling. Our discussion and references have been limited to volatile anesthetic agents since most of the studies at this symposium address the actions of this class of agents. Metabolic and signal transduction systems have not been considered and they may be equally important targets for anesthetic action. It is clearly established that skeletal muscle myofiber type (*i.e.*, type I, slow, *vs.* type II, fast twitch) can be very important in the response to anesthetic action[10] and this fact should always be considered when planning experiments. There are many targets for anesthetic action in the skeletal muscle cell and the advance of technology provides us with new and exciting tools for obtaining new pieces to this very complex puzzle.

REFERENCES

1. B. E. Waud and D. R. Waud, Effects of volatile anesthetics on directly and indirectly stimulated skeletal muscle, *Anesthesiology* 50:103-110 (1979).
2. J. J. Kendig and J. P. Bunker, Alterations in muscle resting potentials and electrolytes during halothane and cyclopropane anesthesia, *Anesthesiology* 36:128-131 (1972).
3. S. Kurihoro, M. Konishi, T. Myagishima and T. Sakai, Effects of enflurane on excitation-contraction coupling in frog skeletal muscle fibers, *Pflugers Arch* 402:345-352 (1984).
4. E. M. Gallant, Porcine malignant hyperthermia: no role for plasmalemmal depolarization, *Muscle and Nerve* 11:785-786 (1988).
5. B. P. Bean, P. Shoryer and D. A. Goldstein, Modification of sodium and potassium channel gating kinetics by ether and halothane, *J Gen Physiol* 77:233-253 (1981).
6. J. P. Ruppersberg and R. Rudel, Differential effects of halothane on adult and juvenile sodium channels in human muscle, *Pflugers Arch* 412:17-21 (1988).
7. F. A. Lai, H. P. Erickson, E. Rousseau, O.-Y. Liu, G. Meissner, Purification and reconstitution of the calcium release channel from skeletal muscle, *Nature* 331:315-319 (1988).
8. M. Fill, E. Stefani, T. E. Nelson, Abnormal human sarcoplasmic reticulum Ca^{2+} release channels in malignant hyperthermia skeletal muscle, *Biophys J* (In Press)
9. M. Fill, R. Coronado, J. R. Mickelson, J. Vilven, J. Mo, B. A. Jacobson and C. F. Louis, Abnormal ryanodine receptor channels in malignant hyperthermia, *Biophys J* 57:471-476 (1990).
10. P. A. Deuster, E. L. Blackman and S. M. Muldoon, In vitro responses of cat skeletal muscle to halothane and caffeine, *J Appl Physiol* 58:521-527 (1985).

2

Effects of Lipid-Soluble Agents on Sodium Channel Function in Normal and MH-Susceptible Skeletal Muscle Cultures

Steven J. Wieland, Jeffrey E. Fletcher, Qi-hua Gong, Henry Rosenberg

INTRODUCTION

Some skeletal muscle abnormalities, including susceptibility to malignant hyperthermia (MH), may be traced to alterations of specific membrane functions. These in turn may be caused directly by mutations in the specific proteins subserving these functions, or by mutations which indirectly alter properties of membrane-bound proteins. Growth of human skeletal muscle cells in primary culture allows pharmacological and physiological exploration of potential regulatory mechanisms which cannot be studied by other means. Subtle, persistent differences in sodium channel function are present in MH-susceptible cells compared to non-susceptible cells. At the moment we are examining the hypothesis that these differences may be due to abnormal lipid metabolism which alters the processing or final environment of certain membrane proteins.

Malignant hyperthermia is a potentially fatal disorder of skeletal muscle which can be induced by potent volatile general anesthetics in susceptible individuals. It is a syndrome which presents multiple clinical and tissue abnormalities. An MH episode may include one or more of the following: hyperthermia, cardiac arhythmia, metabolic acidosis, and muscle rigidity.[1,2] MH susceptibility in humans is considered to have an autosomal dominant pattern of inheritance with a locus for susceptibility near 19q12-13.2,[3] although there may be heterogeneous loci. A widely used diagnostic test for MH susceptibility is the *in vitro* challenge of muscle fiber bundles with halothane and caffeine.[4] In this diagnostic test, muscle from MH-susceptible individuals exhibits a lower threshold of contracture in response to halothane or caffeine than normal muscle. However, since the mechanism of the MH lesion is unknown, the basis for the contracture test remains empirical and subject to misapplication and misinterpretation. The *in vitro* contracture sensitivity[5] and *in vivo* metabolic events in MH have led to a model that postulates greater sensitivity to triggering agents of the Ca^{2+}-releasing mechanism of the sarcoplasmic reticulum (SR) as a possible cause of MH

STEVEN J. WIELAND, QI-HUA GONG, Department of Anatomy, JEFFREY E. FLETCHER, HENRY ROSENBERG, Department of Anesthesiology, Hahnemann University, Philadelphia, Pennsylvania 19102.

Mechanisms of Anesthetic Action in Skeletal, Cardiac, and Smooth Muscle
Edited by T.J.J. Blanck and D.M. Wheeler, Plenum Press, New York, 1991

susceptibility.[6] Since other cellular components also show altered functions, the Ca^{2+}-release mechanism may not be the primary defect. Non-anesthetic-induced abnormalities, including alterations in isolated organelles,[7,8,9,10] suggest a possible membrane defect with more general consequences. Porcine MH-susceptible muscle shows an altered response to Ba^{2+} exposure[11] and a halothane-induced abnormality in electrical excitability[12] even in the absence of anesthetic challenge, and cultured human MH-susceptible myocytes exhibit altered Na^+ currents.[13] Thus, a single altered function in MH-susceptible cells may lead to altered functioning of several membrane-associated systems, including voltage-activated sodium channels and the halothane-sensitivity of SR Ca^{2+} release. A possible key to this spectrum of observations comes from the recent finding that the Ca^{2+}-release sensitivity of sarcoplasmic reticulum can be altered by changes in the concentration of free unsaturated fatty acids.[14]

Free fatty acids are released from skeletal muscle homogenates by the activity of endogenous lipase, and this release is elevated in muscle homogenates from MH-susceptible pigs and humans.[15,16] Intact MH-susceptible muscle does not show overtly altered metabolism unless exposed to triggering agents;[17] the ensuing elevation of free fatty acids is capable of sensitizing the release of Ca^{2+} from SR.[14,18] A cause of the altered lipid metabolism in MH may be abnormal function of the hormone-sensitive lipase, the gene for which is located near the currently proposed locus of MH susceptibility.[3,19,20]

Cultured cells can provide a relatively stable source with which to study inherited physiological lesions, although examination of lesions is dependent on their expression in culture. We recorded plasma membrane ion currents in "myoballs" derived from primary cultures of human skeletal muscle to examine the degree to which muscle-type ion currents would be expressed. We also included cells from MH-susceptible muscle to test the possibility that voltage-activated membrane functions might be measurably altered.

METHODS

In Vitro Contracture Tests

A small strip of vastus lateralis muscle was removed under local anesthesia, and halothane and caffeine contracture tests were performed as previously described.[4] The patient was diagnosed as MH susceptible (MH+) if 1 or more of 6 muscle fiber bundles exhibited a contracture > 0.7 gm within 5 minutes of 3% halothane (bubbled in the gas phase) exposure. Two additional fiber bundles were exposed to increasing caffeine concentrations. Patients were diagnosed as MH+ if either bundle exhibited a contracture > 0.3 gm at a 2 mM concentration. Patients not exhibiting contractures meeting the criteria for MH susceptibility were designated nonsusceptible by diagnostic test (MH−). A small sample (100-200 mg) of muscle was removed from the biopsy before diagnostic tests were done.

Cell Culture and Preparation

Myoblast cultures were prepared from skeletal muscle biopsies by seeding culture flasks as previously described.[13] When cells reached confluence, the growth medium was replaced with fusion-promoting medium. Myoblasts ceased dividing and fused to form multinucleated "myotubes" in 3 to 7 days after exposure to fusion medium.

One to six hours before voltage clamping, cells were released from the substrate by treatment with 0.5 mg/ml trypsin (Sigma Type III) and 0.5 mg/ml collagenase (Sigma Type V) in Dulbecco's phosphate-buffered saline for 3 to 5 min. Myoballs were formed by gently pipetting 5 times with a wide bore (2 mm) pipette. After dispersal, cells were suspended in a 3-fold excess of growth medium and centrifuged at 150 × g for 5 min. The supernatant was discarded and cells were resuspended in growth medium and plated onto gelatin-coated glass cover-slip pieces. After a minimum of 1 hour, cover slips were placed into a 200 μl recording chamber mounted on an inverted microscope, and viewed by phase contrast. Myoballs formed in this manner can be used for recordings up to 24 hours after plating.

Current Recording

Cells were voltage clamped using the whole-cell variant of the patch-clamp method of Hamill et al.,[21] essentially as previously reported.[13] The external recording medium for most recordings was: NaCl 137 mM, KCl 5 mM, $CaCl_2$ 2 mM, glucose 5 mM, and HEPES 10 mM adjusted to pH 7.30. Pipettes were pulled from Kimax tubing; the resistance of the fire-polished tips was in the range 1.2 to 2.5 × 10^6 ohms. The potassium-containing medium used inside the pipettes included; KCl 140 mM, EGTA 2 mM, HEPES 10 mM, pH 7.30.

Fatty Acid Modulation of Whole Cell Ion Currents

Intracellular application of fatty acids to myocyte cytosol was done by including 1-50 μM of polyunsaturated, unsaturated, and saturated fatty acids in recording pipettes during whole-cell voltage clamp recordings. Recordings were made immediately upon establishment of cytoplasmic communication, and at fixed 1 to 3 minute intervals, until the membrane currents stabilized. Cell currents were also measured in parallel under baseline conditions in standard extracellular recording medium, and during extracellular application of the same lipids. Arachidonic acid and other unsaturated fatty acids and derivatives are susceptible to oxidation; therefore, care was taken to prepare fresh stock solutions, keep them under inert gas, and dilute immediately before use.

RESULTS

Membrane Currents in Muscle Cells from MH+ and MH− Patients

Cultured human skeletal muscle cells exhibit both a fast transient inward current and a delayed outward current during voltage clamp steps.[13,22,23] The voltage-dependent activation and inactivation of the fast transient inward current were similar to the properties of this current in rat myoballs.[24] Representative currents of human cells are shown in figures 1A and B. The peak inward currents of cells from patients diagnosed as non-MH-susceptible (MH−) were 2.42 ± 0.50 nA (n = 19), and were 2.57 ± 0.35 nA (n = 35) for cells from MH+ patients.[13] The myoball inward currents ranged from 0.2 to 5 nA; thus, studies which depend on quantitative comparison of absolute magnitudes would require large numbers of cells. However, examination of voltage-dependent functions, kinetics, and modulation of channels is quite feasible.

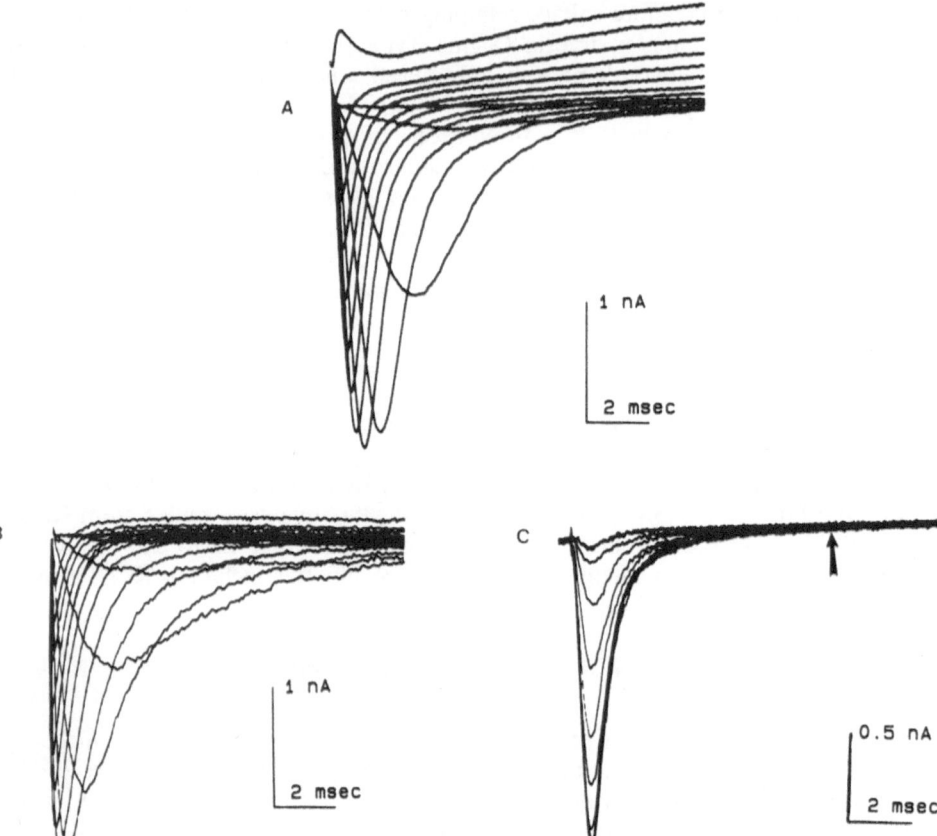

FIGURE 1. Membrane currents of cultured human skeletal muscle. Representative data from 3 different patients. A: Transient inward and delayed outward currents measured in the presence of 140 mM K^+ in recording pipette. The cell was held at −100 mV, and depolarized for 12 msec to test potentials ranging from −60 mV to +70 mV, in 10 mV increments. Test pulses were separated by 2 sec intervals. The resulting 14 current recordings are superimposed. The smallest depolarizations showed the slowest activation of the inward current; successively larger depolarizations showed faster activation. Inward current is graphed downward. Scale bars for A and B: vertical, 1nA; horizontal, 2 msec. B: Inward currents recorded with 140 mM Cs^+ in the pipette. Protocol was the same as in panel A. C: Voltage-dependent inactivation of transient inward current. The cell was held at −100 mV. A 70 msec prepulse preceded each test pulse; prepulses ranged from −130 mV to −30 mV. Test pulses were to −10 mV; currents were measured during each test pulse. Scale bars for panel C: vertical, 0.5 nA; horizontal, 2 msec. Reprinted with permission of the American Physiological Society from Wieland, Fletcher, Rosenberg and Gong, *Am J Physiol* 257:C759-C765 (1989).

Visual inspection of inward currents in cells from MH+ patients frequently showed a slower rate of inactivation. An example of such cells in shown in figure 2B; the presence of residual current (arrow) at 8-12 msec after depolarization to −10 mV can be compared to the lack of current in an MH− cell recorded under the same conditions in figure 1C. Approximately 60% of cells from MH+ cultures had measurable residual inward currents at the end of 12 msec depolarizations to −10 mV, while only 10% of cells from MH− cultures showed large residual current.[13]

Inactivation of the fast component occurred at more negative potentials than those at which the slow component inactivated. This slower component was sodium dependent and blockable with tetrodotoxin (TTX).[13] Thus, it appears to be a slower component of fast inactivation of the sodium current. When the time course of fast inactivation of myoball currents was modeled with exponential decay curves, both MH− and MH+ currents were best fit by two exponential components with few exceptions. However, the maximum magnitude of the "slow" component in MH− cells was still 3-fold smaller than in MH+ cells.[13]

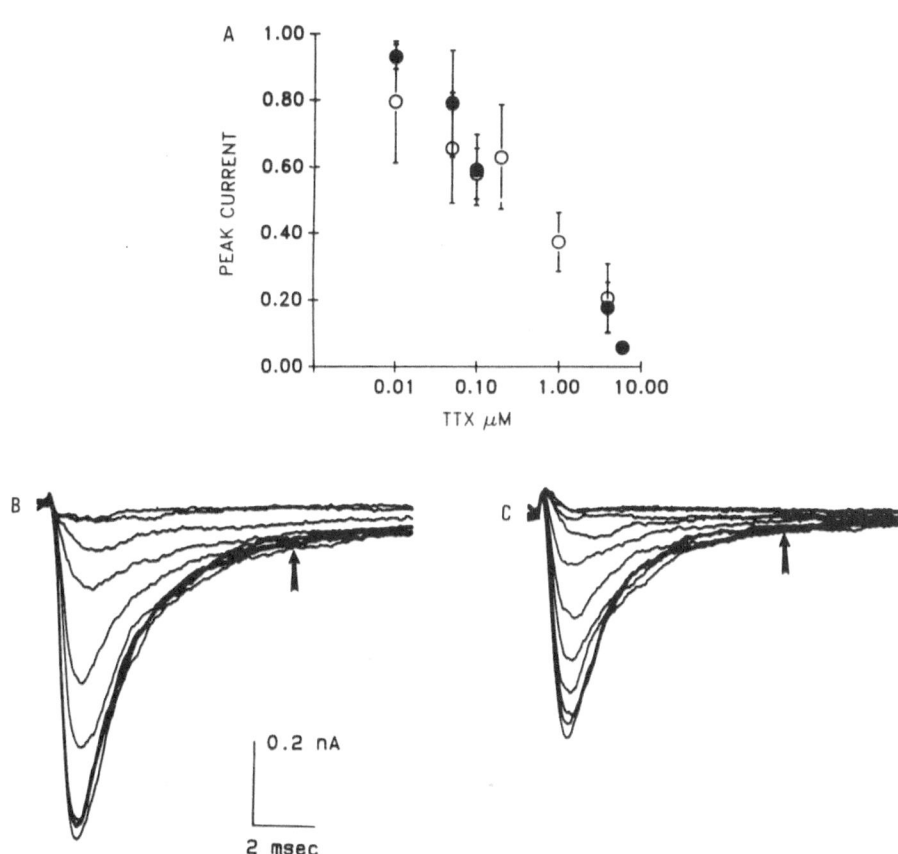

FIGURE 2. Characterization of fast and "slow" components of inward currents. A: The concentration dependence of tetrodotoxin (TTX) block of peak inward currents in MH− and MH+ cells. Cells were held at −100 mV, and the peak inward currents which were elicited by test pulses to −10 mV were compared before and during exposure to varying concentrations of TTX. Data shown are means and SEMs for 2 to 5 independent measurements on different cells. Symbols: o = MH− cells; ● = MH+ cells. B: The voltage-dependent inactivation of inward currents in an MH+ cell showing both fast and slow components. The inactivation protocol was the same as that used in figure 1C. Note the residual current due to the slower-inactivating component (arrow at 8 msec). C: The partial block of both the fast and slow components by 0.1 μM TTX. The inactivation protocol was repeated on same cell as in panel B during exposure to 0.1 μM TTX. Note that both peak current and residual current (arrow) are decreased in magnitude. For a prepulse of −100 mV, the fast inward component was reduced by 42% compared to panel B. The slow component, estimated by the inward current at the arrow, was reduced by 72% compared to the comparable point in panel B. Reprinted with the permission of the American Physiological Society from Wieland, Fletcher, Rosenberg and Gong, *Am J Physiol* 257:C759-C765 (1989).

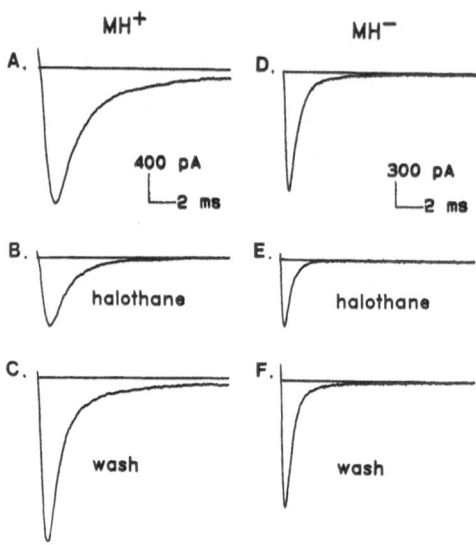

FIGURE 3. Halothane reduces sodium currents in cultured muscle. Cells were held at −100 mV and voltage-activated currents were recorded in response to 15 msec step depolarizations to −10 mV. Solid horizontal lines represent 0 transmembrane current. Panels A, B, and C were recorded sequentially from a single MH+ cell; panels D, E, and F were recorded sequentially from a single MH− cell. Panels A and D are baseline recordings in standard recording medium. Recordings in panels B and E were made while cells were bathed in standard recording medium which had been equilibrated by bubbling with 3% halothane in air for 60 minutes. The recording chamber (volume 0.15 ml) was superfused at a rate of 1 ml/min. Test recordings were made at the end of one minute; further recordings showed no further changes in currents. Panels C and F were recorded after 6 minutes of superfusion with standard recording medium at a flow rate of 0.3-0.5 ml/min.

Effect of Halothane on Inactivation

Myoballs exposed to halothane during recording show inward currents of reduced magnitude and faster inactivation (figure 3), consistent with reports on crayfish giant axon[25] and normal human muscle.[26] Both fast and "slow" components of fast inactivation were reduced in the presence of halothane. Halothane did not consistently alter the relative magnitudes of either component of fast inactivation in either MH+ or MH− cells, although the rates of both were accelerated. Halothane may also alter the number of activatable channels by affecting the voltage dependence of slow inactivation.[25,27]

Effect of Arachidonic Acid on Skeletal Muscle Sodium Current

Several ion conductances have been shown to be modulated by lipids.[28,29,30] Therefore, the effect of altering the concentrations of normal endogenous lipid components was examined in cultured muscle cells. Human muscle cells were voltage clamped in whole-cell recording mode using standard recording medium with 5-20 μM arachidonic acid added to the standard intracellular solution in the recording pipette (figure 4A). Voltage-activated inward currents increased in magnitude for several minutes, stabilizing by 15 minutes at levels 30-50% higher than at the start of recording (figure 4B). Occasional cells showed responses which were 1-2 fold greater than initial values. Cells recorded with no additions to the pipette showed no increase in inward currents. Twenty-two cells followed for 15 minute showed final peak currents which were only 1 ± 5% greater than their initial values.

Cells show profound but reversible inhibition of inward currents after incubation in extracellular medium containing 5-15 μM arachidonic acid (figure 4C). Although extracellular arachidonic acid activates outward rectifying K channels in cardiac and smooth muscle,[28,29] it inhibits sodium currents in skeletal muscle (figure 4C) and neuroblastoma cells.[30] Thus, although increased extracellular free fatty acids may activate some membrane processes, they may serve to reduce membrane electrical excitability by activating K channels and inhibiting Na channels.

Oleic Acid Produces Similar Intracellular and Extracellular Effects

Arachidonic acid might act *via* its metabolites in the lipoxygenase or cyclooxygenase pathways.[29] Oleic acid, an unsaturated 18 carbon fatty acid, is not a substrate for the lipoxygenase or cyclooxygenase enzymes. Oleic acid was approximately as effective as arachidonic acid in both intracellular activation and extracellular inhibition of skeletal muscle sodium currents. However, the saturated 18 carbon stearic acid was ineffective as a modulator by either route.

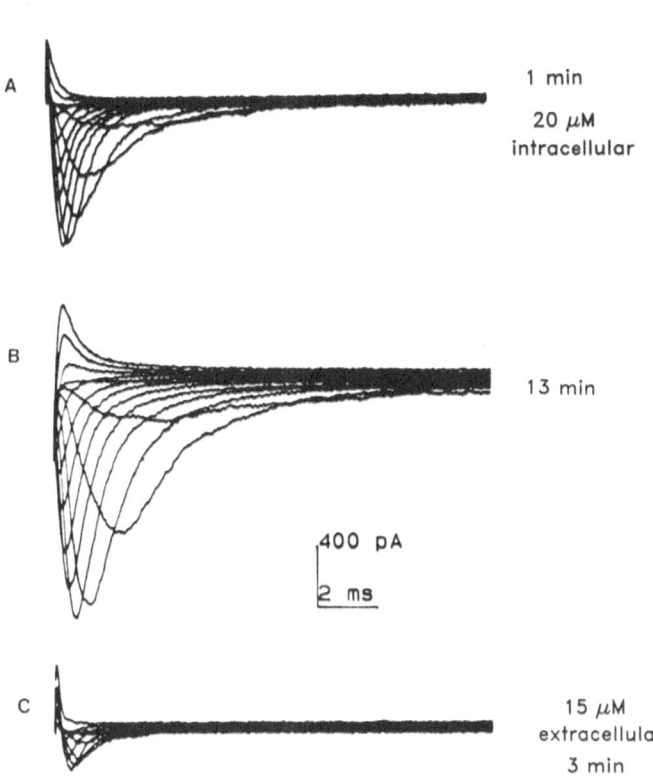

FIGURE 4. Arachidonic acid modulates skeletal muscle sodium currents. A: Currents recorded from a normal cell with 20 μM arachidonic acid in the pipette at the initiation of pipette contact with the cell interior. The pipette contained 120 mM Cs+ to block outward K+ currents. B: Same cell after 13 minutes of exposure. C: Same cell after an additional 3 min incubation with 15 μM extracellular arachidonic acid. For all panels: holding potential −100 mV; test pulses ranged from −60 to +70 mV.

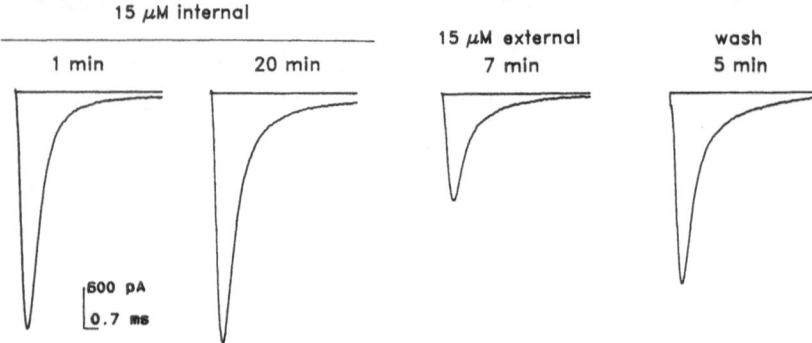

15 µM internal

1 min   20 min

15 µM external
7 min

wash
5 min

600 pA
0.7 ms

FIGURE 5. Reduced sensitivity of an MH+ cell to arachidonic acid exposure. Records from a single MH+ cell during intracellular exposure to 15 µM arachidonic acid. The external recording medium initially had no additions, and holding potential was −100 mV. Shown is the voltage-activated inward current in response to a step to −10 mV shortly after initiation of pipette contact with cytoplasm and again, 20 minutes later. The external recording medium was then replaced with medium containing 15 µM arachidonic acid for 7 minutes, followed by a third recording of inward currents. A final recording was made after a 5 minute wash of the chamber with 3 ml of external recording medium.

Arachidonic Acid Response May Be Reduced in MH+ Cells

An example of response by an MH+ cell to intracellular and extracellular exposure to arachidonic acid is shown in figure 5. Only a modest increase in the peak inward current occurred over 20 minutes of monitoring, although inhibition by extracellular exposure was strong and reversible. A gradual increase in the magnitude of the slower component of inactivation was also apparent over the entire sequence. Groups of cells were continuously monitored under whole-cell recording conditions with or without 15 µM arachidonic acid in the pipette. After 15 minutes of exposure to arachidonic acid, the peak inward currents in cells from normal patients (MH−) had increased in magnitude above their initial magnitude by 41 ± 10% (n = 7). Cells from MH-susceptible patients (MH+) showed a smaller increase of 22 ± 6% (n = 15).

DISCUSSION

"Myoballs" can be easily produced from cultured human myotubes and myoblasts to provide the proper geometry and size for whole-cell tight seal recording. The transient inward currents (TTX blockable) and delayed outward currents (Cs^+ sensitive) which were recorded under these conditions are similar to those reported in rat[24] and human[23] muscle cultures.

A slowly inactivating component of inward current which could be separated from the larger, faster inactivating current by pre-pulse inactivation was detected in these cultured skeletal muscle cells. Although variable in magnitude, this component was significantly larger in MH+ cells than in MH− cells. The voltage gating and kinetic properties of sodium channels can be modified by temperature,[33,34] deoxycholate treatment[35] and several lipid-soluble and polypeptide neurotoxins (reviewed in ref. 36). Thus, this second, slowly inactivating population of channels in skeletal myoballs may not be genetically different from the primary, faster channels. The data from halothane and fatty acid exposure of human myoballs show that the magnitude and kinetics of sodium current components can be broadly manipulated in both MH− and MH+ cells.

One possible cause for the larger magnitude of the slowly inactivating component of Na$^+$ current in cultured MH+ cells may be an alteration of the membrane environment with a resultant change in the gating properties of the channels. Other functions which appear abnormal in MH+ tissues include uncoupling of Ca^{2+}-stimulated succinate oxidation[7] and oligomycin-insensitive Ca^{2+},Mg^{2+}-ATPase activity in mitochondria,[15] increased sarcolemmal acetylcholinesterase activity[37] and ATP-dependent Ca^{2+} transport,[8] altered ryanodine binding by the sarcoplasmic reticulum,[32] altered triglyceride metabolism,[17] and altered inositol 1,4,5-trisphosphate phosphatase.[31] MH+ cells might be more susceptible to proteolytic damage, resulting in abnormal gating properties of some sodium channels. There are multiple mechanisms which could covalently damage ion channels; these might underlie the reported abnormalities of sodium currents in several myopathies.[38] On the other hand, the reversible changes in magnitude and kinetics which are produced in both MH− and MH+ cells by halothane and fatty acid exposure argue that sodium channels are subject to lipid-based modulatory mechanisms. The measurement of an altered inward current in cultured cells from MH-susceptible patients shows that a heritable abnormality, perhaps in a modulatory pathway, is expressed in these cells *in vitro*.

The activation and inactivation properties of voltage-dependent ion channels are modulated by both covalent and non-covalent mechanisms. Experimental addition of specific free fatty acids can activate potassium channels in cardiac[28] and smooth muscle[29] cells, and inhibit sodium and calcium channels in neuroblastoma cells.[30] Channel inhibition in neuroblastoma cells has been proposed to be due to activation of protein kinase C subtypes,[30] although the data were not fully consistent with this mechanism. The existence of non-covalent binding sites for specific lipid-soluble and water-soluble exogenous toxins suggests that related sites may exist for endogenously produced regulatory molecules, including fatty acids and their metabolites.

Intracellular versus extracellular application of unsaturated fatty acids produces divergent actions on voltage-activated sodium currents of human skeletal muscle cells. This indicates that at least two sites of action exist in these cells. The inhibitory action of extracellular fatty acids is consistent with the results of Linden and Routtenberg[30] on neuroblastoma cells. However, the facilitation due to intracellular application of fatty acids correlates more with the effects of fatty acids on smooth muscle potassium channels.[29] Arachidonic acid may act through metabolic conversion to prostaglandins *via* the cyclooxygenase pathway, or through conversion to leukotrienes *via* the lipoxygenase pathway.[39] If one of these pathways were required, then oleic acid would not be an active substitute. Since oleic acid does show modulatory activity in human skeletal muscle and in smooth muscle,[29] unmetabolized arachidonic acid may act directly as a modulatory ligand. The next question is the character of the modulatory sites of action, particularly whether they are direct, part of the channel complexes, or *via* intermediate steps.

These data suggest that the actions of lipid-soluble agents are sensitive to membrane sidedness. They also show that sodium channels may be targets of lipid-based signalling pathways; therefore, muscle relaxants as well as muscle stimulants could be designed for intervention in these pathways. Finally, the data from cultured cells of malignant-hyperthermia-susceptible individuals suggest differences in the magnitude of sodium channel modulation in these cells compared to those from nonsusceptible individuals. We believe that the variant regulatory mechanisms themselves, not their effects on *in vitro* sodium channels, may be linked to the mechanism of susceptibility.

ACKNOWLEDGEMENTS

This work was supported in part by the Muscular Dystrophy Association.

REFERENCES

1. G. A. Gronert, Malignant hyperthermia, *Anesthesiology* 53:395-423 (1980).
2. H. Rosenberg and J. E. Fletcher, Malignant hyperthermia, *in:* "Muscle Relaxants: Side Effects and Rational Approach to Selection," I. Azar, ed., Marcel Dekker, New York (1987) pp. 115-148.
3. T. V. McCarthy, J. M. S. Healy, J. J. A. Heffron, M. Lehane, T. Deufel, F. Lehmann-Horn, M. Farrall and K. Johnson, Localisation of the malignant hyperthermia susceptibility locus to human chromosome 19q12-13.2, *Nature* 343:562-564 (1990).
4. J. E. Fletcher and H. Rosenberg, Laboratory methods for malignant hyperthermia diagnosis, *in:* "Experimental Malignant Hyperthermia," C. H. Williams, ed., Springer-Verlag, New York (1988) pp. 121-140.
5. J. E. Fletcher and H. Rosenberg, *In vitro* interaction between halothane and succinylcholine in human skeletal muscle: Implications for malignant hyperthermia and masseter muscle rigidity, *Anesthesiology* 63:190-194 (1985).
6. J. R. Mickelson, J. A. Ross, B. K. Reed and C. F. Louis, Enhanced Ca^{2+}-induced calcium release by isolated sarcoplasmic reticulum vesicles from malignant hyperthermia susceptible pig muscle, *Biochim Biophys Acta* 862:318 (1986).
7. K. S. Cheah and A. M. Cheah, Mitochondrial calcium transport and calcium-activated phospholipase in porcine malignant hyperthermia, *Biochim Biophys Acta* 634:70-84 (1981).
8. J. R. Mickelson, J. A. Ross, R. J. Hyslop, E. M. Gallant and C. F. Louis, Skeletal muscle sarcolemma in malignant hyperthermia: Evidence for a defect in calcium regulation, *Biochim Biophys Acta* 897:364-376 (1987).
9. T. E. Nelson, Abnormality in clacium release from skeletal sarcoplasmic reticulum of pigs susceptible to malignant hyperthermia, *J Clin Invest* 72:862-870 (1983).
10. T. E. Nelson, E. H. Flewellen and D. W. Arnet, Prolonged electromechanical coupling time in skeletal muscle of pigs susceptible to malignant hyperthermia, *Muscle and Nerve* 6:263-268 (1983).
11. E. M. Gallant, Porcine malignant hyperthermia: No role for plasmalemmal depolarization (letter), *Muscle and Nerve* 11:785-786 (1988).
12. P. A. Iaizzo, F. Lehmann-Horn, S. R. Taylor and E. M. Gallant, Malignant hyperthermia: Effects of halothane on the surface membrane, *Muscle and Nerve* 12:178-183 (1989).
13. S. J. Wieland, J. E. Fletcher, H. Rosenberg and Q. H. Gong, Malignant hyperthermia: Slow sodium current in cultured human muscle cells, *Am J Physiol* 257:C759-C765 (1989).
14. J. E. Fletcher, L. Tripolitis, K. Erwin, S. Hanson, H. Rosenberg, P. A. Conti, J. Beech, Fatty acids modulate calcium-induced calcium release from skeletal muscle heavy sarcoplasmic reticulum fractions: Implications for malignant hyperthermia, *Biochem Cell Biol* 68:1195-1201 (1990).
15. K. S. Cheah and A. M. Cheah, Skeletal muscle mitochondrial phospholipase A_2 and the interaction of mitochondria and sarcoplasmic reticulum in porcine malignant hyperthermia, *Biochim Biophys Acta* 638:40-49 (1981).
16. J. E. Fletcher and H. Rosenberg, *In vitro* muscle contractures induced by halothane and suxamethonium: II. Human skeletal muscle from normal and malignant hyperthermia susceptible patients, *Br J Anaesth* 58:1433-1439 (1986).
17. J. E. Fletcher, H. Rosenberg, K. Michaux, L. Tripolitis and F. H. Lizzo, Triglycerides, not phospholipids, are the source of elevated free fatty acids in muscle from patients susceptible to malignant hyperthermia, *Eur J Anesth* 6:355-362 (1989).
18. S. C. Chow and M. Jondal, Polyunsaturated free fatty acids stimulate an increase in cytolsolic Ca^{2+} by mobilizing the inositol 1,4,5-trisphospahte-sensitive Ca^{2+} pool in T cells through a mechanism independent of phosphoinositide turnover, *J Biol Chem* 265:902-907 (1990).
19. C. Holm, T. G. Kirchgessner, K. L. Svenson, G. Fredrikson, S. Nilsson, C. G. Miller, J. E. Shively, C. Heinzmann, R. S. Sparkes, T. Mohanda, A. J. Lusis, P. Belfrage and M. C. Schotz, Hormone-sensitive lipase: Sequence, expression and chromosomal localization to 19 cent-q13.3, *Science* 241:1503-1506 (1988).
20. R. C. Levitt, V. A. McKusick, J. E. Fletcher and H. Rosenberg, Gene candidate (letter), *Nature* 345:297-298 (1990).

21. O. P. Hamill, A. Mary, E. Neher and B. Sakmann, Improved patch-clamp techniques for high-resolution current recording from cells and cell-free membrane patches, *Pflugers Arch* 391:85-100 (1981).

22. T. Probstle, R. Rudel and J. P. Ruppersberg, Hodgkin-Huxley parameters of the sodium channels in human myoballs, *Pflugers Arch* 412:264-269 (1988).

23. A. Trautmann, C. Delaporte and A. Marty, Voltage-dependent channels of human muscle cultures, *Pflugers Arch* 406:163-172 (1986).

24. C. Frelin, H. P. M. Vijverberg, G. Romey, P. Vigne and M. Lazdunski, Different functional states of tetrodotoxin sensitive and tetrodotoxin resistant Na^+ channels occur during the *in vitro* development of rat skeletal muscle, *Pflugers Arch* 402:121-128 (1984).

25. B. P. Bean, P. Shrager and D. A. Goldstein, Modification of sodium and potassium channel gating kinetics by ether and halothane, *J Gen Physiol* 77:233-253 (1981).

26. J. P. Ruppersberg and R. Rudel, Differential effects of halothane on adult and juvenile sodium channels in human muscle, *Pflugers Arch* 412:17-21 (1988).

27. R. L. Ruff, L. Simoncini and W. Stuhmer, Slow sodium channel inactivation in mammalian muscle: A possible role in regulating excitability, *Muscle and Nerve* 11:502-510 (1988).

28. D. Kim and D. E. Clapham, Potassium channels in cardiac cells activated by arachidonic acid and phospholipids, *Science* 244:1174-1176 (1989).

29. R. W. Ordway, J. V. Walsh, J. J. Singer, Arachidonic acid and other fatty acids directly activate potassium channels in smooth muscle cells, *Science* 244:1176-1179 (1989).

30. D. J. Linden and A. Routtenberg, Cis-fatty acids, which activate protein kinase C, attenuate Na^+ and Ca^{2+} currents in mouse neuroblastoma cells, *J Physiol (Lond)* 419:95-119 (1989).

31. P. S. Foster, E. Gesini, C. Claudianos, K. C. Hopkinson and M. A. Denborough, Inositol 1,4,5-trisphosphate phosphatase deficiency and malignant hyperpyrexia in swine, *Lancet* July 15:124-127 (1989).

32. J. R. Mickelson, E. M. Gallant, L. A. Litterer, K. M. Johnson, W. W. Rempel and C. F. Louis, Abnormal sarcoplamic reticulum ryanodine receptor in malignant hyperthermia, *J Biol Chem* 263:9310-9315 (1988).

33. E. Benoit, A. Corbier and J. M. Dubois, Evidence for two transient sodium currents in the frog node of Ranvier, *J Physiol (Lond)* 361:339-360 (1985).

34. D. R. Matteson and C. M. Armstrong, Evidence for a population of sleepy sodium channels in squid axon at low temperature, *J Gen Physiol* 79:739-758 (1982).

35. D. H. Wu, P. J. Sides and R. Narahashi, Interaction of deoxycholate with the sodium channel of squid axon membranes, *J Gen Physiol* 76:355-379 (1980).

36. B. Hille, "Ionic Channels of Excitable Membranes," Sinauer Associates, Sunderland, MA (1984) pp. 96-116,303-328.

37. J. R. Mickelson, H. S. Thatte, T. M. Beaudry, E. M. Gallant and C. F. Louis, Increased skeletal muscle acetylcholinesterase activity in porcine malignant hyperthermia, *Muscle and Nerve* 10:723:727 (1987).

38. R. Rudel, J. P. Ruppersberg and W. Spittelmeister, Abnormalities of the fast sodium current in myotonic dystrophy, recessive generalized myotonia, and adynamica episodica, *Muscle and Nerve* 12:281-287 (1989).

39. Y. Kurachi, H. Ito, T. Sugimoto, T. Shimizu, I. Miki and M. Ui, Arachidonic acid metabolites as intracellular modulators of the G protein-gated cardiac K^+ channel, *Nature* 337:555-557 (1989).

3

Effect of Halothane on Human Skeletal Muscle Sarcoplasmic Reticulum Calcium-Release Channel

Thomas E. Nelson

INTRODUCTION

In order to gain an understanding of how anesthetics alter muscle function, we have utilized the planar lipid bilayer technique for recording the conductance and gating properties of a calcium-release channel molecule derived from native skeletal muscle sarcoplasmic reticulum membranes. The incorporation of this protein molecule into an artificial membrane simplifies investigation, limiting anesthetic action to the bilayer itself and/or the single protein molecule. Thus, any effect the anesthetic has on the protein's conductance and gating functions will be a consequence of the anesthetic's action directly on the protein and/or on the lipid bilayer in such a way as to alter the protein's function. Our attempts to understand a piece of the puzzle of anesthetic action on skeletal muscle do, in fact, begin at the molecular, single protein level of attack and even so, this experimental model is far from simple, and the possible ways by which volatile anesthetics could alter its function represent a challenge. Our piece of the puzzle involves a large protein molecule located in the structure bridging the transverse tubule and the terminal cisternae of the sarcoplasmic reticulum (see Chapter 1, figure 1). This protein, sometimes referred to as the ryanodine receptor protein, is thought to play a significant role in excitation-contraction coupling of skeletal muscle. In our present studies, we have incorporated this protein into a planar lipid bilayer and have recorded and measured its properties as a cation-conducting channel. These studies parallel those on the pharmacogenetic disease malignant hyperthermia (MH) and illustrate how such a disease may lead to a better understanding of normal muscle response to volatile anesthetics.

In this chapter, preliminary results are presented which, it is hoped, will provide a stimulus for similar investigations into the effects of anesthetics on this calcium release system in other organs, especially cardiac and vascular smooth muscle.

METHODS AND MATERIALS

Sarcoplasmic Reticulum Membrane Isolation

Vastus lateralis muscle biopsies were obtained from patients referred to our Malignant Hyperthermia Diagnostic and Research Unit because of a history for this

THOMAS E. NELSON, Department of Anesthesiology, University of Texas Health Science Center, Houston, Texas 77030.

Mechanisms of Anesthetic Action in Skeletal, Cardiac, and Smooth Muscle
Edited by T.J.J. Blanck and D.M. Wheeler, Plenum Press, New York, 1991

21

genetic disease. The diagnostic contracture test protocol of the North American MH Diagnostic Group was followed.[1] Patients received either an epidural or general anesthetic, the latter utilizing thiopental, N_2O, and narcotic. Extra muscle for the isolation of sarcoplasmic reticulum membranes was obtained after informed consent, and the protocol was approved by the University Committee for the Protection of Human Subjects. The extra muscle, 1-5 g, was placed in ice cold homogenization solution immediately after excision and chopped into small pieces with scissors. Within an hour after excision, the muscle was homogenized 4 times for 15 sec each with a Polytron homogenizer. The homogenization solution (HEPES/PI) contained HEPES, 20 mM (pH 6.8) and protease inhibitors: PMSF, 200 μM; aminobenzamidine, 200 μM; aprotinin, 2 μg/ml; pepstatin, 2 μg/ml; leupeptin, 2 μg/ml; soybean trypsin inhibitors, 10 μg/ml; and iodoacetamide, 20 μM. The homogenate was centrifuged at 8,000 rpm in a Beckman JA-20 rotor for 30 min and the supernatant decanted. The myofibrillar pellet was rehomogenized in 5 volumes HEPES/PI and the initial centrifugation step repeated. The supernatants from these initial steps were combined and concentrated KCl added to produce a final concentration of 0.6 M in order to solubilize actomyosin. The KCl extract was centrifuged for 30 min at 33,000 rpm in a Beckman TI 45 rotor and the resultant pellet suspended in a solution of HEPES, 20 mM (pH 6.8); KCl, 150 mM; trehalose, 0.3 M and the protease inhibitors listed above. This mixture was centrifuged again for 30 min at 33,000 rpm and the final pellet transferred to a vial and stored at −85°C. This SR pellet is similar to what is commonly referred to as a heavy SR fraction containing a high percentage of terminal cisternae membranes and by SDS gel electrophoresis, the largest amount of the ryanodine binding protein. The pellets served as the source of SR membrane used in the single channel measurements detailed below.

Single Channel Reconstitution and Analysis

Details of the planar bilayer equipment and methodology have been reported previously.[2] Planar bilayers were formed across a 228 μm diameter aperture of a Delrin cup by applying a 7:3 mixture (50 mg/ml in decane) of palmitoyl-oleoyl-phosphatidylethanolamine: palmitoyl-oleoyl-phosphatidylcholine. Salt-agar electrodes electrically bridged the cis and trans sides of the bilayer to $Ag/AgCl_2$ electrodes attached to the amplifier head stage. SR membranes were applied to the cis side of the bilayer and this contained 3.5 ml of 250 mM $CsCH_3SO_4$, 10 mM CsHEPES (pH 7.4), CaEGTA, pCa = 5.2. The trans chamber contained 0.5 ml of the same solution except $CsCH_3SO_4$ was 50 mM. A membrane-bilayer fusion voltage pulse protocol (0 to −50 to +50 to 0 mV) was applied to enhance fusion of SR membranes into the bilayer. Membrane potentials are referenced to the trans side of the bilayer. After a Ca-release channel was observed, the incorporation protocol was changed to a data sampling protocol in which single channel data was collected at positive potentials (50 mV) during short (100 ms) repetitive pulses from 0 mV (0.5 to 0.05 Hz). Data acquisition software and hardware (pClamp, Tl-1 interface, Axon Instruments, Burlingame, CA) were computer interfaced. Data were digitized at 5-10 kHz, filtered at 3-4 kHz, and stored on floppy disk. Analysis software was provided by Dr. T. VanDongen and Dr. A. Brown (Baylor College of Medicine, Houston, TX) and supplemented by commercial programs (pClamp, Axon Instruments, Burlingame, CA). Drugs were added to the cis chamber and mixed by a magnetic stir bar. Halothane was prepared as a 28 mM stock solution in ethanol and added to the 3.5 ml cis chamber by 0.5 μl additions, each producing a concentration of 4 μM. After each addition of halothane, the cis chamber was covered and stirred for 1 min and measurements followed. Dead space volume for the covered cis chamber was 1.7 ml

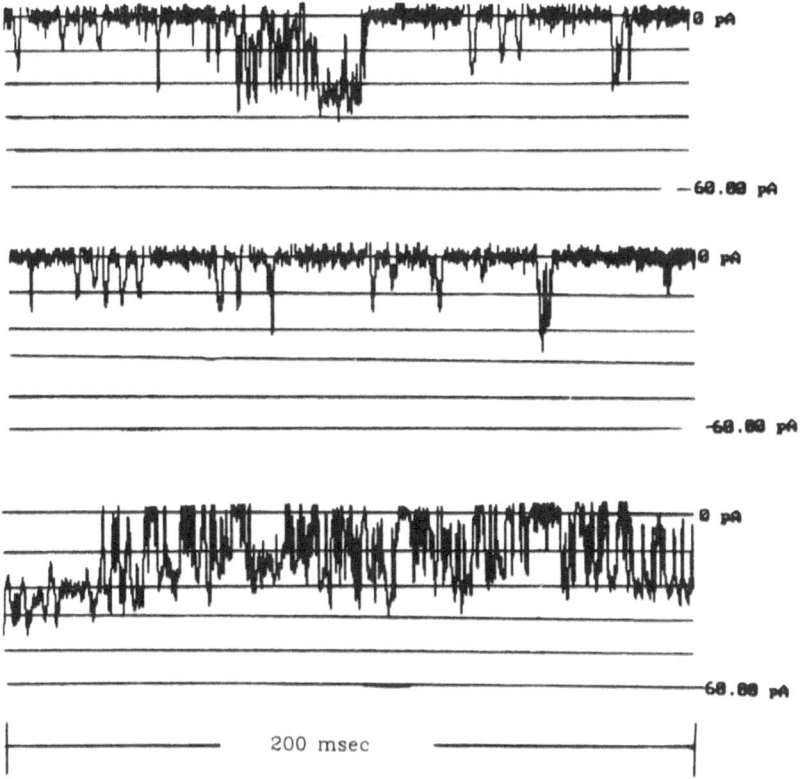

FIGURE 1. Open state levels of conductance through the human skeletal muscle calcium release channel. The figure is a series of traces selected from a single channel to illustrate the method of measuring open states. The top of each trace represents the channel in a closed state (0 pA) and when the channel opens, ion conductance produces a downward trace. The thin longitudinal lines are spaced at 12 pA intervals and open state levels can be observed near the 12, 24 and 36 pA lines.

and partition coefficients for halothane of 0.96 for the buffer/gas and of 100 for buffer/bilayer phases were assumed.[3] Each addition of halothane (4 μM initial) would, at equilibrium, produce approximately 2 μM halothane in the gas phase and this represents about 0.003 of MAC, the minimum alveolar concentration of halothane at which 50% of stimulated subjects do not respond.

Computer software developed by the Department of Molecular Physiology and Biophysics at the Baylor College of Medicine has provided us with the capabilities of detecting, measuring and analyzing the effects of halothane on the different open state levels of the Ca^{2+}-release channel. We selected 5 different open state levels at intervals of 12 pA each, providing data for analysis at openings from 0 to 12 pA, 0 to 24 pA, 0 to 36 pA, 0 to 48 pA and 0 to 60 pA. A portion of a channel record showing these open state levels is depicted in figure 1. For each of these open states, values for open probability (Po), integral, mean open and closed times with appropriate histograms, mean amplitude with histogram, and the number of observations (events) could be obtained. Once a channel was observed we usually obtained 2 control measurements after which halothane, 4 μM final concentration, was added to the cis chamber, the chamber covered and contents stirred for 1 min and then additional recordings were obtained.

Table 1. Descriptive Statistics for Properties of the Calcium Release Channel
from MH-Nonsusceptible (MHN) and MH-Susceptible (MHS) Human Skeletal Muscle

	Open Probability (%)	Slope Conductance for Cesium (pS)	Mean Open Time (msec)	Mean Closed Time (msec)
MHN	3.8 ± 0.9	480 ± 7	0.39 ± 0.04	11.0 ± 1.6
MHS	5.7 ± 1.7	467 ± 14	0.52 ± 0.11	11.1 ± 2.0

Values are means ± SEM. Data for MHN derives from 4 patients and 6 channels; data for MHS from 4 patients and 8 channels.

RESULTS

The human Ca^{2+}-release channel is similar, if not identical, in conductance and gating properties to those reported for the Ca^{2+}-release channels isolated from rabbit[4] and pig[5] skeletal muscle. The slope conductance is 480 ± 24 pS, and two time constants are fitted for each of the open and closed states.[2] The human Ca^{2+}-release channel, as well as those described in other species, is blocked by ruthenium red, $pCa^{2+} > 3$ and millimolar concentrations of Mg^{2+}, and is activated by ATP and caffeine, while ryanodine produces a fixed substate opening.[2] These conductance, gating and pharmacologic properties of the human Ca^{2+}-release channel associate it with the homo-tetramer protein complex commonly referred to as the ryanodine receptor.[6] We found no statistically significant differences between the properties of Ca^{2+}-release channels from MH-susceptible (MHS) and MH-nonsusceptible (MHN) human skeletal muscle (table 1). As illustrated in figure 2, skeletal muscle biopsies from these patients will produce *in vitro* isometric contractures when exposed to

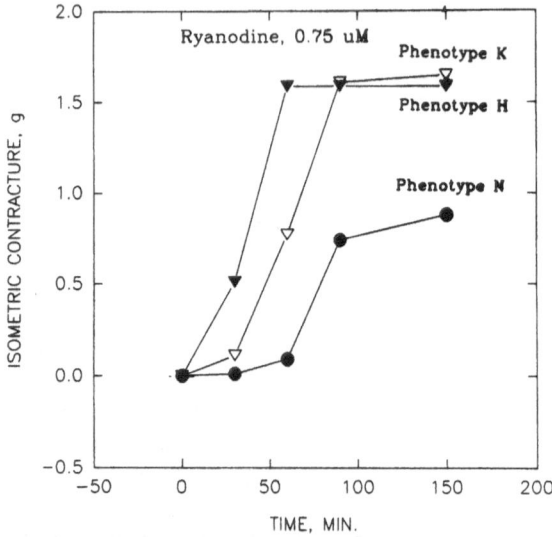

FIGURE 2. Ryanodine-induced contracture in malignant hyperthermia susceptible (MHS) and nonsusceptible (MHN) skeletal muscle. Biopsies of vastus lateralis muscle were obtained during an elective MH diagnostic procedure. Ryanodine 0.75 µM was added to the muscle bath (37°C), and the isometric contracture measured with a force transducer.

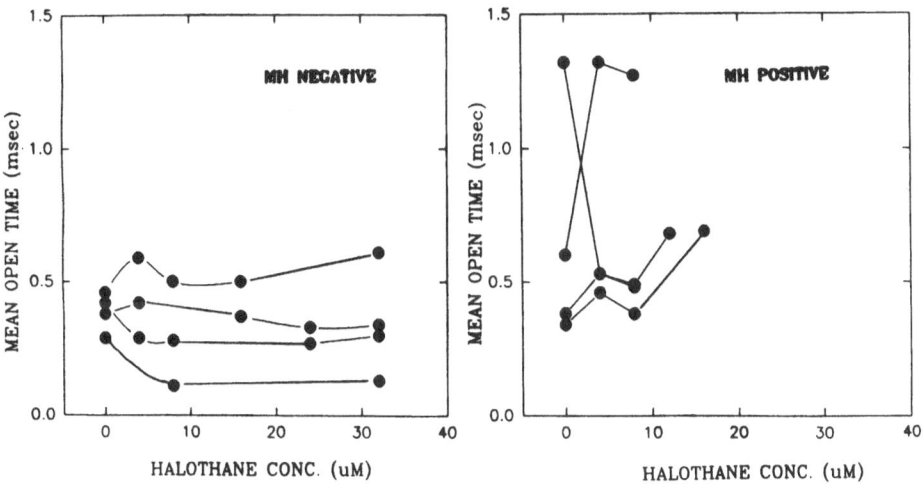

FIGURE 3. Halothane effect on the mean open time of the human calcium release channel. Each line represents a single channel treated with serial increases of the concentrations of halothane indicated.

ryanodine, and the MHS muscle is more sensitive to these contracture effects of ryanodine. If it is assumed that the contracture-producing effects of ryanodine are mediated by the ryanodine-binding Ca^{2+}-release channel, then this would suggest an abnormality for this protein in human MHS muscle. In the clinical arena, volatile anesthetics such as halothane, but not caffeine or ryanodine, are responsible for producing the MH syndrome. Therefore, if abnormal effects of halothane on the Ca^{2+}-release channel are observed, such results could be informative for how volatile anesthetics alter muscle function in healthy and diseased states.

The probability of the Ca^{2+} channel to reside in an open state, Po, was significantly increased by halothane (4-16 μM) in Ca^{2+}-release channels from MHS, but not from MHN muscle. This halothane-induced increase in Po of the MHS

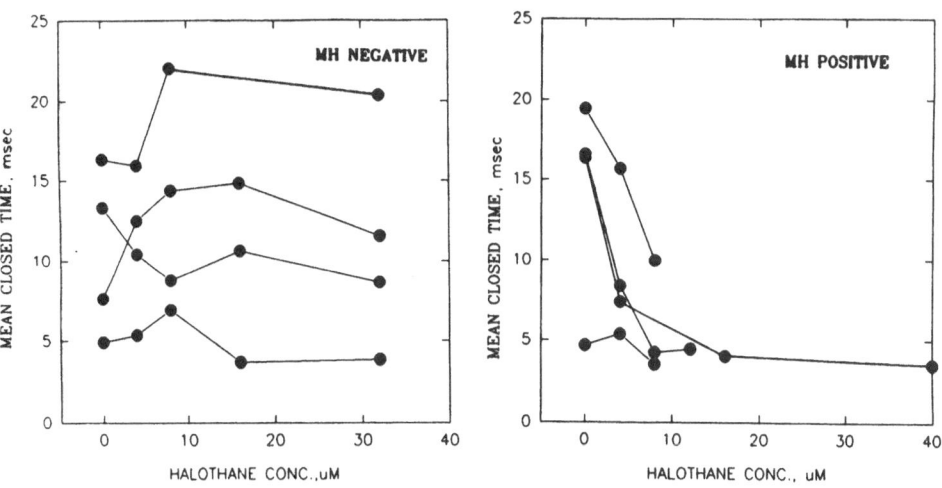

FIGURE 4. Halothane effect on mean closed time of single human calcium release channels. Data obtained and analyzed from the same channels depicted in figure 3.

25

Ca^{2+}-release channel was associated with an increase in the mean open time (figure 3) and a decrease in mean closed time (figure 4). Halothane did not significantly change either the mean open time (figure 3) or the mean closed time (figure 4) in the Ca^{2+}-release channel from MHN muscle. The effect of higher concentrations of halothane has not been tested on normal human Ca^{2+}-release channels, and it may be that similar effects would be observed in MHN channels treated with higher concentrations. The effect of halothane to decrease mean closed time and increase the mean open time and the Po is probably an effect on the gating properties of the MHS channel to produce more ion flux per channel, and if such an effect occurs in intact muscle, an increased myoplasmic Ca^{2+} would be expected. Representative single channel recordings from MHS and MHN Ca^{2+}-release channels in the presence and absence of halothane illustrate these halothane effects (figures 5 and 6). The single Ca^{2+}-release channel can be analyzed by selecting different open state level thresholds and measuring the frequency of their occurrence. Five arbitrary open state levels, each separated by 12 pA, were measured for each channel in the presence and absence of halothane (figure 1). In the absence of halothane, single MHN channels were

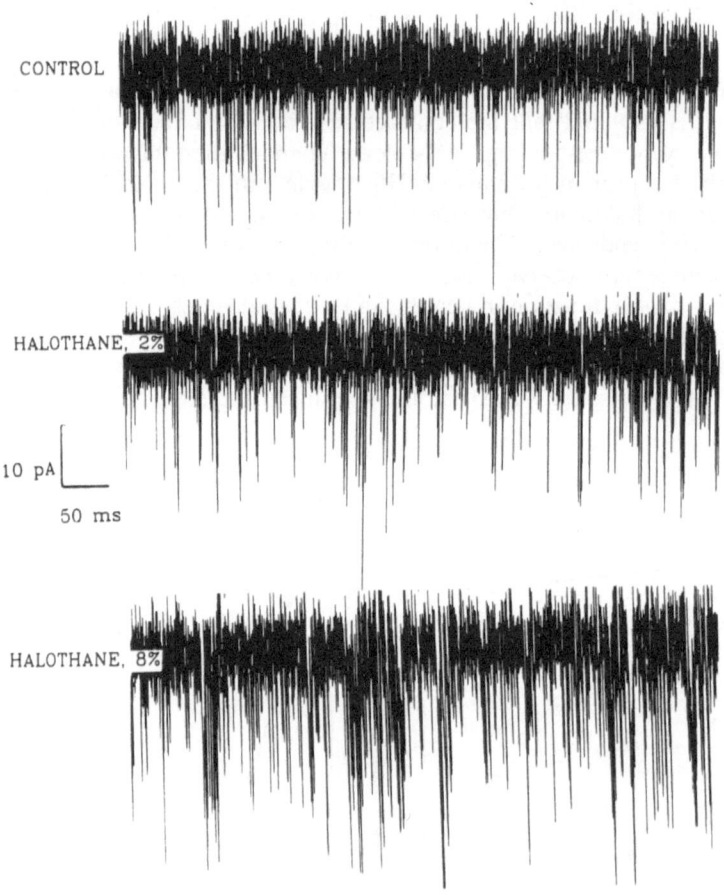

FIGURE 5. Halothane-treated MHN human calcium release channel. Each trace was recorded from the same channel prior to (control) and after treatment with halothane. The upper thick part of each trace is the zero level of conductance and the downward spikes represent rapid channel events of opening and closing. The values corresponding to 2% and 8% represent 8 and 16 μM concentrations of halothane in solution at equilibrium. No discernable halothane effects were observed.

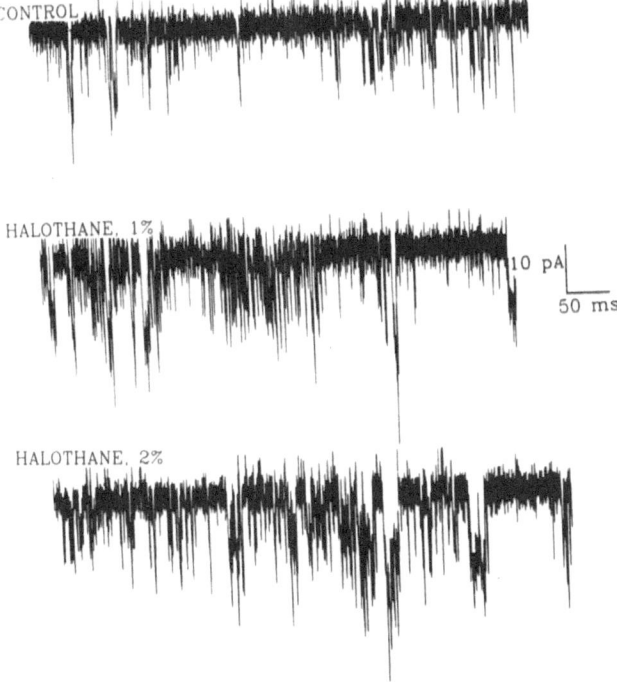

FIGURE 6. Halothane-treated MHS human calcium release channel. See figure 5 for explanation of traces. Halothane, 1% (2 μM final) and 2% (4 μM final) increases the probability of this channel to reside in an open state by increasing the number of channel openings (decreased mean closed state time) and the length of time the channel remains in the open state. Gaps in the upper baseline (lower trace) represent prolonged open state dwell times produced by halothane.

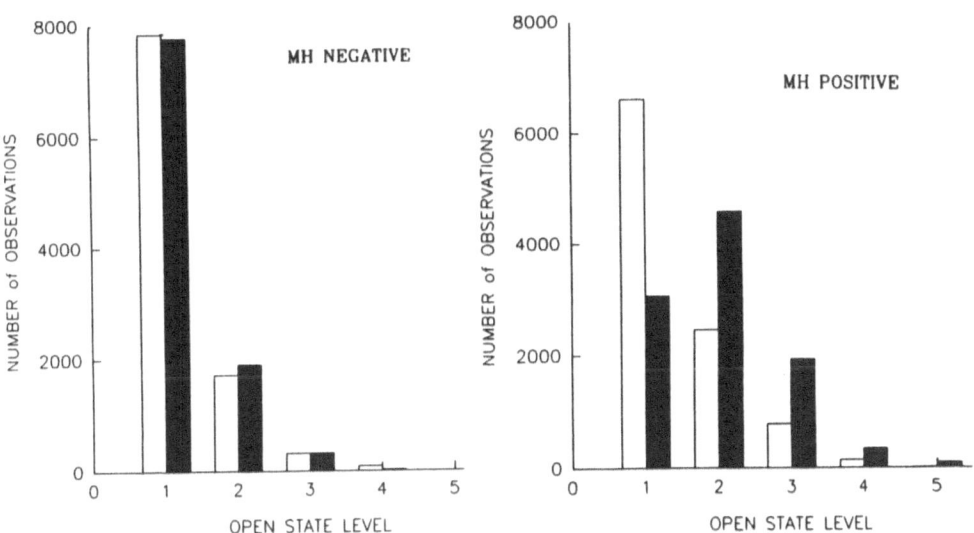

FIGURE 7. Halothane effect on open state levels of MHN and MHS human calcium release channels. Compared to MHN channels, those from MHS muscle have more openings to higher levels of conductance, and halothane (filled bars) produced an increase in the number of these larger conductance levels. Halothane (filled bars) did not alter the conductance levels in MHN channels. Total observations (MHN = 8552, MHS = 9417) were normalized to 10,000 for each diagnostic group and for each there were 6 channels from 3 different patients.

27

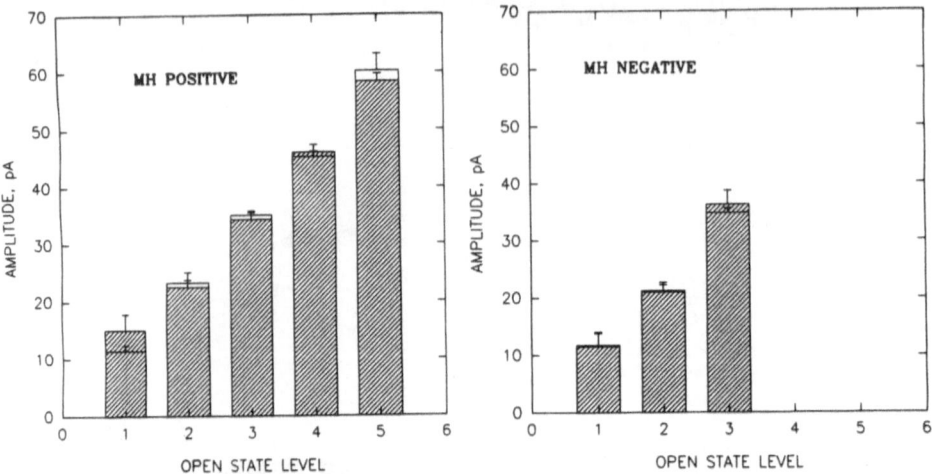

FIGURE 8. Halothane effect on the mean amplitude of each open state level of MHN and MHS channels. The values were measured in the same channels depicted in figure 7. Measured amplitudes in the presence of halothane (shaded bars) did not differ from those in the absence of halothane (open bars overlapping) in calcium release channels from MHS and MHN human skeletal muscle. Means for MHS and MHN represent for each, 6 channels from 3 different patients.

observed with a distribution of 78.4%, 17.2%, 3.2%, 0.9% and 0.3% for open state levels 1, 2, 3, 4 and 5 respectively. As shown in figure 7, halothane, 4 μM, had no effect on the distribution of these open state levels in Ca^{2+}-release channels from MHN muscle. In the absence of halothane, MHS Ca^{2+}-release channels were observed at 66.0%, 24.6%, 7.8%, 1.4% and 0.1% for open state levels 1, 2, 3, 4 and 5, respectively. Compared to channels from MHN human muscle, those from MHS muscle had more events at levels 2, 3 and 4 and fewer at levels 1 and 5 (figure 7). Halothane, 4 μM, markedly decreased the number of events observed at open state level 1 (-53.6%) while increasing the number of events measured at open state levels 2 ($+85.9\%$), 3 ($+147\%$), 4 ($+137\%$) and 5 ($+850\%$). The numbers in parentheses represent percent change from control values. The amplitude for each open state level for MHS and MHN channels was not affected by halothane (figure 8).

DISCUSSION

Halothane, enflurane and isoflurane concentrations equal to or less than 0.01 MAC were shown in our previous study to alter the Ca^{2+}-release channel in an intact, isolated SR membrane vesicle preparation.[7] The incorporation of single protein molecules of this Ca^{2+}-release channel into a planar lipid bilayer and measurement of their channel conductance properties show a similar, low-anesthetic-concentration effect with the protein from malignant hyperthermia susceptible patients. Most studies on anesthetic action have used concentrations of anesthetics close to or in excess of MAC values and it is exceptional to find such dramatic effects of an anesthetic at values far below the effective anesthetic concentration. The sarcoplasmic reticulum membrane is an excellent model for studying anesthetic action because within the same membrane, functional effects of anesthetics on several different channels (*i.e.*, Ca^{2+}, Na^+, K^+, Cl^-) and on the Ca^{2+}-Mg^{2+}-ATPase calcium pump can be measured. The concentration dependence (ED_{50}) of these volatile anesthetics on the Ca^{2+}-Mg^{2+}-

ATPase calcium pump protein was shown to be about 5 times greater than the ED_{50} effect for the Ca^{2+}-release channel.[7] This isolated membrane model suggests that anesthetic action on Ca^{2+} regulation in muscle cells of a patient undergoing anesthesia may change during induction and after the anesthetic is discontinued. In these circumstances when anesthetic concentration is changing, systems affected by low (subMAC) concentrations may be expressed only on induction of, or long after discontinuing the anesthetic. When the desired level of anesthesia is obtained, systems affected by MAC and subMAC anesthetic concentrations may be expressed. Thus, multiple end organ sites of anesthetic action may have different anesthetic concentration dependencies and the net effect observed may differ at different levels of anesthetic concentration in the patient. Thus, anesthetic action in skeletal muscle may not be the continuation of a single process changing with the concentration of anesthetic, but it may reflect anesthetic action on several different processes, each having a different anesthetic concentration dependency.

The Ca^{2+}-release channel that we have recorded in this study was only recently discovered and little is known about its functional properties. It is a receptor-regulated, as opposed to voltage-regulated, channel and several regulatory ligands have been identified. Calmodulin,[8,9] Mg^{2+},[2] ruthenium red,[2] and tetracaine[10] each reduces the conductance of the Ca^{2+}-release channel. Ryanodine produces a prolonged substate level opening,[11] while ATP[11] and caffeine[2] produce marked increases in conductance of the Ca^{2+}-release channel. The ryanodine receptor protein has been cloned[12] and based on its amino acid sequence and on hydrophobicity assessments, transmembrane segments and sidedness for regulatory sites have been postulated.[12] Each of these regulatory sites could represent potential targets for the action of anesthetics. The halothane effect(s) that we observed on this Ca^{2+}-release channel may be limited to a defect in the MH channel protein or it could be a change in sensitivity to a halothane effect in the normal Ca^{2+}-release channel protein complex. The answer to this question must await future experiments in which we plan to test higher concentrations of halothane on the Ca^{2+}-release channel from normal human muscle.

Accepting the possibility that the effects of halothane on MH Ca^{2+}-release channels may reflect effects on a normal Ca^{2+}-release channel, what are the possible mechanisms involved? Two halothane effects on this channel are obvious. First, the number of events (openings and closures) is markedly increased by halothane, and this is associated with an increase in the open state time and a decrease in the time the channel is in a closed state. Since this channel is considered to be a major source of Ca^{2+} for contraction, the net effect would be to increase contractility of skeletal muscle. It is interesting to note that biopsied human skeletal muscle twitch tension is increased by *in vitro* exposure of the muscle to halothane[13] and that MH human muscle undergoes a contracture when exposed to halothane.[14] Associated with the halothane-induced increased activity of the Ca^{2+} channel is a change in the conductance state level of the channel. In the absence of halothane, the channel conducts at lower state amplitudes, with very few large open state levels occurring. After exposure to halothane, there is a shift from lower to higher conductance levels and the most probable explanation for this is for halothane to affect the gating mechanism of the channel. Basically, it appears that halothane produces an effect that favors greater conductance levels by increasing how widely the gate is opened. Since the amplitude for each open state level was not changed by halothane, the most likely explanation is a shift from lower to higher conductance states by opening the channel gate to a more opened, wider state.

These data are very preliminary and a considerable amount of speculation about their possible meaning has been exercised. The new technological advances in single channel recording and analysis provide many exciting opportunities for testing hypotheses regarding anesthetic action.

ACKNOWLEDGEMENTS

The single channel recordings were made possible through the patience, guidance and generosities of Drs. Enrico Stefani and Mike Fill. Marina Lin provided technical help and Margie Choate helped in manuscript preparation. This study was supported by the Department of Anesthesiology, University of Texas Health Sciences Center, Houston.

REFERENCES

1. M. G. Larach, Standardization of the caffeine halothane muscle contracture test, *Anesth Analg* 69:511-515 (1989).
2. M. Fill, E. Stefani and T. E. Nelson, Abnormal human sarcoplasmic reticulum Ca^{2+} release channels in malignant hyperthermia skeletal muscle, *Biophys J* (in press).
3. L. L. Firestone, J. C. Miller and K. W. Miller, Appendix: Tables of physcial and pharmacological properties of anesthetics, *in:* "Molecular and Cellular Mechanisms of Anesthetics," S. H. Roth and K. W. Miller, eds., Plenum Publishing, New York (1986) p. 455.
4. J. S. Smith, T. Imagawa, J. J. Ma, M. Fill, K. Campbell and R. Coronado, Purified ryanodine receptor from rabbit skeletal muscle is the calcium release channel of the sarcoplasmic reticulum, *J Gen Physiol* 92:1-26 (1988).
5. M. Fill, J. R. Coronado, J. Mickelson, J. M. Vilven, B. A. Jacobson and C. F. Louis, Abnormal ryanodine receptor channels in malignant hyperthermia, *Biophys J* 57:471-476 (1990).
6. A. Saito, M. Inui, M. Radenmacher, J. Frank and S. Fleischer, Ultrastructure of the calcium release channel of sarcoplasmic reticulum, *J Cell Biol* 107:211-219 (1988).
7. T. E. Nelson and T. Sweo, Ca^{2+} uptake and Ca^{2+} release by skeletal muscle sarcoplasmic reticulum: Differing sensitivity to inhalational anesthetics. *Anesthesiology* 69:571-577 (1988).
8. T. E. Nelson, Effect of calmodulin on calcium pulse-induced calcium release from fragmented skeletal sarcoplasmic reticulum, *Fed Proc* 43:498 (1984).
9. G. Meissner, Evidence for a role for calmodulin in the regulation of calcium release from skeletal muscle sarcoplasmic reticulum, *Biochemistry* 25:244-250 (1986).
10. S. T. Ohnishi, Calcium-induced calcium release from fragmented sarcoplasmic reticulum, *J Biochem* 86:1147-1150 (1979).
11. G. Meissner, Adenine nucleotide stimulation of Ca^{2+}-induced Ca^{2+} release in sarcoplasmic reticulum, *J Biol Chem* 259:2365-2374 (1984).
12. H. Takeshima, S. Nishimura, T. Matsumoto, H. Ishida, K. Kangawa, N. Minamino, N. Mastuo, M. Ueda, M. Hanaoka, T. Hirose and S. Numa, Primary structure and expression from complementary DNA of skeletal muscle ryanodine receptor, *Nature* 339:439-445 (1989).
13. T. E. Nelson and M. A. Denborough, Studies on normal human skeletal muscle in relation to the pathopharmacology of malignant hyperpyrexia, *Clin Exp Pharm Physiol* 4:315-322 (1977.)
14. F. R. Ellis, D. G. F. Harriman, N. P. Keany, Kyei-Mensah and J. H. Tyrrell, Halothane-induced muscle contracture as a cause of hyperpyrexia. *Brit J Anaesth* 43:721-722 (1971).

4

Ca Release from Skeletal Muscle SR
Effects of Volatile Anesthetics

G. Salviati, S. Ceoldo, G. Fachechi-Cassano, R. Betto

INTRODUCTION

The effects of volatile anesthetics on the physiology of skeletal muscle have been extensively studied. However, conflicting results have been presented, probably due to different experimental models and different concentrations of the drugs used. Often these studies address volatile anesthetic effects at concentrations well above those required for clinical anesthesia.

A great number of studies have focused on the effects of halothane on the sarcoplasmic reticulum (SR) from both skeletal and cardiac muscle. Halothane can release calcium from the sarcoplasmic reticulum,[1-3] possibly by potentiating calcium-induced calcium release.[4-7] An inhibitory effect of halothane on SR calcium transport has been reported,[8] probably due to inhibition of the SR calcium-activated ATPase,[9] although in other studies halothane increased the activity of this enzyme.[10] Because of these conflicting results, we have re-examined the mechanism of action of halothane and other general anesthetics on the SR, using chemically skinned rabbit muscle fibers with the aim to identify the target of these drugs on the SR.

It is now generally accepted that the calcium efflux channel of the SR is formed by the ryanodine receptor, which is located on the junctional face of the terminal cisternae of the SR where it forms the structures called "feet." Smith et al.[11] reported that the activity of the channel is modulated by calmodulin, in addition to Mg^{2+} and ATP. Our results indicate that halothane activates calcium efflux from the SR by interacting with calmodulin.

MATERIALS AND METHODS

SR Ca Uptake and Release in Skinned Fibers

Chemically skinned fibers were obtained from a fast-twitch muscle, the psoas muscle of New Zealand rabbits. Chemical skinning was carried out according to the methods of Wood et al.[12] Single fibers were then isolated with the help of a dissecting

G. SALVIATI, S. CEOLDO, G. FACHECHI-CASSANO, R. BETTO, Istituto di Patologia generale, Universita' di Padova and NRC Unit for Muscle Biology and Physiopathology, Via Trieste 75, Padova, Italy

Mechanisms of Anesthetic Action in Skeletal, Cardiac, and Smooth Muscle
Edited by T.J.J. Blanck and D.M. Wheeler, Plenum Press, New York, 1991

microscope. Fibers were transferred to a chamber containing 1.0 ml of a "relaxing" (R) solution (0.17 M K^+, 2.5 mM Mg^{2+}, 5 mM ATP, 5 mM EGTA, 10 mM imidazole, pH 7.0) and were attached to two clamps. The fibers were then incubated with a Ca-loading solution (composition given below). Ca-loading activity of sarcoplasmic reticulum was measured by following the increase in light scattering after the addition of 10 mM K pyrophosphate or 5 mM oxalate (pH 7.0) to the Ca-loading solution. In the presence of either of these two precipitating anions, active Ca transport leads to the formation of Ca-anion crystals in the SR lumen, and to a progressive increase in the light scattering of the fiber. The change in light scattering is proportional to the increase in calcium content.[13] The experimental setup has been previously described.[14] The Ca-loading solutions were modified R solutions (0.17 M K^+, 2.5 mM Mg^{2+}, 5.0 mM ATP, 5 mM EGTA, 10 mM imidazole, pH 7.0) containing 5 mM oxalate and total Ca 3.27 mM or 10 mM K pyrophosphate with different total Ca concentrations (1.60-4.89 mM). The concentration of Mg-ATP was 0.65 mM. Free Mg^{2+} was 90 μM in the experiments where oxalate was present and 30 μM in those where K pyrophosphate was present. The free pCa range was 6.4 to 5.0. Apparent dissociation constants were from Orentlicher et al.[15]

Calcium efflux from SR was measured by following the decrease of light scattering of fibers preloaded with Ca^{2+} in the presence of oxalate. Ca-loading was carried out by incubating fibers at pCa 6.0 in the presence of 5 mM oxalate. When the light scattering signal attained a plateau level, calcium efflux was initiated by incubating in pCa 7.0 or 5.6 solutions which did not contain Mg^{2+} or ATP. The absence of these substrates also prevented contraction of the fiber, which would have interfered with light scattering measurements.

Tension Measurements

For tension measurements, single fibers were isolated and transferred to a chamber containing 1.0 ml of the "relaxing" (R) solution described above. The fibers were attached to two clamps, one of which was connected to a tension transducer as described elsewhere.[18] Fibers were exposed to Ca^{2+}-loading solution (pCa 6.6) for 30 sec, then incubated in a modified R solution without EGTA and challenged with stepwise increasing concentrations of caffeine until tension development was observed (caffeine threshold). When the fiber relaxed, maximum tension was measured by adding 20 mM caffeine. All activities were measured at room temperature (23-25°C).

Other Measurements

The activity of SR Ca^{2+}-ATPase was measured spectrophotometrically with the enzyme-coupled assay of Warren et al.[16] These experiments were performed at 25°C on purified SR membrane fractions derived from longitudinal tubules,[17] in the presence of 1.5 μM calcium ionophore A23187 to abolish back inhibition on the Ca pump by the accumulated calcium.

SDS-gel electrophoresis was carried out on the discontinuous gel system of Laemmli.[19] The separating gel was a 10-20% polyacrylamide linear gradient gel. After electrophoresis, gels were stained with silver.

Protein was measured with the method of Lowry et al.[20] using bovine serum albumin as standard. ATP, EGTA, and imidazole were purchased from Sigma Chemical Co., St Louis, MO. Calmodulin was from Boehringer, Mannheim, W. Germany. Thymol-free halothane was a generous gift of Dr. T. J. J. Blanck. All other chemicals were analytical grade.

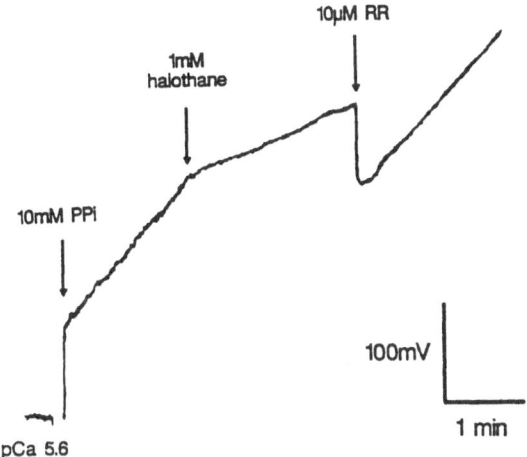

FIGURE 1. Halothane inhibition of the rate of SR calcium-loading activity supported by pyrophosphate (PPi) and its reversal by ruthenium red (RR). The light scattering by a chemically skinned single rabbit skeletal muscle fiber was measured at pCa 5.6 in the presence of 10 mM K pyrophosphate. The decrease in light scattering after the addition of ruthenium red is due to the absorbance of light by this compound.

RESULTS

Effects of Halothane on SR Calcium Loading

The effect of halothane on the SR Ca-loading activity of chemically skinned rabbit fast-twitch fibers was tested during the linear phase of Ca-loading in the presence of 10 mM K pyrophosphate (figure 1). The addition of 1 mM halothane inhibited the calcium loading rate by about 50 per cent. This inhibitory effect could be due to the inhibition of the Ca pump of the SR. This possibility was tested by measuring the effects of halothane on the SR calcium-activated ATPase. Halothane, at the same concentration (1 mM), did not inhibit the activity of the calcium ATPase when measured on an SR longitudinal tubule preparation, which lacks the calcium release channel. The enzyme activities were 3.30 and 3.42 μmol ADP/mg protein/min in the absence and in the presence of 1 mM halothane, respectively.

Since light scattering measures net calcium accumulation into the SR, *i.e.*, the difference between the total Ca^{2+} transported into the SR lumen and the Ca^{2+} released, the above results suggest that halothane reduces the net Ca loading of the SR not by inhibiting the calcium pump but by increasing Ca^{2+} efflux from the SR. This conclusion is further supported by the interaction of ruthenium red (RR) and halothane as shown in figure 1. The inhibitory effect of halothane on the rate of calcium loading was completely counteracted by the addition of 10 μM ruthenium red. Since RR is a well known inhibitor of the Ca-efflux channel of SR terminal cisternae,[21] these results indicate that 1 mM halothane neither perturbs the lipid bilayer nor activates non-specific calcium efflux pathways, but rather acts at the efflux channel. The inhibitory effect of halothane is dose-dependent (figure 2), a small inhibitory effect being already evident at 250 μM halothane. However, halothane did not abolish the calcium-loading activity of the SR. In all experiments maximum inhibition was about 50%. This limited effect could be due in part to the fact that in the medium containing 10 mM K pyrophospate, the concentration of free Mg^{2+} was very low (30 μM). It is known that Mg^{2+} has a closing effect on the channel.[21] Therefore, halothane could have a limited effect because the channel was already partially open. This was tested by assaying the effect of halothane on the SR calcium loading activity supported by 5 mM oxalate. Under these conditions, the concentration of free Mg^{2+} was higher (0.1 mM) than in the experiments using pyrophosphate. Halothane was more potent with oxalate as the precipitating anion; inhibition of Ca uptake was apparent at halothane concentrations as low as 100 μM. However, the maximum inhibition was again not more than 50%. Thus, the limitation on the halothane effect is not likely related to Mg^{2+}. Other general anesthetics such as isoflurane and enflurane showed effects similar to those of halothane under the same experimental conditions. However, enflurane was the less effective. In the presence of 10 mM K pyrophosphate, the inhibition by 500 μM was only 8%.

Calcium Dependence of the Effects of Halothane

It has been reported that the halothane effect on SR calcium efflux is Ca-dependent.[4,7] We indirectly analyzed the effect of halothane at different calcium concentrations by measuring the effect of RR on the rate of calcium loading in the presence of K pyrophosphate. The rationale for this experiment was the following: the activation of the rate of calcium loading by RR is maximal when the channel is fully open whereas no activation by RR can be observed when the channel is closed (see for example ref. 18). As shown in figure 3, RR did not modify the rate of calcium loading at pCa's higher than 6.0. The activating effect of RR was maximum at pCa 5.4-5.2. Halothane shifted the curve to the left, the maximum being achieved at pCa 5.6. Similar results were obtained by using isoflurane (not shown). These results agree with previous results[4,7] in showing that halothane and isoflurane activate the calcium-induced calcium release mechanism of the SR.

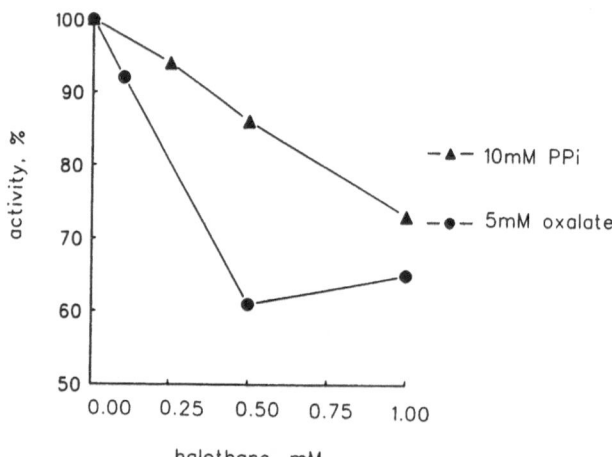

FIGURE 2. Dose-dependence of halothane inhibition of the rate of anion-supported SR calcium loading. The rate of calcium loading was measured by light scattering at pCa 5.6 in the presence of 10 mM K pyrophosphate (▲) or 5 mM K oxalate (•).

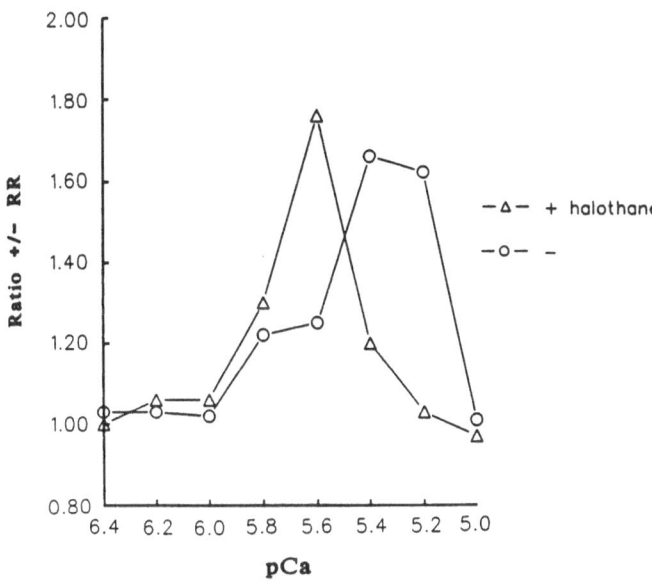

FIGURE 3. Calcium dependence of halothane effects on the rate of SR calcium loading. Calcium loading rate was measured by light scattering in the presence of 10 mM K pyrophosphate in the absence (○) and in the presence (△) of 500 μM halothane. When the rate of increase of light scattering was linear, 10 μM ruthenium red was added and the activation of the rate of calcium loading was measured. Data are expressed as the ratio between the rate measured in the presence and in the absence of RR.

Effects of Halothane on SR Calcium Release

The results so far obtained suggest that the main effect of halothane on skeletal muscle SR is that of activating the release of calcium. However, the experiments provided only indirect evidence of this phenomenon. To test directly the effects of halothane on SR calcium release, skinned muscle fibers were allowed to accumulate calcium oxalate in the SR lumen (figure 4). When a plateau level of Ca-loading was attained, the loading solution was rapidly exchanged with a releasing solution (a pCa 7.0 solution that contained no Mg^{2+} or ATP). Under these conditions there was a release of calcium which exhibited an exponential decay and was completely abolished by the addition of RR (figure 4). As shown in table 1 and figure 4B, addition of 1 mM

Table 1. Halothane Activates the Rate of SR Calcium Efflux

pCa	Rate Constant (min⁻¹)		Initial Efflux Rate (nmol/sec/g muscle)	
	Control	Halothane (1 mM)	Control	Halothane (1 mM)
7.0	0.26	0.54	92	258
5.6	0.48	0.93	127	641

Calcium efflux was measured as described in figure 4.

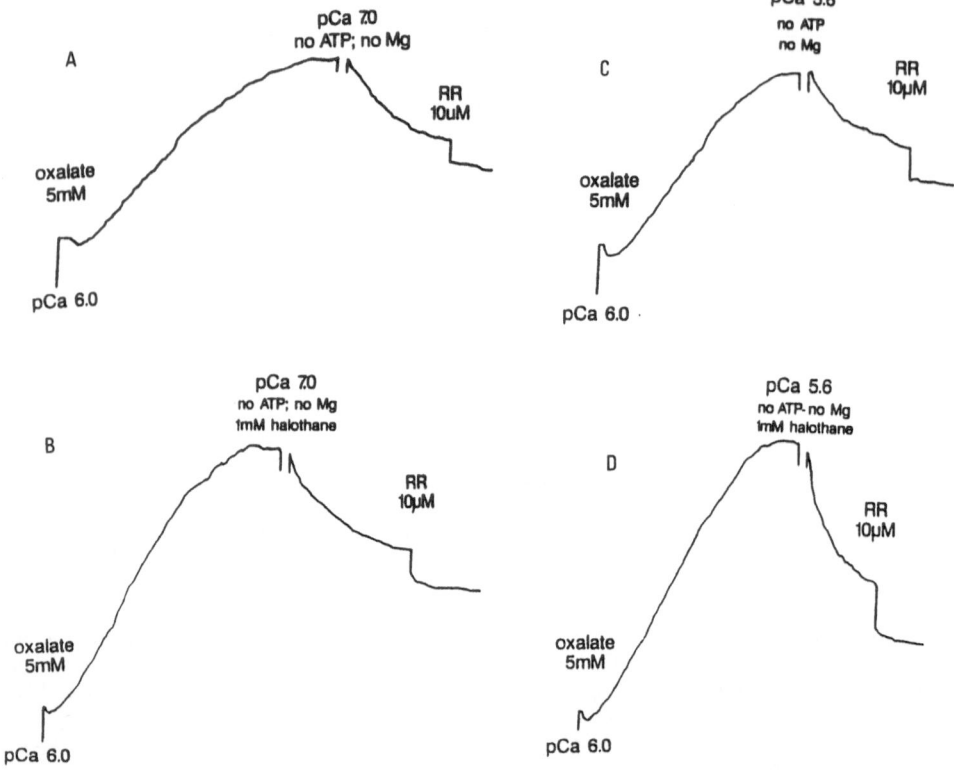

FIGURE 4. Effects of halothane on the rate of SR calcium release. Active SR calcium loading and calcium release were measured by light scattering. Calcium loading was carried out by incubating at pCa 6.0 in the presence of 5 mM K oxalate. When the plateau level was attained, the calcium loading solution was rapidly exchanged with a calcium releasing solution (pCa 7.0) that did not contain ATP and Mg^{2+}. A: pCa 7.0, no halothane; B: pCa 7.0, 1 mM halothane; C: pCa 5.6, no halothane; D: pCa 5.6, 1 mM halothane.

halothane to the release medium increased by two to three times the first order rate constant of calcium efflux. The release of calcium was again inhibited by RR. We also tested the calcium dependence of the rate constant of release. Figure 4 and table 1 show that the rate constant of release at pCa 5.6 was higher than that at pCa 7.0 both in the absence (figure 4C) and in the presence (figure 4D) of 1 mM halothane. Table 1 shows the calcium dependence of the halothane effect on the initial rate of calcium release. Halothane increased the initial rate of calcium release both at pCa 7.0 and 5.6. However, at pCa 5.6 the effect of the drug was twice that at pCa 7.0.

The Effects of Halothane Are Not Completely Reversible

A skinned fiber, after loading the SR with calcium by incubating in the absence of precipitating anions at pCa 6.4 for 30 sec, was challenged with caffeine, a well known calcium releasing agent. Calcium release, as indicated by tension development, was obtained when the caffeine concentration was raised to 3 mM. The addition of 20 mM caffeine released all the loaded calcium. The same fiber was then exposed to 1 mM halothane in relaxing solution. After 2 min, halothane was removed by washing with relaxing solution not containing the drug, and the fiber was again challenged with

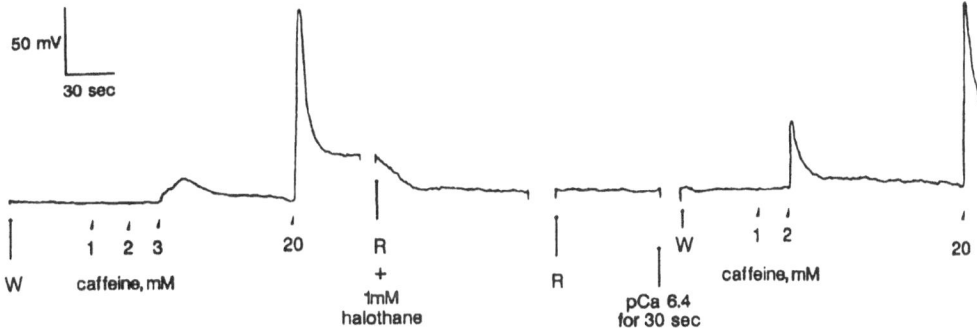

FIGURE 5. Pretreatment with halothane increases the sensitivity to caffeine of the SR of rabbit skinned muscle fibers. Caffeine-induced calcium release from the SR was measured indirectly by following tension development, after loading the SR by incubating at pCa 6.4 for 30 sec as described in Materials and Methods. "R" represents incubation in the relaxing solution described in the text, and "W" is a modified relaxing solution, containing no EGTA.

caffeine after loading the SR with calcium. As shown in figure 5, preincubation with halothane decreased the concentration of caffeine necessary to cause the release of calcium. Furthermore, more calcium was released by the threshold concentration of caffeine. On the other hand, preincubation with halothane did not change the amount of calcium loaded into the SR, as indicated by the tension developed by 20 mM caffeine. Preincubation with 1 mM halothane in relaxing solution increased the activating effects of RR on the rate of calcium loading in the presence of K pyrophosphate (not shown). These results suggest that halothane may affect the calcium release mechanism of the SR by interacting with some modulator(s) of the activity of the calcium channel.

Calmodulin Counteracts the Effects of Halothane

It has been shown that calmodulin modulates the activity of the calcium release channel of the SR of skeletal and heart muscle.[11] Calmodulin decreases the open probability without affecting the ion conductance.[11] Figure 6 shows that the addition of 2 μM calmodulin to the calcium release solution at pCa 5.6 decreased the rate of efflux. In agreement with the results of Meissner,[22] the first order rate constant (k) was decreased by two-fold ($k = 0.25$ min^{-1} vs. a control k of 0.48 min^{-1}). In the presence of 1 mM halothane, calmodulin was even more effective in decreasing the rate of calcium efflux ($k = 0.36$ min^{-1} vs. a k in halothane alone of 0.93 min^{-1}). Accordingly, the initial rates of calcium release were also decreased by calmodulin to 63 and 89 nmoles calcium/sec/g muscle, in the absence and in the presence of halothane, respectively (i.e., two and seven times lower than those in the absence of calmodulin; see table 1). These results suggest that halothane may interact with endogenous calmodulin bound to the SR[22] and remove the modulator effect of the protein on the SR calcium-release channel. This hypothesis was tested in the following experiment. A purified membrane preparation derived from the SR terminal cisternae[17] was incubated with 1 mM halothane at 4°C. After 30 min, the membrane were pelleted by centrifugation. Both the supernatant and the pellet were analyzed by SDS gel electrophoresis. As shown in figure 7, membrane purified from SR terminal cisternae contained a protein band of an estimated molecular weight of 17,000 Daltons, co-migrating with purified calmodulin. Membranes sedimented after

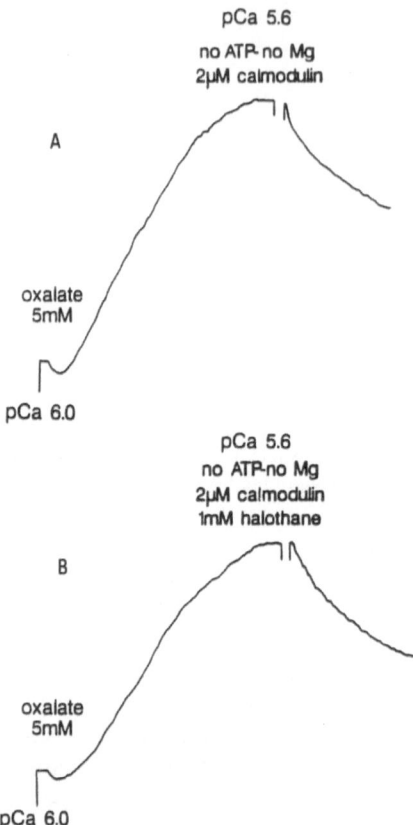

FIGURE 6. Calmodulin reverses the activating effects of halothane on SR calcium release. Calcium release from the SR was measured as described in figure 4. The calcium releasing solution (pCa 5.6) contained 2 μM calmodulin. A: no halothane; B: 1 mM halothane.

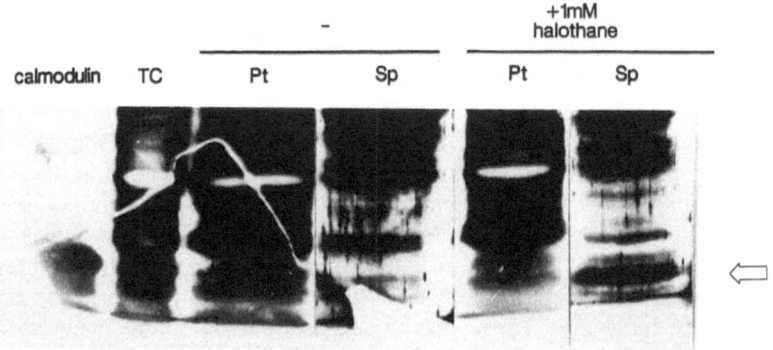

FIGURE 7. Halothane dissociates calmodulin from the SR membranes. Purified SR terminal cisternae membrane preparation (1 mg protein/ml) was incubated at 4°C with 1 mM halothane in R solution (final volume was 0.2 ml). After 30 min the membrane were sedimented by centrifuging at 100,000 g for 10 min in a Beckman Airfuge. The pellet was resuspended in 0.2 ml of 10 mM imidazole buffer. Total pellet protein and total supernatant protein were analyzed by SDS gel electrophoresis on a 10-20% polyacrylamide linear gradient. After electrophoresis the gel was stained with silver. Only the low molecular weight region is shown. Key: calmodulin, purified calmodulin; TC, untreated terminal cisternae preparation; Pt, pellet; Sp, supernatant.

incubation with 1 mM halothane contained a reduced amount of this protein which, on the other hand, was enriched in the supernatant. These results indicate that halothane released calmodulin from the SR membranes. Similar results were obtained by incubating the terminal cisternae membrane preparation with 10 μM trifluoperazin, a calmodulin antagonist (not shown).

DISCUSSION

The results indicate that halothane, at clinical concentrations, releases calcium from the sarcoplasmic reticulum of chemically skinned rabbit skeletal muscle fibers by activating the calcium-induced calcium release mechanism. This effect is mediated at least in part by the interaction of halothane with calmodulin and the dissociation of calmodulin from the SR membranes.

Our results show that halothane inhibits the anion-supported calcium loading activity of the SR. An inhibition of the calcium loading rate of the SR could be explained by: 1) the inhibition of the calcium pump activity; 2) an activation of calcium efflux from the SR through the calcium release channel; 3) an increase of the permeability of the SR membranes. Our results demonstrate that halothane, at concentrations up to 1 mM, does not inhibit the calcium ATPase of the SR. In these experiments we used highly purified membrane preparations derived from longitudinal tubules of the SR. These membranes contain a high density of the calcium pump protein, but are almost completely devoid of the calcium release channel (the ryanodine receptor) which is localized to the junctional membrane of the SR terminal cisternae.[23] Furthermore, the assay was carried out in the presence of the calcium ionophore A23187 to prevent the inhibitory effect on the calcium pump by the calcium transported into the SR lumen. Under these conditions, no effects of halothane on the activity of the calcium pump were observed. These results can explain the divergent results already published and suggest the need of very controlled experimental conditions, including the source of membrane preparations, for this type of study. Halothane inhibition of the rate of anion-supported calcium loading of the SR was completely reversed by ruthenium red. Since RR is an inhibitor of the SR calcium release channel,[21] these results suggest that the main target of halothane on the SR membranes is the physiological calcium release pathway. On the other hand, these results argue strongly against the hypothesis that clinical concentrations of halothane may affect the non-specific permeability characteristics of the SR membrane.

Our results show that the maximum inhibition by halothane of the rate of calcium loading of the SR of skinned rabbit fibers is only about 50%. This result can be explained by taking into account the fact that in skinned fibers the SR is intact[24] and composed by longitudinal tubules and terminal cisternae. Since the target of halothane, the calcium release channel, is located on the terminal cisternae, it is likely that the calcium loading activity of longitudinal tubules is not affected. It has been shown that the volume of the longitudinal tubules of the SR of fast-twitch muscle of guinea pig[25] is about twice that of terminal cisternae. If one assumes that the rate of calcium transport is the same in longitudinal tubules and in terminal cisternae and that halothane, by opening the calcium release channel, completely abolishes the net accumulation of calcium into the terminal cisternae without affecting that of longitudinal tubules, then the maximum inhibition should not be greater than 30-50%.

In agreement with previous results,[4,7] our results show that the effect of halothane is calcium dependent and that halothane activates the calcium-induced calcium release mechanism by increasing the affinity for calcium of the gating system of the channel. This suggests that halothane may interact directly with the calcium release channel of the SR. However, our results also demonstrate that halothane treatment causes permanent changes in the behavior of the channel of skinned fibers. Therefore, it is likely that halothane may interact with some modulator(s) of the activity of the channel. Our results demonstrate that calmodulin counteracts almost completely the calcium releasing effects of halothane on the SR. It has been shown that calmodulin inhibits the release of calcium from isolated terminal cisternae preparations.[22] The inhibition is not mediated by the phosphorylation of the channel by a calmodulin-dependent protein kinase, since the calmodulin effect can be demonstrated in the absence of ATP (ref. 22, and our results). Calmodulin decreases two- to three-fold the first order rate constant of calcium efflux by decreasing the opening probability of the channel.[11]

Our results suggest that one mechanism by which halothane induces release of calcium from the SR is that of removing calmodulin bound to the SR membrane. The molecular structure of calmodulin is similar to a dumbbell with a central portion consisting of an alpha-helix structure connecting the two lobes where the calcium binding sites are located.[27] The calcium-induced conformational changes allow this portion of the molecule to interact with enzymes, peptides and pharmacological agents such as phenothiazines. In sarcoplasmic reticulum preparations from heart muscle, calmodulin has been shown to bind to the ryanodine receptor.[28] Halothane may bind to this domain of calmodulin, thus inhibiting the interaction of the protein with the ryanodine receptor. Once calmodulin is removed, the calcium channel open time would increase with a consequent increase of calcium efflux from the SR.

It is interesting that the effects of calmodulin are greater in the cardiac SR.[11] This can be explained by assuming that heart and skeletal muscle express different isoforms of the ryanodine receptor and that the two isoforms have different affinity for calmodulin. Recent results support such an interpretation by showing that cardiac and skeletal muscle ryanodine receptors are not identical and are coded by separate genes.[26] These difference may represent the molecular basis of a greater sensitivity of heart muscle to halothane. It is very likely that the effects we found *in vitro* would occur also *in vivo* with a consequent increase in the myoplasmic concentration of Ca^{2+}. Interestingly, it has been reported that treatment with micromolar concentrations of calmodulin antagonists induced contracture in both normal and malignant hyperthermia pig muscles and pre-treatment with these drugs potentiated the response to halothane.[29]

REFERENCES

1. A. Takagi, Abnormaltiy of sarcoplasmic reticulum in malignant hyperpyrexia, *Adv Neurol Res* 20:107-113 (1976).
2. Y. Ogawa and N. Kurebayashi, The Ca-releasing action of halothane on fragmented sarcoplasmic reticulum, *J Biochem* 92:899-905 (1982).
3. T. Beeler and K. Gable, Effect of halothane on Ca^{2+}-induced Ca^{2+} release from sarcoplasmic reticulum vesicles isolated from rat skeletal muscle, *Biochim Biophys Acta* 821:142-152 (1985).
4. M. Endo, S. Yagi, T. Ishizuka, K. Horiuti, Y. Koga, and K. Amaha, Changes in the Ca^{2+}-induced Ca^{2+} release mechanism in the sarcoplasmic reticulum of the muscle from a patient with malignant hyperthermia, *Biomed Res* 4:83-92 (1983).
5. D. H. Kim, F. A. Sreter, S. T. Ohnishi, J. F. Ryan, J. Roberts, P. D. Allen, L. G. Meszaros, B. Antoniu, and N. Ikemoto, Kinetic studies of Ca^{2+} release from sarcoplasmic reticulum of normal and malignant hyperthermia susceptible pig muscles, *Biochim Biophys Acta* 755:320-324 (1984).
6. J. R. Mickelson, J. A. Ross, B. K. Reed, and C. F. Louis, Enhanced Ca^{2+}-induced Ca^{2+} release by isolated sarcoplasmic reticulum vesicles from malignant hyperthermia susceptible pig muscle, *Biochim Biophys Acta* 862: 318-328 (1986).

7. L. Carrier and M. Villaz, Effects of halothane on calcium release from sarcoplasmic reticulum of rabbit psoas and semitendinosus skinned muscle fibers, *Biochem Pharmacol* 39:145-149 (1990).
8. J. J. A. Heffron and G. A. Gronert, Effect of halothane (2-bromo-2-chloro-1,1,1-trifluoroethane) on calcium binding and release by sarcoplasmic reticulum, *Biochem Soc Trans* 7:44-47 (1979).
9. N. Kurebayashi and Y. Ogawa, Effect of halothane on the calcium activated ATPase reaction of fragmented sarcoplasmic reticulum in reference to the Ca releasing action, *J Biochem* 92:907-913 (1982).
10. E. M. Diamond and M. C. Berman, Effect of halothane on the stability of Ca^{2+} transport activity of isolated fragmented sarcoplasmic reticulum, *Biochem Pharmacol* 29:375-381 (1980).
11. J. S. Smith, E. Rousseau and G. Meissner, Calmodulin modulation of single sarcoplasmic reticulum Ca^{2+}-release channels from cardiac and skeletal muscle, *Circ Res* 64:352-359 (1989).
12. D. S. Wood, J. R. Zollman, J. P. Reuben, and P. W. Brandt, Human skeletal muscle: Properties of chemically skinned fibers, *Science* 187:1075-1076 (1975).
13. M. M. Sorenson, J. P. Reuben, A. B. Eastwood, M. Orentlicher and G. M. Katz, Functional heterogeneity of the sarcoplasmic reticulum within sarcomeres of skinned muscle fibers, *J Membrane Biol* 53:1-17 (1980).
14. G. Salviati, M. M. Sorenson and A. B. Eastwood, Calcium accumulation by the sarcoplasmic reticulum in two populations of chemically skinned human muscle fibers: Effects of calcium and cyclic AMP, *J Gen Physiol* 79:603-632 (1982).
15. M. Orentlicher, P. W. Brandt and J. P. Reuben, Regulation of tension in skinned muscle fibers: Effect of high concentrations of MgATP, *Am J Physiol* 233:C127-C1374 (1977).
16. G. B. Warren, P. A. Toon, N. J. M. Birdsall, A. G. Lee and J. C. Metcalfe, Reconstitutiion of a calcium pump using defined membrane components, *Proc Nat Acad Sci* (USA) 71:622-626 (1974).
17. A. Saito, S. Seiler, A. Chu and S. Fleischer, Preparation and morphology of sarcoplasmic reticulum terminal cisternae from rabbit skeletal muscle, *J Cell Biol* 99:875-885 (1984).
18. G. Salviati and P. Volpe, Ca^{2+} release from sarcoplasmic reticulum of skinned fast- and slow-twitch muscle fibers, *Am J Physiol* 254:C459-C465 (1988).
19. U. K. Laemmli, Cleavage of structural proteins during assembly of head of bacteriophage T4, *Nature* 227:680-685, 1970).
20. O. H. Lowry, N. J. Rosebrough, A. L. Farr and R. J. Randall, Protein measurement with the folin phenol reagent, *J Biol Chem* 193:265-275 (1951).
21. J. R. Smith, R. Coronado and G. Meissner, Sarcoplasmic reticulum contains adenine nucleotide-activated calcium channels, *Nature* 316:446-449 (1985).
22. G. Meissner, Evidence of a role for calmodulin in the regulation of calcium release from skeletal muscle sarcoplasmic reticulum, *Biochemistry* 25:244-251 (1986).
23. S. Fleischer, E. M. Ogumbunmi, M. C. Dixon and E. A. M. Fleer, Localization of Ca^{2+} release channels with ryanodine in junctional terminal cisternae of sarcoplasmic reticulum in fast skeletal muscle, *Proc Natl Acad Sci* (USA) 82:7256-7259 (1985).
24. A. B. Eastwood, D. S. Wood, K. R. Bock and M. M. Sorenson, Chemically skinned mammalian skeletal muscle: I. Structure of skinned rabbit psoas, *Tissue & Cell* 11:553-566 (1979).
25. B. R. Eisenberg and A. M. Kuda, Stereological analysis of mammalian skeletal muscle: II. White vastus muscle of adult guinea pig, *J Ultrastructure Res* 51:176-187 (1975).
26. K. Otsu, H. F. Willard V. K. Khanna, F. Zorzato, N. M. Green and D. H. MacLennan, Molecular cloning of cDNA encoding the Ca^{2+} release channel (ryanodine receptor) of rabbit cardiac sarcoplasmic reticulum, *J Biol Chem* 15:13472-13483 (1990).
27. Y. S. Babu, J. S. Sack, T. J. Greenhough, C. E. Bugg, A. R. Means and W. J. Cook, 3-dimensional structure of calmodulin, *Nature* 315:37-40 (1985).
28. S. Seiler, A. D. Wegener, D. H. Wang, D. R. Hathaway and L. R. Jones, High molecular weight proteins in cardiac and skeletal muscle junctional sarcoplasmic reticulum vesicles bind calmodulin, are phosporylated, and are degraded by Ca^{2+}-activated protease, *J Biol Chem* 259:8550-8557 (1984).
29. S. P. Collins, M. D. White and M. A. Demborough, Calmodulin antagonist drugs and porcine malignant hyperpyrexia, *Clin Exp Pharmacol Physiol* 15:473-477 (1988).

5

Halothane-Cooling Contractures and Regulation of the Myoplasmic Ca^{2+} Concentration in Skeletal Muscle

Roberto Takashi Sudo, Guilherme Suarez-Kurtz

INTRODUCTION

Malignant hyperthermia (MH) is a pharmacogenetic syndrome which may be triggered in susceptible individuals by drugs, such as halothane and succinylcholine, commonly used by anesthesiologists.[1] Two standard procedures are presently used for investigating susceptibility to MH (MHS) in humans[2,3] and swine.[4] Both of these procedures rely on the increased sensitivity of muscle fragments obtained from biopsies to the contractile effects of halothane and/or caffeine. Caffeine- and halothane-induced contractures of isolated skeletal muscles are highly temperature-dependent. Reducing the temperature of the bathing medium from 37°C to 25°C reduces the ability of halothane to elicit tension in MHS swine muscles[5] and decreases caffeine-induced tension in both normal or MHS human muscles.[6]

In contrast, Sakai and his coworkers,[7,8] have shown that rapid cooling to below 10°C potentiated the caffeine-induced tension of isolated frog or mammalian muscles. These "rapid-cooling contractures" (RCC) were attributed to a combined inhibitory effect of caffeine and cooling on Ca^{2+} sequestration by the sarcoplasmic reticulum (SR). Results from our laboratory[9] revealed that cooling (2-4°C) elicited contractures in frog muscles exposed *in vitro* to halothane concentrations comparable to those present clinically. This observation raised our interest in the possibility of using these "halothane-cooling contractures" (HCCs) as a paradigm for investigating the effects of halothane on the regulation of the myoplasmic Ca^{2+} concentration and muscle contractility. Because halothane is frequently the triggering agent of MH episodes in susceptible individuals, and because the halothane concentrations required for the HCC are comparable to those present clinically, we suggested[10,11] that the HCC could provide a simple, reproducible experimental model for investigating MH susceptibility. In addition, the HCC may prove to be useful for studying drugs of potential therapeutic value in the management of the MH episodes.

In this article we review our data on the HCCs of different muscle preparations *in vitro*, including intact and denervated extensor digitorum longus (EDL) and soleus muscles of mouse, fragments of human muscle biopsies and chemically-skinned[12]

ROBERTO TAKASHI SUDO, GUILHERME SUAREZ-KURTZ, Departamento de Farmacologia Basica e Clinica, Universidade Federal do Rio de Janeiro, Rio de Janeiro, RJ - 21941, Brazil.

Mechanisms of Anesthetic Action in Skeletal, Cardiac, and Smooth Muscle
Edited by T.J.J. Blanck and D.M. Wheeler, Plenum Press, New York, 1991

innervated and denervated rabbit muscle fibers. The experimental procedures used in these studies were described in detail in the the original publications.[9-11,13-15]

HCC IN MAMMALIAN EDL AND SOLEUS MUSCLES

Figure 1 shows the protocol used for eliciting HCCs in isolated muscle preparations. Cooling from 23°C to 2°C had no effect on the resting tension of mouse EDL muscles superfused with control saline, but elicited a small ($2.0 \pm 0.4\%$ of P_o, the maximum tetanic tension; $N = 23$), transient contracture (cooling-induced contracture, CC) in the soleus muscle (figure 1A). Preequilibration of the muscles with halothane (0.3%) at room temperature markedly potentiated the CC of the soleus and endowed the EDL with the ability to contract upon cooling. The HCCs of both muscles were transient, and tension returned to baseline during the cooling period (figure 1B). The HCCs could be reversed upon superfusion of the muscles with control saline for 15 to 30 min (figure 1C).

The HCCs are highly reproducible and can be repeated 8-10 times in the same muscles, provided that 15-20 min are allowed between successive challenges.[10] The amplitude of the HCCs depends on several factors, of which the most important are the halothane concentration, the final cooling temperature and the muscle fiber type. Muscles containing predominantly slow-twitch fibers, such as the mouse soleus, develop larger HCCs for any given halothane concentration (0.3 - 2.0%) and temperature

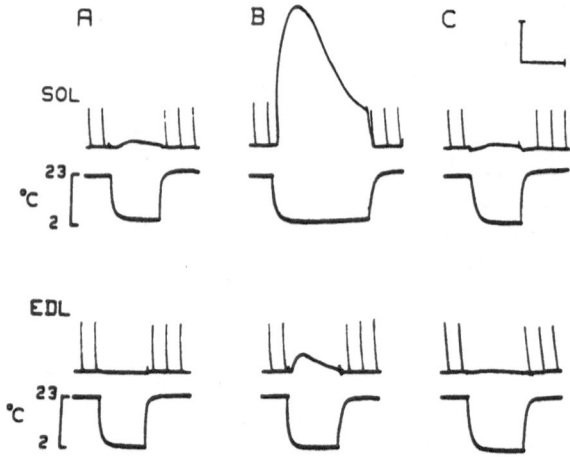

FIGURE 1. Cooling-induced contractures (CC) in soleus (SOL) and EDL muscles. In all panels, the upper trace is the isometric tension recording and the lower trace is the temperature of the bathing medium. The muscles were cooled from 23 to 2°C before the exposure to halothane (A), after 10 min of equilibration with 0.7% halothane (B), and 30 min after the return to control saline (C). The procedure used for applying halothane to the muscle in this and all subsequent figures was the following (cf. ref. 10): a mixture of halothane (0.1-2.0%) and oxygen obtained from a calibrated vaporizer was bubbled at a flow rate of 1.5 l/min into a stoppered flask (200-300 ml) with 80% of its capacity filled with the appropriate saline solution. After 10 min of equilibration with the anesthetic mixture and as bubbling continued, this solution was pumped into the muscle chamber at constant rate for 10 min. The muscle was then rapidly cooled as described in the text. The procedures described in this legend for cooling and for equilibrating the muscles with different concentrations of halothane were also used in the experiments depicted in figures 2-4 and 8-10. Reprinted from R. T. Sudo, M. R. B. Souza, G. Zapata and G. Suarez-Kurtz, *J Pharmacol Exp Ther* 237:600-697 (1986) by permission of copyright holder and authors.

(2-10°C) than muscles, such as the EDL, in which the fast-twitch fibers predominate.[10,11] The greater susceptibility of the slow-twitch fibers to the CC (figure 1A) and the HCC (figure 1B) may be related to their lower threshold for activation by Ca^{2+} at 3-5°C,[16] and/or to intrinsic differences in SR function between fast and slow mammalian fiber types.[17,18]

In both EDL and soleus muscles, the amplitude of the HCC is dose-dependent, threshold responses being observed with 0.3% halothane and maximal contractures occurring after equilibration with 1.5 - 2.0% halothane (figure 2A). The HCC_{50} of both soleus and EDL muscles requires pre-equilibration at room temperature with 0.7% halothane, which corresponds closely to the MAC for this general anesthetic in humans. The amplitude of the HCC_{50} is equivalent to that of the caffeine-cooling

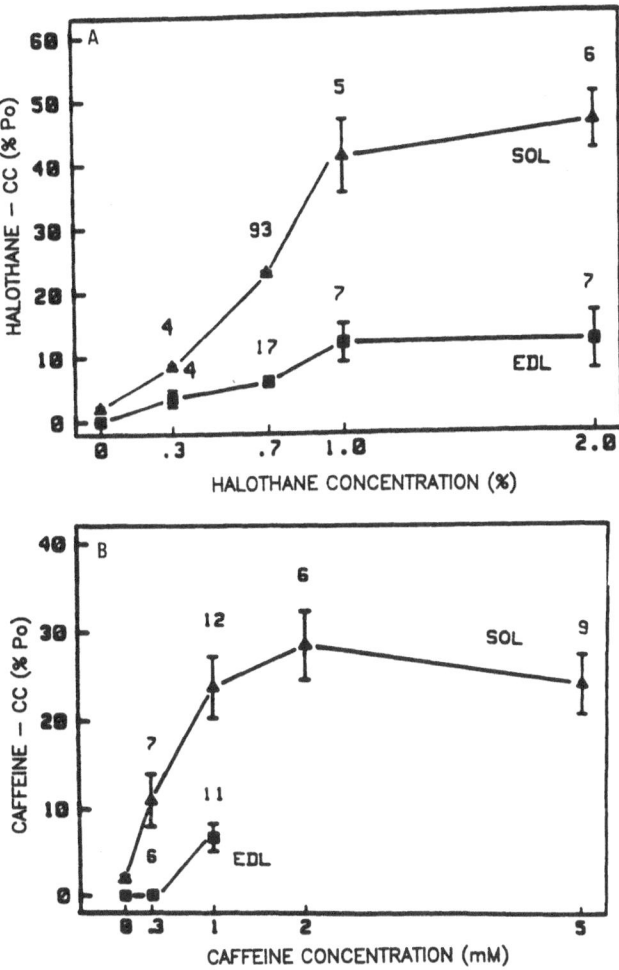

FIGURE 2. Dose-response curves for the HCC and the caffeine-cooling contractures (CAFF-CC) of mouse EDL and soleus muscles. The contractures were elicited by cooling to 3°C after the muscles had been equilibrated for 10 min with increasing concentrations of halothane or caffeine. The amplitude of the contractures is expressed (mean ± SEM) relative to the maximum tetanic tension (P_o) for each muscle. The number of muscles for each point is indicated above the error bars. Reprinted from R. T. Sudo, M. R. B. Souza, G. Zapata and G. Suarez-Kurtz, *J Pharmacol Exp Ther* 237:600-607 (1986) by permission of copyright holder and authors.

contractures of mouse EDL or soleus muscles equilibrated with 1 mM caffeine (figure 2B).

HCC IN HUMAN MUSCLE BIOPSIES

We have previously reported[11] that HCCs can be elicited in muscles biopsies obtained from patients with no muscle disease, undergoing elective surgery under spinal anesthesia. These results have been recently confirmed in muscle fragments obtained from patients being tested for MH. Figure 3 shows data from three patients, who had a negative response to the standard North American caffeine-halothane muscle contracture test. The upper part of figure 3 shows the muscle twitches at 22°C followed by several contractures elicited during cooling to 2°C in the absence or in the presence of increasing concentrations (0.2 - 3.0%) of halothane. A plot of the amplitude of the HCCs versus the halothane concentration is shown in the lower part of figure 3. It is significant that the halothane concentrations required for just-detectable HCCs (0.2%) and for maximum HCCs (1 - 2%) in the human biopsies corresponded closely to those reported for comparable effects on mouse soleus and EDL muscles (figure 2).

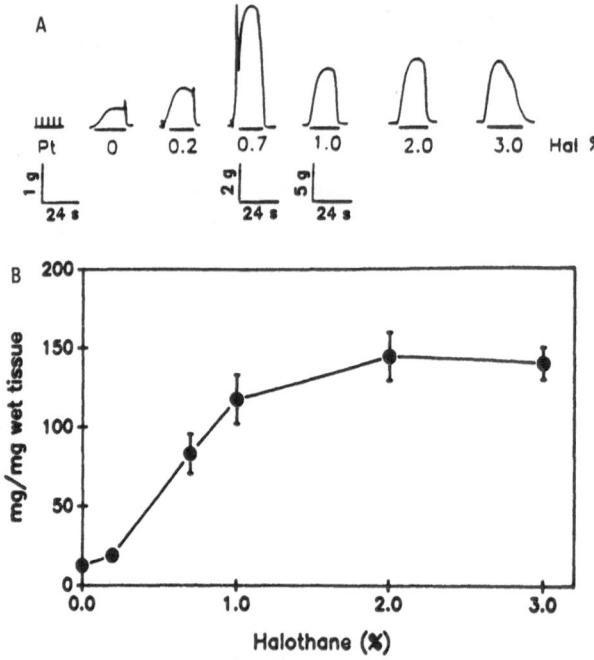

FIGURE 3. HCC in human vastus lateralis muscle. The experiment was performed at 22°C, except during the cooling periods. A: isometric tension recordings from a muscle strip. The first panel shows the electrically-evoked twitches at 22°C. The subsequent panels show contractures induced by cooling from 22°C to 2°C in the absence (second panel) or in the presence of increasing concentrations of halothane. The cooling period in this and in subsequent figures is indicated by the horizontal bars under the tension tracings. B: the amplitude of the HCC (expressed in mg of tension per mg of wet weight of the muscle strip) *versus* the halothane concentration in the superfusate. Data are expressed as means ± SEM for six muscle strips from three MH normal patients.

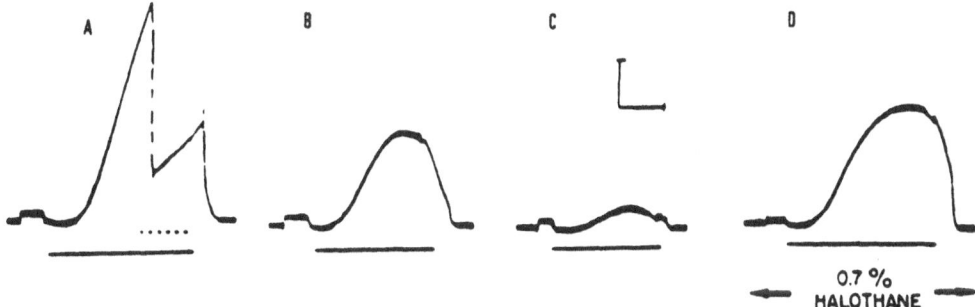

FIGURE 4. CC of human muscle. Recordings from a fascicle of rectus abdominis muscle removed from a pacient under general anesthesia with halothane. A was recorded *ca.* 45 min after the muscle biopsy; the large CC saturated the tension recording system, and its sensitivity was reduced 2.5 times during the period indicated by the dotted line. B and C were recorded after perfusion of the fascicle in vitro with the control saline for additional 30 and 60 min periods, respectively. Note the progressive decline in the amplitude of the CC. D shows the HCC recorded after 10 min equilibration of the fascicle *in vitro* with 0.7% halothane. The experiment was performed at 22°C except during the cooling periods. Reprinted from R. T. Sudo, G. Zapata and G. Suarez-Kurtz, *Can J Physiol Pharmacol* 65:697-703 (1987) by permission of copyright holder and authors.

The experiment shown in figure 4, performed on a muscle fascicle obtained from a patient under halothane anesthesia for programmed surgery, provides direct evidence in support of our contention that the HCCs can be elicited in the presence of clinically relevant concentrations of halothane. Each panel in figure 4 shows contractures elicited by cooling to 3°C. Panels A, B and C were recorded 45, 75 and 105 min, respectively, after removal of the muscle fascicle from the patient, and during superfusion with control (halothane-free) saline. Lowering the temperature of the saline to 2°C elicited contractures, the amplitude of which declined progressively between A and C. The large contracture in figure 4A can be described as an HCC, due to the presence in the muscle fibers of the halothane administered *in vivo* to the patient. The progressive decline in amplitude between A and C is attributed to the washout of halothane from the muscle during the superfusion with control saline *in vitro*. After recording panel C, in which the amplitude of the contracture elicited by cooling had declined to *ca.* 6% of its amplitude in A, halothane (0.7%) was added to the superfusion saline, and 10 min later the muscle was again cooled to 2°C (figure 4D). The amplitude of the ensuing HCC was smaller than that recorded in figure 4A, strongly suggesting that the halothane concentration required for the HCC_{50} in mammalian muscles (0.7%, see above) are less than those present in skeletal muscle during general anesthesia with halothane.

Experiments are now in progress to explore the possibility of using the HCC to detect MH susceptiblity in humans. Because the SR of MHS muscles is more sensitive to the Ca^{2+} releasing effects of halothane than normal muscle,[19] and because cooling enhances markedly this effect of halothane, we would predict that lower concentrations of halothane, possibly equivalent to those present clinically, will elicit HCCs in MHS muscle, as compared to normal muscle.[15] The possibility of using the HCC for the detection of MH susceptibility is of interest for several reasons: First, in contrast to the currently used tests for MH susceptibility, HCCs can be elicited in muscles exposed to halothane concentrations comparable to those present clinically. Second, HCCs are completely reversible, allowing comparison of the HCC data with other contracture tests for MH susceptibility in the same muscle fragment. Third, the HCCs are highly reproducible over several hours of experimentation at 22°C, and dose-response curves can be readily constructed.

MECHANISM OF THE HCC

Experiments with intact muscles provided important information regarding the mechanism of generation of the HCC. These contractures could be elicited in muscles rendered inexcitable by prolonged membrane depolarization with high-KCl solutions or by exposure to the local anesthetic lidocaine;[10,11] actually, lidocaine potentiated the HCC of frog and mammalian muscles[9,10] (see below). We concluded from these results that the HCC does not require functional integrity of the excitation-contraction coupling process. Since the ability of mammalian muscles to develop HCCs was not affected by brief (2 min) exposures to Ca^{2+}-free saline containing EGTA[10] or by the Ca^{2+}-entry blocker D600 (Sudo, unpublished results), we concluded that the HCC does not involve activation of L-type Ca^{2+} channels and/or increased influx of extracellular Ca^{2+} across the sarcolemma. However, depletion of intracellular Ca^{2+} stores by prolonged equilibration (30 min) of mammalian muscles with Ca^{2+}-free saline decreased both the HCC and the twitch tension.[10] This suggests that intracellular Ca^{2+} stores, in particular the sarcoplasmic reticulum (SR) are mobilized during the HCC. This suggestion is consistent also with our observations that procaine abolished the HCCs of frog and mammalian muscle, whereas lidocaine potentiated these contractures. We attributed these opposing effects of the two local anesthetics to the fact that procaine, but not lidocaine, inhibits the release of SR-stored Ca^{2+}.[20]

Chemically-skinned mammalian muscle fibers[12] were used to investigate further the mechanisms involved in the generation of the HCC.[15] With this technique it is possible to examine directly the effects of halothane or cooling on the contractile proteins and on Ca^{2+} uptake and release by the SR. The data showed that HCCs can be elicited in Ca^{2+}-loaded, skinned fibers from the EDL muscle of adult rabbits when they are exposed to clinically relevant halothane concentrations. The protocol used in

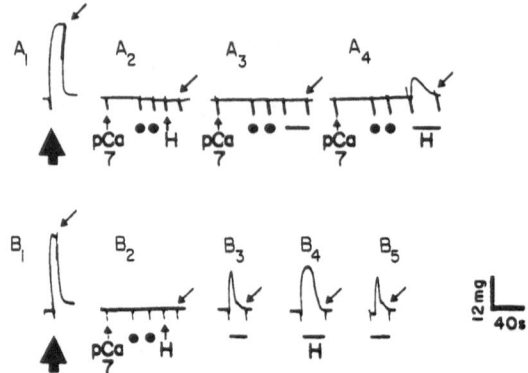

FIGURE 5. CC and HCC of chemically-skinned fibers of rabbit EDL muscle, prepared as described by Wood et al.[12] and studied at 22°C, except during the cooling periods. A_1 and B_1 show the maximal contractile response of the fiber (P_0) elicited by 0.5 mM $CaCl_2$ (large arrows). After relaxation with an EGTA-containing relaxing solution (solution R; slanted arrows) and release of the Ca^{2+} remaining in the SR by exposure to 20 mM caffeine in solution R, the fibers were submitted to successive cycles of Ca^{2+} loading into the SR and Ca^{2+} release (A_{2-4}; B_{2-5}). Ca^{2+} loading was performed by soaking the fiber in a solution of buffered Ca^{2+} (pCa 7.0) for 1 min, followed by two washes with control saline (washing solution, indicated by the dots). The loading step is not shown in B_{3-5}. Halothane (0.65 mM) and cooling (12°C in A and 8°C in B) was used to induce release of SR-stored Ca^{2+}. At the end of each loading-release cycle, the fiber was exposed to solution R containing 20 mM caffeine (not shown). Further description in the text. Reprinted from R. T. Sudo, G. Zapata-Sudo and G. Suarez-Kurtz, *Anesthesiology* 73:958-963 (1990) by permission of copyright holder and authors.

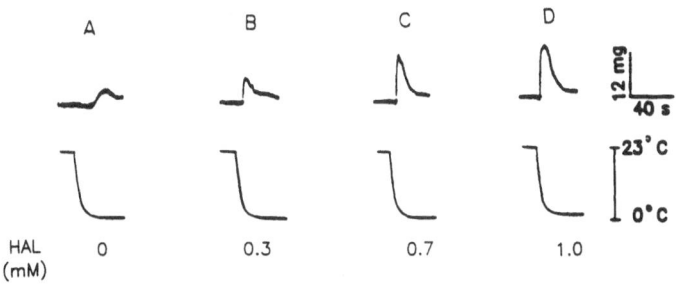

FIGURE 6. Dose-dependence of the HCC in a skinned fiber of rabbit EDL muscle, studied at 23°C, except during the cooling periods. The upper tracings in A-D show tensions induced in a Ca^{2+}-loaded fiber by cooling to 2-3°C in the absence (A) or the presence of halothane in increasing concentrations (B-D). The procedures used for eliciting the CC and the HCCs were those described in the legend of figure 5.

these experiments and representative recordings of HCCs are shown in figure 5. Fiber A developed no tension when challenged with either halothane (0.65 mM) at 22°C (figure 5A$_2$) or when cooled to 12°C (figure 5A$_3$); however, a transient HCC was recorded when the fiber was cooled to 12°C in the presence of halothane (figure 5A$_4$). Fiber B also failed to contract in the presence of 0.65 mM halothane at room temperature (figure 5B$_2$), but developed a transient CC when cooled to 8°C in the absence of halothane (figure 5B$_3$). When the cooling step was performed in the presence of halothane (figure 5B$_4$), the resulting HCC had a larger amplitude and longer duration than the CC recorded in the absence of halothane. The HCCs could be repeated several times in the same skinned fibers provided that the SR was loaded with Ca^{2+} between successive challenges with halothane at low temperature. Ca^{2+}-depleted fibers did not develop HCCs or CC's, which indicates that these tensions depend on the Ca^{2+} release from the SR. We have previously reported[15] that inhibition of SR Ca^{2+} uptake plays no significant role in the generation of the HCCs in skinned mammalian muscle fibers.

Figure 6 shows that the HCCs of skinned muscle fibers, like those of intact muscles, are dose-dependent. In this experiment a skinned fiber was cooled to 2-3°C in the absence or in the presence of halothane in increasing concentrations. Notice that despite the increase in amplitude of the HCC with the increase in halothane concentration, the tensions were always transient and the fiber relaxed to nearly baseline during the cooling period.

To investigate whether the transient nature of the HCCs resulted from depletion of halothane-releasable Ca^{2+} in the SR,[10,11] we loaded Ca^{2+} into the SR in the presence of oxalate. Because the SR in fibers loaded with Ca oxalate has a much higher Ca content than in skinned fibers loaded without oxalate,[21] Ca^{2+} depletion is unlikely to occur during the HCCs. Figure 7 shows that the HCCs of fibers loaded with Ca oxalate were sustained throughout the cooling period but relaxed upon removal of halothane and addition of the relaxing solution. Another important observation made on the experiment of figure 7 was that Ca-oxalate-loaded fibers were capable of developing successive HCCs without the need of interposing Ca^{2+} loading periods after each contracture (figure 5).

The HCCs of skinned fibers, like those of intact mammalian and frog muscles, were inhibited by procaine and potentiated by lidocaine.[15] Procaine is a potent inhibitor of the Ca^{2+}-induced-Ca^{2+}-release mechanism,[22] and this effect is thought to account for the block of CC in the absence of halothane[23] and for the inhibition of the halothane-induced tensions at room temperature.[24] It is possible that a similar

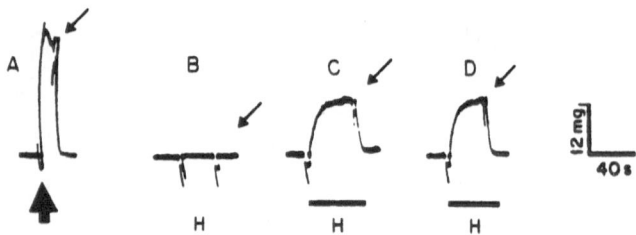

FIGURE 7. HCCs in a skinned fiber loaded with Ca oxalate. A: Po elicited by 0.5 mM CaCl$_2$. Solution R was used to relax the fiber (slanted arrows). Between A and B, the fiber was soaked for 30 min in a solution of pCa 7.0, containing 5 mM K oxalate. B-D: effects of 0.93 mM halothane (H) at 22°C (B) or at 12°C (C,D); the cooling period is indicated by the horizontal bars. Reprinted from R. T. Sudo, G. Zapata-Sudo and G. Suarez-Kurtz, *Anesthesiology* 73:958-963 (1990) by permission of copyright holder and authors.

mechanism acounts for the blockade of the HCCs by procaine. Accordingly, lidocaine, which does not mimic the inhibitory effects of procaine on Ca^{2+}-induced Ca^{2+} release potentiated rather than depressed the HCCs. We emphasize that the concentrations of procaine and lidocaine required for modifying the HCCs of both intact and skinned muscle fibers are considerably higher than those present clinically.

Taken together, our results with both intact and skinned muscle fibers provide strong evidence in support of the proposal[10,11,15] that the HCC results from a synergistic interaction between halothane and cooling, promoting release of SR-stored Ca^{2+}, possibly by facilitating the Ca^{2+}-induced-Ca^{2+}-release mechanism. Our observations[15] that halothane has no significant effect on the amplitude or time course of sub-maximal Ca^{2+}-induced tensions studied at 8-14°C, whereas cooling *per se* inhibited these tensions, indicate that the HCC cannot be ascribed to increased response of the contractile proteins to Ca^{2+}.

EFFECTS OF SALICYLATE ANALOGS ON THE HCC

The HCC provides a convenient experimental model for the investigation of drug interactions involving the effects of halothane on myoplasmic Ca^{2+} homeostasis and contractility in skeletal muscle fibers. Potentially, these studies might reveal drugs of interest for the management of halothane-triggered episodes of MH. The original observation[25] that sodium salicylate inhibited the caffeine-cooling contractures of skeletal muscles led us to investigate its influence on the HCC. The data showed that the HCCs and the electrically-evoked twitches of mammalian muscles were depressed by salicylate in a dose-dependent and reversible manner.[9,11] These effects are illustrated in figure 8. Recently, Zapata-Sudo[26] extended these observations to several benzoate analogs. Her data revealed that the ability of these compounds to inhibit the HCC is markedly affected by the introduction of substituents in position 5 of the aromatic ring. Analogs with reduced liposolubility, such as 5-amino salicylic acid and gentisic acid were ineffective, whereas the lipophilic derivatives 5-Cl- and 5-Br-salicylic acid were *ca.* 10 fold more potent than sodium salicylate in inhibiting the HCC. When 5-Cl- and 5-Br-salicylic acids were tested in skinned mammalian fibers[26] it was observed that they did not inhibit the halothane-induced release of SR-stored Ca^{2+} ions; indeed, the HCCs of skinned fibers were potentiated, rather than inhibited by the 5-Cl and 5-Br analogs of salicylic acid. These observations led to the suggestion of two alternative mechanisms to explain the blockade of the HCC in intact muscle by

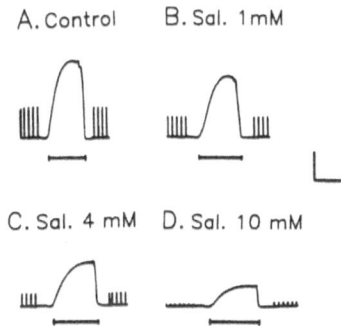

FIGURE 8. Inhibition of the HCC by sodium salicylate. A-D show tension recordings from mouse SOL muscle equilibrated with 0.7% halothane throughout the experiment. In each panel, 3-5 electrically evoked twitches, recorded at 23°C, precede and follow an HCC induced by cooling to 3°C. In B-D, the muscle was treated with increasing concentrations of sodium salicylate for 15 min before the cooling step. Notice the dose-dependent blockade of the HCC and the twitch tension by the salicylate. Calibration bars: horizontal, 2 min (twiches) or 24 sec (HCCs); vertical, 1g. From G. Zapata-Sudo (ref. 26) with permission from the author.

salicylate analogs. First, acidification of the myoplasm due to the influx of the salicylates across the sarcolemma, leading to reduction in the Ca^{2+} affinity for the contractile proteins.[27,28,29] Second, interference with the excitation-contraction coupling process due to salycilate-induced changes in surface charges at the sarcolemma.[30,31]

HCC IN DENERVATED MUSCLES

Chronic denervation affects several morphological, biochemical and mechanical characteristics of mammalian skeletal muscle fibers.[14] Caffeine-induced contractures of mammalian muscle can be potentiated or inhibited, depending on the muscle type.[32] Trachez et al.[14] investigated the influence of 2 to 90 days denervation on the HCCs of mouse EDL and soleus muscles, and observed potentiation of the HCCs (and also the CC) in both muscles, the effects being more striking in the EDL. Figure 9 shows tension recordings from paired EDL and soleus muscles from the same animal, one

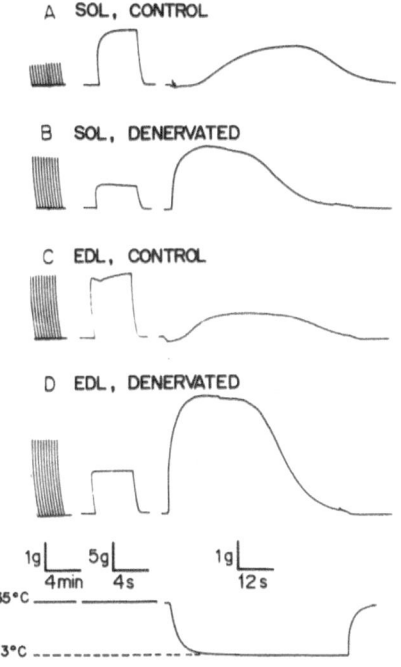

FIGURE 9. Effects of denervation on the twitches, tetanic tension and HCC of SOL (A,B) and EDL (C,D). The tension recordings are from 21-day denervated SOL and EDL and of the contralateral control muscles, normally innervated. The first two panels in each recording show the twitches elicited at 0.05 Hz and the maximal tetanic tension in the absence of halothane. The muscles were then exposed to halothane (0.7%) for 10 min (not shown) and subsequently cooled to 3°C to elicit the HCCs (third panel in each recording). Note the different calibration bars for each segment of the four tension recordings. The tracing at the bottom of the figure indicates the temperature of the bathing medium. Reprinted from M. M. Trachez, R. T. Sudo and G. Suarez-Kurtz, *Can J Physiol Pharmacol* 68:1207-1213 (1990) by permission of copyright holder and authors.

muscle of each pair having being denervated for 21 days. Comparison of the paired recordings reveals that denervation reduced the latency, increased the rate of tension development and the peak amplitude of the HCC (0.7% halothane) of both muscles. Application of the same protocol to EDL and soleus muscles denervated for 2 to 90 days, revealed significant increases in the peak amplitude and the rate of tension development of the HCC within 2 days of denervation, and maximum potentiation between 14 and 21 days. Dose-response curves for the HCC in control and in muscles denervated for 14 days are presented in figure 10. It is apparent that the denervation potentiated the HCCs at all halothane concentrations tested including 1.0%, which induces nearly maximum HCCs in control EDL and soleus muscles.[10,11]

For each concentration of halothane, potentiation of the HCC was larger in the EDL, as compared to the soleus muscle. This was ascribed[14] to differences in the fiber type composition of the two muscles. Denervation converts some properties of the SR of fast-twitch fibers, which predominate in the EDL, to the slow-twitch type.[33] In denervated muscles, the difference in maximum amplitude of the HCCs between EDL and soleus muscles was eliminated. A similar situation was reported for the caffeine-induced contractures of denervated rat muscles.[32]

Chemically-skinned fibers from the rabbit EDL were used by Trachez et al.[13] to investigate the mechanism of the denervation-induced potentiation of the HCCs. Their data revealed that, compared to control fibers, skinned fibers obtained from 14-day denervated muscles had significantly higher rates of net ATP-dependent Ca^{2+}-uptake by the SR and significantly less spontaneous release ("leakage") of the Ca^{2+} stored in the SR. The resulting increase in Ca^{2+} accumulation in the lumen of the SR was proposed as the principal mechanism responsible for the potentiation of the HCCs in denervated muscles. Trachez et al.[13] suggested that changes in the interaction of halothane with the SR membranes might also contribute to the potentiation of the HCCs in denervated muscles. Their data, reproduced in figure 11, showed that the dose-response curve for halothane-induced tensions of Ca^{2+}-loaded fibers studied at 22°C was displaced to the left after denervation. A striking observation was that 0.7 mM halothane, a concentration that had no effect on the

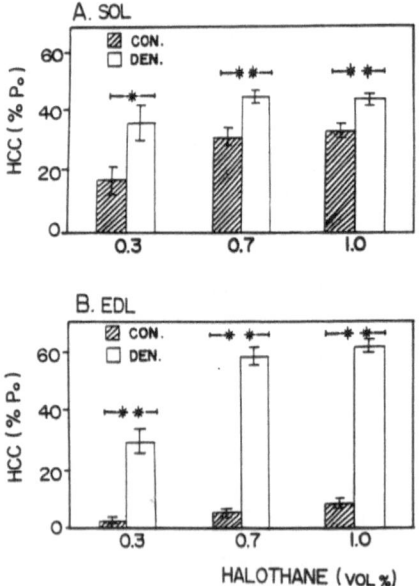

FIGURE 10. Effects of denervation on the dose-response relationship for the HCC of SOL (A) and EDL (B). The peak amplitudes of the of innervated (shaded bars) and denervated muscles (white bars) are plotted against the concentrations of halothane. The data are expressed as means ± S.E.M. for 6-10 paired muscles. * P < 0.05, ** P < 0.005, for the differences between the HCC of control and denervated muscles. Reprinted from M. M. Trachez, R. T. Sudo and G. Suarez-Kurtz, *Can J Physiol Pharmacol* 68:1207-1213 (1990) by permission of copyright holder and authors.

FIGURE 11. Effects of denervation on the halothane-induced tensions of skinned fibers from rabbit EDL muscle. Data from 10 control fibers and from 10 fibers obtained from 14-day denervated muscles were used to construct the plot. The ordinate represent percentage of Ca^{2+}-loaded fibers responding with a detectable tension ($>5\%$ of P_o) when challenged with the halothane concentrations indicated on the abcissa. The experiments were performed at 23°C using the procedures described in the legend of figure 5 for recording P_o, for loading Ca^{2+} into the SR and for releasing Ca^{2+} from the SR with halothane. Reprinted from M. M. Trachez, R. T. Sudo and G. Suarez-Kurtz, *Am J Physiol* 259:C503-C506 (1990) by permission of copyright holder and authors.

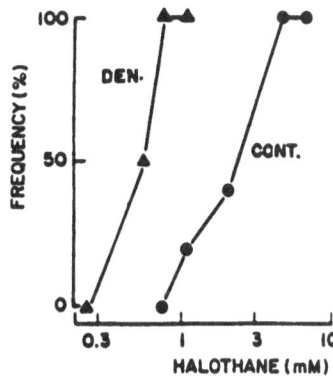

control fibers (n = 10), induced contratures in the 10 denervated fibers tested. Zorzato et al.[32] have recently shown that denervation increases the number of Ca^{2+}-release sites in the terminal cisternae of the SR. This finding might be related to the increased sensitivity of the SR of denervated muscles to the Ca^{2+}-releasing effects of halothane.[13]

This latter observation is reminiscent of the increased effectiveness of halothane in inducing Ca^{2+} release from the SR in another condition associated with functional alterations of the SR, namely MH. Thus, Ohnishi et al.[34] observed that, for a given level of Ca^{2+} in the SR, halothane is more effective in causing Ca^{2+} release in MHS, as compared to control muscles.

CONCLUSION

The data reviewed in this article showed that the HCC is a powerful experimental paradigm for investigating the effects of halothane on intracellular Ca^{2+} homeostasis and contractility in skeletal muscles. The HCCs are dose-dependent, readily reproducible and reversible in intact muscles, muscle strips obtained from human biopsies, and skinned fibers obtained from normal and denervated mammalian muscles. We are presently exploring the usefulness of the HCCs for detecting MH susceptibility in humans and for investigating drugs of potential value in the management and/or prevention of halothane-triggered episodes of MH. Because denervation affects markedly the HCCs of both intact and skinned mammalian muscle fibers, their study in muscle biopsies from patients with neurogenic myopathies could provide important information on the homeostasis of myoplasmic Ca^{2+} concentration in the skeletal muscles of these patients.

ACKNOWLEDGEMENTS

We are grateful to Drs. Henry Rosenberg and Jeffrey E. Fletcher from Hahnemann University for providing the human muscle biopsies, laboratory supplies and equipment used in the experiments shown in figure 3. This work was supported by the Conselho Nacional de Desenvolvimento Científico e Tecnológico (CNPq), the Financiadora de Estudos e Projetos (FINEP), and the Coordenaça~o de Aperfeiçoamento de Pessoal de Nível Superior (CAPES).

REFERENCES

1. B. A. Britt and W. Kalow, Malignant hyperthermia: a statistical review, *Canad Anaesth Soc J* 17:293 (1970).
2. The European Malignant Hyperpyrexia Group, A protocol for the investigation of malignant hyperpyrexia (MH) susceptibility, *Br J Anaesth* 56:1267 (1984).
3. North American Malignant Hyperthermia Group, Standardization of the caffeine halothane muscle contracture test, *Anesth Analg* 69:511 (1989).
4. G. C. Allen, J. E. Fletcher, F. J. Huggins, P. A. Conti and H. Rosenberg, Caffeine and halothane contracture testing in swine using the recommendations of the North American Malignant Hyperthermia Group, *Anesthesiology* 72:71 (1990).
5. T. E. Nelson, D. M. Bedell and E. W. Jones, Porcine Malignant Hyperthermia: effects of temperature and extracellular calcium concentration on halothane-induced contracture of susceptible skeletal muscle, *Anesthesiology* 42:301 (1975).
6. B. A. Britt, L. Endrenyi, E. Scott and W. Frodis, Effect of temperature, time and fascicle size on the caffeine contracture test, *Can Anaesth Soc J* 27:1 (1980).
7. T. Sakai, The effect of temperature and caffeine on activation of the contractile mechanism in the striated muscle fibers, *Jikeikai Med J* 12:88 (1965).
8. T. Sakai and S. Kurihara, A study of rapid cooling contracture from the viewpoint of excitation-contraction coupling, *Jikeikai Med J* 21:47 (1974).
9. G. Suarez-Kurtz and R. T. Sudo, The temperature dependence of halothane-induced contractures of skeletal muscle, *Muscle & Nerve* 9:47 (1986).
10. R. T. Sudo, M. R. B. Souza, G. Zapata and G. Suarez-Kurtz, Halothane-cooling contractures of mammalian muscles, *J Pharmacol Exp Ther* 237:600 (1986).
11. R. T. Sudo, G. Zapata and G. Suarez-Kurtz, Studies of the halothane-cooling contractures of skeletal muscle, *Can J Physiol Pharmacol* 65:697 (1987).
12. D. S. Wood, J. P. Zollman, J. P. Reuben and P. W. Brandt, Human skeletal muscle: properties of the "chemically skinned" fiber, *Science* (Wash. DC) 187:1075 (1975).
13. M. M. Trachez, R. T. Sudo and G. Suarez-Kurtz, Alterations in the functional properties of skinned fibers from denervated rabbit skeletal muscle, *Am J Physiol* 259:C503 (1990).
14. M. M. Trachez, R. T. Sudo and G. Suarez-Kurtz, Potentiation of the halothane-cooling contractures of mammalian muscles by denervation, *Can J Physiol Pharmacol* 68:1207 (1990).
15. R. T. Sudo, G. Zapata-Sudo and G. Suarez-Kurtz, Halothane cooling contractures of skinned mammalian muscle fibers, *Anesthesiology* 73:958 (1990).
16. D. G. Stephenson and D. A. Williams, Calcium-activated force responses in fast- and slow-twitch skinned muscle fibres of the rat at different temperatures, *J Physiol* (Lond) 317:281 (1981).
17. F. N. Briggs, J. L. Poland and R. J. Solaro, Relative capabilities of sarcoplasmic reticulum in fast and slow mammalian skeletal muscles, *J Physiol* (Lond) 266:587 (1977).
18. G. Salviati, M. M. Sorenson and A. B. Eastwood, Calcium accumulation by the sarcoplasmic reticulum in two populations of chemically skinned human muscle fibers: Effects of calcium and cyclic AMP, *J Gen Physiol* 79:603 (1982).
19. M. Endo, S. Yagi, T. Ishizuka, K. Horiuti, Y. Koga and K. Amaha, Changes in the Ca-induced Ca release mechanism in the sarcoplasmic reticulum of the muscle from a patient with malignant hyperthermia, *Biomed Res* 4:83 (1983).
20. C. P. Bianchi and T. C. Bolton, Action of local anesthetics on coupling systems in muscle, *J Pharmacol Exp Ther* 157:388 (1967).
21. M. M. Sorenson, H. S. L. Coelho and J. P. Reuben, Caffeine inhibition of calcium accumulation by the sarcoplasmic reticulum in mammalian skinned fibers, *J Membrane Biol* 90:219 (1986).
22. M. Endo, Calcium release from the sarcoplasmic reticulum, *Pharmacol Rev* 57:71 (1977).
23. K. Horiuti, Mechanism of contracture on cooling of caffeine-treated frog skeletal muscle fibres, *J Physiol* (Lond) 398:131 (1988).
24. A. Takagi, H. Sugita, Y. Toyokura and M. Endo, Malignant hyperpyrexia: Effects of halothane on single skinned muscle fibers, *Proc Jpn Acad* 52:603 (1976).
25. G. Suarez-Kurtz, M. J. B. Costa and S. Coutinho, The inhibitory effects of salicylate on contractility in skeletal muscle, *J Pharmacol Exp Ther* 230:478 (1984).
26. G. Zapata-Sudo, Efeitos de derivados salicilicos nas contraturas induzidas por resfriamento em musculos esqueleticos tratados com halotano, M.Sc. Dissertation presented to Universidade Federal do Rio de Janeiro, Rio de Janeiro, 1989.
27. T. M. Nosek, K. Y. Fender and R. E. Godt, It is diprotonated inorganic phosphate that depresses forces in skinned muscle fibers, *Science* 236:191 (1987).

28. A. Fabiato and F. Fabiato, Effects of pH on the myofilaments and the sarcoplasmic reticulum of skinned cells from cardiac and skeletal muscles, *J Physiol* (Lond) 276:233 (1978).

29. Y. Nakamaru and A. Schwartz, Possible control of intracellular calcium metabolism by $[H^+]$: Sarcoplasmic reticulum of skeletal and cardiac muscle, *Biochem Biophys Res Commun* 41:830 (1970).

30. S. McLaughlin, Salicylates and phospholipid bilayer membranes, *Nature* 243:234 (1973).

31. F. Ricciopo Neto and T. Narahashi, Ionic mechanism of the salicylate block of nerve conduction, *J Pharmacol Exp Ther* 199:454 (1976).

32. F. Zorzato, P. Volpe, E. Damiani, D. Quaglino Jr. and A. Margreth, Terminal cisternae of denervated rabbit skeletal muscle: Alterations of functional properties of Ca^{2+} release channels, *Am J Physiol* 26:C504 (1989).

33. A. Margreth, G. Salviati, S. Di Mauro and G. Turati, Early biochemical consequences of denervation in fast and slow skeletal muscle and their relationship to neural control over muscle differentiation, *Biochem J* 126:1099 (1972).

34. S. T. Ohnishi, S. Taylor and G. A. Gronert, Calcium-induced Ca^{2+} from sarcoplasmic reticulum of pigs susceptible to malignant hyperthermia, *FEBS Lett* 161:103 (1983).

6

Interactions of Fatty Acids with the Calcium Release Channel in Malignant Hyperthermia

Jeffrey E. Fletcher, Henry Rosenberg, Jill Beech

INTRODUCTION

The anesthesia-induced malignant hyperthermia (MH) syndrome has been suggested to be a consequence of a halothane-sensitive defect in Ca^{2+} regulation, based on muscle rigidity during the syndrome and the increase in Ca^{2+} in isolated fiber bundles exposed to halothane. Additionally, the threshold of Ca^{2+}-induced Ca^{2+} release (TCICR) is lower than normal in isolated fractions of heavy sarcoplasmic reticulum (HSR) from porcine MH muscle.[1,2] However, the defect need not reside in the Ca^{2+}-release channel protein, as there are reports of nonrigid MH in humans[3] and loss of Ca^{2+} regulation could be the result of a disturbance in fatty acid metabolism.[4] The present study examines fatty acid metabolism and the influence of fatty acids on various aspects of Ca^{2+} regulation and on caffeine, halothane and succinylcholine action in normal and MH muscle. Additionally, since phenytoin has been suggested to antagonize MH,[5] its effects on Ca^{2+} regulation and fatty acid metabolism have been examined.

METHODS

Subjects

Humans were referred for diagnostic testing for MH. Horses were either controls or those referred for chronic intermittent rhabdomyolysis or hyperkalemic periodic paralysis. The swine (Yorkshire × Duroc cross) were obtained from Iowa State University through Biomedical Alternatives (Raleigh, NC) and were positive for MH susceptibility by the barnyard challenge (halothane 6%), serum CK values, H blood typing and the halothane and caffeine contracture tests.[6] At the completion of the muscle biopsies some swine were challenged *in vivo* with halothane 3% for 5 min followed by halothane 2% for an additional 43 min. Twenty min after reducing the halothane concentration to 2%, succinylcholine (1 mg/kg per dose) was administered

JEFFREY E. FLETCHER[a,b] and HENRY ROSENBERG[a], Departments of [a]Anesthesiology and [b]Biochemistry, Hahnemann University, Philadelphia, Pennsylvania 19102-1192. JILL BEECH, Department of Clinical Studies, University of Pennsylvania School of Veterinary Medicine, New Bolton Center, Kennett Square, Pennsylvania 19348.

Mechanisms of Anesthetic Action in Skeletal, Cardiac, and Smooth Muscle
Edited by T.J.J. Blanck and D.M. Wheeler, Plenum Press, New York, 1991

five times at 5 min intervals. Blood was drawn for analysis before the halothane administration and 3 min after the last succinylcholine administration.

In one experiment, phenytoin was administered orally to a horse (12-14 mg/kg, twice daily). Muscle biopsies (semimembranosus) were taken immediately before beginning phenytoin and after 1 weeks treatment. Halothane and caffeine contracture tests for MH were conducted using standard methods.[6,7]

Approval was obtained from the Hahnemann University Animal Welfare Committee and the University of Pennsylvania Institutional Animal Care and Use Committee for these studies.

Muscle Contracture Studies

Caffeine-induced (8 mM) contractures were examined in an *in vitro* tissue bath in fiber bundles from human (vastus lateralis) or equine (semimembranosus) biopsies.[8,9] Dantrolene (10 μM) was added 5 min before phospholipase A_2 (PLA_2) or caffeine. Bee venom PLA_2 (1 μM; Sigma Chemical Co.) was added 5 min before caffeine. Isoproterenol (10 μM) was added 10 min before caffeine. Contractures to succinylcholine and halothane were examined in control porcine gracilis muscle.[8-10] Succinylcholine was added 5 min before halothane (3%). Isoproterenol (1 μM) was added 10 min before succinylcholine. Insulin (1 μM) was added 10 min before isoproterenol.

Ca^{2+} Release and Ryanodine Binding in Isolated SR

The following methods were used to measure the effects of inhibition of fatty acid release on the TCICR. HSR fractions were prepared by differential centrifugation (8,000 - 12,000 × g) of homogenates of longissimus dorsi from swine (Yorkshire × Duroc cross) biopsied with nontriggering anesthesia[4] and the TCICR was determined with pyrophosphate,[11] as previously described.[4] Ca^{2+} was added in 10 μM increments to a 1.5 ml volume maintained at 37°C. The biopsy specimen was divided into two samples before homogenization. The lipase inhibitor p-bromophenacyl bromide (100 μM) was added to one sample during homogenization and remained present during subsequent isolation of the HSR fraction.

To determine the effects of fatty acids on ryanodine binding to the Ca^{2+}-release channel, highly enriched terminal cisternae (TC) fractions from porcine longissimus dorsi biopsied with nontriggering anesthesia were isolated on a discontinuous sucrose gradient.[12] ^{3}H-ryanodine (New England Nuclear; 99% purity, 95 μCi/pmol) binding was determined under the following conditions:[13] 37°C, 90 min, pH 7.0, 0.2 mg TC protein/ml, with a PIPES buffer containing PIPES (10 mM) and KCl (100 mM) adjusted for a final Ca^{2+} concentration of 6 μM with a solution of $CaCl_2$ (2 mM), EGTA (3.7 mM) and nitrilotriacetic acid (3.7 mM). The incubations were stopped with PIPES buffer (4°C), filtered and washed with cold (4°C) buffer.[14] Liquid scintillation counting and Scatchard analysis was performed as previously described.[14] Oleic acid (20 μM final concentration) was added immediately before ryanodine.

Fatty Acid Metabolism and Protein Analysis

Lipase activities were determined using radiolabeled artificial phospholipid substrates as follows. The muscle biopsies were removed with nontriggering anesthesia and frozen in liquid N_2 for later analysis. For the fixed pH and time studies, on the day of the experiment the tissues were thawed and 0.5 g (wet wt) specimens weighed and homogenized (Brinkman Polytron) in 1.2 ml Tris buffer containing: Tris (200 mM),

CaCl$_2$ (2 mM) and fatty-acid-free bovine serum albumin (0.5%) at pH 7.4 (4°C). Aliquots (0.2 ml) of homogenate were added to 0.3 ml substrate buffer [Tris buffer containing ^{14}C-triolein (Amersham; 55 mCi/mol; 10 μM final concentration), or ^{14}C-phosphatidylcholine (Amersham; 58 mCi/mol; 10 μM final concentration)] and incubated at 37°C for 60 min (triglyceride substrate) or 1 min (phospholipid substrate). The incubation was stopped by adding methanol and the lipids further extracted.[15,16] The neutral lipids were separated by one-dimensional (1-D) thin-layer chromatography (TLC) and radioactivity in each lane was quantitated with a radioimaging TLC analyzer.[15,16] The pH profiles were conducted by homogenizing the tissues in distilled water and adding a two-fold concentrated Tris buffer (see above) to the homogenates at a 1:1 ratio to yield the same final concentration as described above. At the lower pH range (pH 3-6) acetate (200 mM final concentration) replaced Tris as the buffering agent. In some cases, CaCl$_2$ (2 mM) was replaced with EGTA (10 mM).

Free fatty acids were determined as previously described for liquid N$_2$ frozen biopsied specimens of whole muscle,[17] or thawed and homogenized whole muscle and whole muscle incubates.[4] The phospholipids, which remain at the origin, were separated from neutral lipids by 1-D TLC[18] and the fatty esters were determined by GC analysis of their methyl esters,[19] as previously described for triglycerides.[4,17,18,20,21]

To measure the stimulation of fatty acid metabolism by isoproterenol in human skeletal muscle, primary skeletal muscle cultures were established.[22] The cells were radiolabeled with ^{14}C-linolenic acid (18:3; 10 μM) for three days, as described for radiolabeling immortalized airway epithelial cells.[15,16] The cultures were then incubated with isoproterenol (10 μM; 20 min; 37°C) in a HEPES buffer containing physiological concentrations of NaCl, KCl, MgCl$_2$, CaCl$_2$ and glucose.[15,16] The lipids were extracted, neutral lipids separated by 1-D TLC and the radioactivity in each fraction determined.[15,16]

Protein was determined by standard methodology.[23] All lipid and Ca^{2+} values are expressed per mg protein.

RESULTS

Contracture Studies in Human and Porcine Muscle

Dantrolene antagonized contractures induced by caffeine in human skeletal muscle (table 1). PLA$_2$ has previously been demonstrated to increase contractures to halothane, but not to succinylcholine.[10] Since caffeine is a diagnostic agent for MH, we tested whether caffeine contractures were increased by PLA$_2$ pretreatment. PLA$_2$

Table 1. Effect of Dantrolene (10 μM) and Bee Venom PLA$_2$ (1 μM)
on Caffeine-Induced (8 mM) Contractures in Normal Human Vastus Lateralis Muscle

Sequence of Addition	N	Contracture (g; mean ± SEM)
Caffeine	3	1.25 ± 0.38
Dantrolene → Caffeine	3	0.13 ± 0.03*
PLA$_2$ → Caffeine	3	2.33 ± 0.35*
Dantrolene → PLA$_2$ → Caffeine	3	0.03 ± 0.03*

Asterisk (*) indicates significant difference ($P < 0.05$) compared to caffeine alone by paired t-test.

alone did not induce contractures in these preparations (data not shown). Contractures induced by caffeine were increased by almost two-fold in PLA$_2$-treated preparations (table 1). These contractures were completely antagonized by dantrolene (table 1), suggesting the effects of the PLA$_2$ were on Ca^{2+} release from the sarcoplasmic reticulum, not on sarcolemmal Ca^{2+} influx.

Succinylcholine pretreatment in control porcine muscle had very little effect on contractures subsequently induced by halothane (figure 1A). Isoproterenol, by acting on the β-adrenergic receptor, phosphorylates and thereby activates hormone sensitive lipase (HSL).[24] Insulin dephosphorylates and inactivates HSL.[24] Pretreating normal

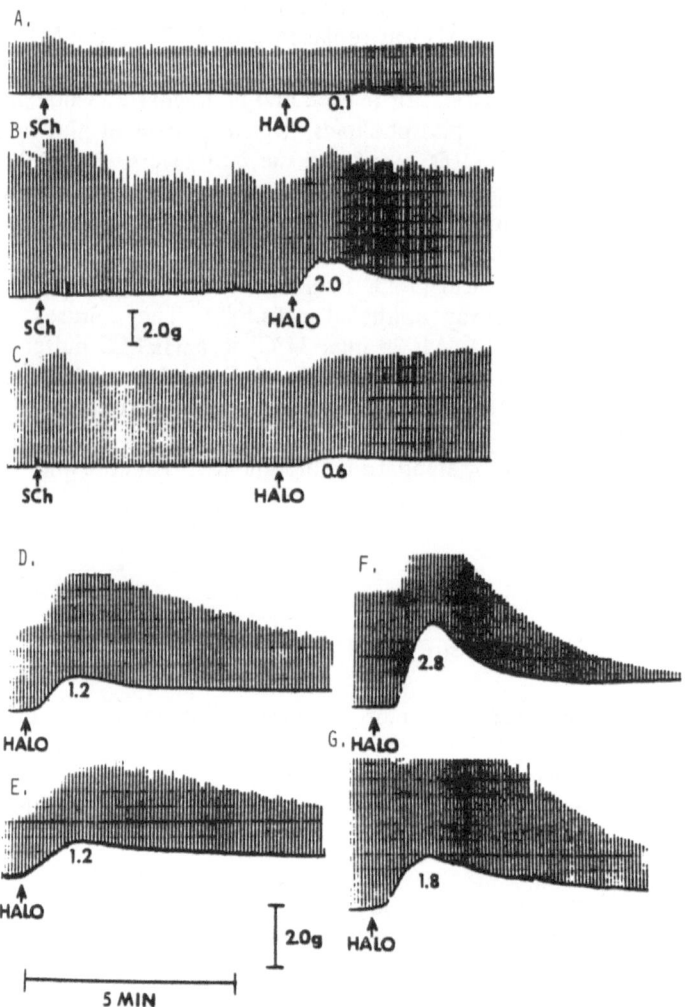

FIGURE 1. Antagonism of isoproterenol-enhanced halothane contractures in normal and MH porcine gracilis muscle by insulin. Panels A-C are muscle strips from a MH− pig. Panels D-G are muscle strips from a MH+ pig. In panels A-C succinylcholine (SCh; 50 mM) was added to the bath 5 min before halothane (HALO; 3%). In Panel B the muscle strip was exposed to isoproterenol (1 μM; 10 min) before SCh. In panel C the muscle strip was exposed to insulin (1 μM; 10 min), then isoproterenol (1 μM; 10 min), then SCh. In Panels D-G halothane (3%) was added to the bath. Panels E and F were pretreated with insulin (1 μM) or isoproterenol (1 μM), respectively, 10 min before halothane. In Panel G the muscle strip was exposed to insulin (1 μM; 10 min), then isoproterenol (1 μM; 10 min), then halothane. The magnitude of each contracture is indicated in grams.

muscle strips with isoproterenol to increase HSL-generated free fatty acid levels results in very large contractures to halothane in the presence of succinylcholine (figure 1B). Insulin antagonizes the effect of isoproterenol (figure 1C). Halothane-induced contractures in porcine MH muscle (figure 1D) are not antagonized by insulin (figure 1E). Unlike human MH muscle in which halothane contractures are unaffected by isoproterenol,[4] the contracture response of porcine MH muscle to halothane is greatly increased by isoproterenol (figure 1F). The isoproterenol-induced increase in response to halothane in porcine MH muscle is antagonized by insulin (figure 1G). Additionally, caffeine (8 mM) contractures in equine semimembranosus muscle were enhanced by isoproterenol to 277 $\pm$ 35% (mean $\pm$ SEM) of those in untreated muscle in 11 biopsies from nine different horses.

Effects of Inhibition of Fatty Acid Release on the TCICR

Unsaturated fatty acids liberated during isolation of HSR fractions have been postulated to lower the TCICR in porcine MH muscle.[4] In the present study treatment with oleic acid (18:1; 10 μM) decreased the TCICR to 52 $\pm$ 12% of untreated HSR preparations (mean $\pm$ SEM; n = 6 pigs). Therefore, the effects of adding a lipase inhibitor (p-bromophenacyl bromide; 100 μM) during homogenization of the muscle and subsequent HSR isolation were examined on the TCICR. Separate samples from the same biopsy specimen were minced and homogenized in a medium without (control) or containing lipase inhibitor. The TCICR in control preparations (5.1 $\pm$ 0.8 μmol Ca^{2+}/mg HSR protein; mean $\pm$ SEM for six different normal swine) was lower than the TCICR determined in HSR from the same pigs prepared in the presence of lipase inhibitor (7.4 $\pm$ 0.6 μmol Ca^{2+}/mg HSR protein; P < 0.01 by paired t-test).

Effects of Fatty Acids on Ryanodine Binding

We recently found that the low K_d of ryanodine binding to TC suggested by other investigators to indicate MH susceptibility was instead related to the use of two different strains of swine for the control and MH groups.[14] However, the MH swine used in our ryanodine binding study were the same as those in which we reported decreased triglyceride levels and an inability to generate excess free fatty acids in skeletal muscle homogenates.[4] Therefore, if free fatty acids regulate ryanodine binding to TC in the same manner as they regulate the TCICR,[4] we would not have observed altered ryanodine binding in our previous study.[14] A typical Scatchard plot for ^{3}H-ryanodine binding to a TC fraction is shown in figure 2A. Addition of 20 μM oleic acid decreased the B_{max} and K_d values by about 20% (figure 2B). Since a 20% change is still within the normal range for the control population,[14] it is unlikely that regulation of ryanodine binding by fatty acids would account for the 3- to 3.5-fold[13,25] differences in K_d reported by other investigators comparing Pietrain MH and Yorkshire normal muscle. Instead, because our strain of swine exhibits a K_d different from either Yorkshire or Pietrain pigs,[13,25] strain differences, not MH susceptibility or an effect of free fatty acids, seem a more likely explanation for the lower K_d in Pietrain swine.

Lipase Activities Using Radiolabeled Artificial Phospholipid Substrates

Both PLA_2[10,26] and triglyceride lipase[20] activities have been suggested to be the source of elevated fatty acids in MH muscle homogenates or mitochondrial fractions. However, the activities of either of these enzymes have not been directly determined in normal or MH porcine muscle. The pH profiles of PLA_2 activity and triglyceride

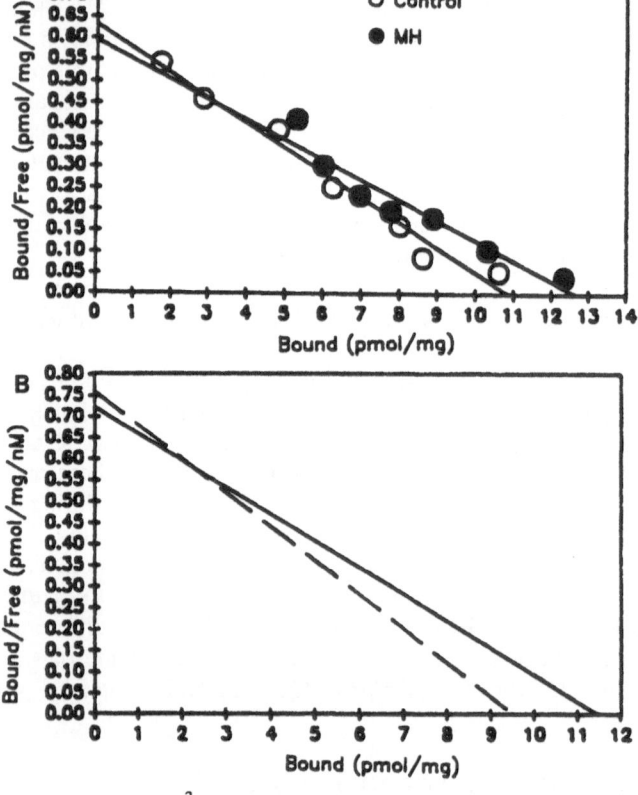

FIGURE 2. Effects of fatty acids on ^{3}H-ryanodine binding to terminal cisternae (TC) from normal and MH swine. A: Typical Scatchard plot of ^{3}H-ryanodine binding to TC. B: Effects of oleic acid (20 μM) on ^{3}H-ryanodine binding in normal porcine TC. In Panel B the solid line is in the absence and the dashed line is in the presence of oleic acid.

lipase activity are shown in figure 3. The triglyceride lipase and PLA_2 enzyme activities were the same whether determined in a Ca^{2+}-containing (2 mM) or Ca^{2+}-free (EGTA 10 mM) medium (data not shown), suggesting these lipase activities are Ca^{2+}-independent under these conditions. The Ca^{2+}-independence of these enzymes was observed in fresh and frozen tissue (data not shown). Directly determining these lipolytic activities on radiolabeled substrates revealed normal PLA_2 and triglyceride lipase activities in MH muscle (table 2).

Lipase Activities and Lipid Analysis in Skeletal Muscle

We had reported that, in British Landrace swine, 18:3 was significantly higher in MH+ than in MH− whole muscle analyzed after freezing in liquid N_2.[17] However, this finding was not verified in the Yorkshire × Duroc strain in the present study (table 3), supporting our original suggestion that this observation was not directly related to MH.[17] Freezing skeletal muscle in liquid N_2 eliminates the elevated FA release in MH muscle (table 3). This treatment also reduces fatty acid release by about 40% compared to values reported for fresh muscle from this strain.[4] This loss is about equal to the Ca^{2+}-dependent component observed in fresh tissue.[4]

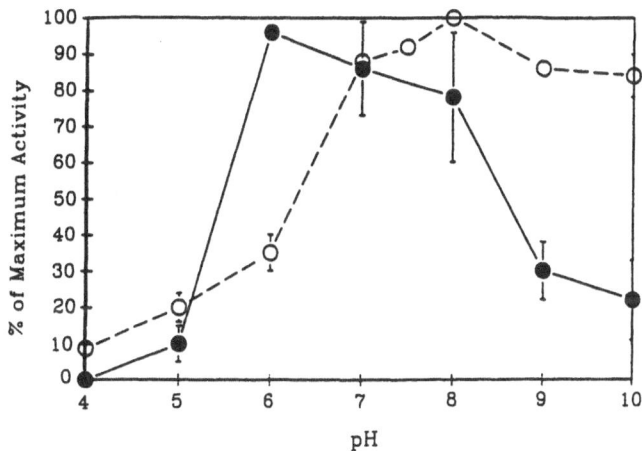

FIGURE 3. The pH dependence of PLA$_2$ and triglyceride lipase activities in normal porcine muscle. The filled circles are triglyceride lipase activity (n = 3 pigs) and the open circles are PLA$_2$ activity (n = 4 pigs). Error bars are shown when the SEM exceeds the size of the symbols.

Our previous studies of lipid analysis in swine did not report fatty esters on phospholipids.[17] We have now analyzed the phospholipid fatty esters and found them to be normal in MH muscle (table 4). However, we still have not looked at individual classes of phospholipids.

Stimulation of Fatty Acid Metabolism by Isoproterenol

Isoproterenol (10 μM, 20 min) stimulation of normal human skeletal muscle causes no change in the ratio of radiolabel in the phospholipid to neutral lipid fractions (96 ± 11% of that with no isoproterenol; mean ± SEM; n = 6 patients). In contrast, isoproterenol stimulation of primary cultures of MH skeletal muscle causes a shift in the ratio of radioactivity in the phospholipid to neutral lipid fractions (176 and 148%; n = 2). This shift most probably reflects the hydrolysis of triglyceride and cholesterol esters by HSL and loss of fatty acids through β-oxidation.

Table 2. PLA$_2$ and Triglyceride Lipase Activities as Determined
on Radiolabeled Substrates, in Longissimus Dorsi from 6 MH− and 6 MH+ Pigs

Susceptibility	Lipolytic Activities (mean ± SEM)	
	PLA$_2$ (pmol FA/mg/min)	Triglyceride Lipase (pmol FA/mg/hr)
MH−	2.7 ± 0.5	6.4 ± 0.7
MH+	2.6 ± 0.5	6.2 ± 0.9

Experimental conditions: pH 7.4, 37°C.

Table 3. Free Fatty Acid Analysis of Liquid N_2 Frozen Specimens, or Homogenates Prepared from Thawed Specimens of Biopsied Skeletal Muscle

Group	Time (hr)	Free Fatty Acids						
		16:0	16:1	18:0	18:1	18:2	18:3	20:4
MH−	N_2	214 ±41[a]	27±4[a]	126 ±21[a]	260 ±48[a]	73 ±14[a]	11±1	11 ±4[a]
	0	304 ±62[b]	41±12	164 ±26[b]	415 ±140	94 ±18[b]	15±2	16 ±3[b]
	2	698 ±77[a,b]	80±10[a]	314 ±26[a,b]	732 ±73[a]	500 ±61[a,b]	23±1	104 ±15[a,b]
MH+	N_2	222 ±45[c]	26±5[b]	162 ±38[c]	276 ±51[b]	73 ±11[c]	15±2[a]	16 ±4[c]
	0	282 ±22[d]	39±4[c]	154 ±9[d]	321 ±21[c]	86 ±9[d]	16±2	20 ±3[d]
	2	676 ±46[c,d]	82 ±11[b,c]	314 ±27[c,d]	747 ±87[b,c]	429 ±30[c,d]	26±3[a]	95 ±11[c,d]

Homogenates of initially liquid-N_2-frozen longissimus dorsi from 5 MH− and 7 MH+ pigs were incubated for 0 or 2 hrs before lipid extraction. The subgroups are: N_2, analysis done on frozen specimens;[17] 0 hr, tissue thawed and subsequently homogenized and extracted; 2 hr, tissue thawed, homogenized, incubated 2 hr then extracted. Values have units of pmol/mg protein (mean ± SEM).

Abbreviations: 16:0, palmitic acid; 16:1, palmitoleic acid; 18:0, stearic acid; 18:1, oleic acid; 18:2, linoleic acid; 18:3, linolenic acid; 20:4, arachidonic acid.

Within each fatty acid, significant ($P < 0.05$) differences as determined by a one-way ANOVA and Scheffe test are indicated by the same superscript. There were no significant difference between the MH− and MH+ groups for any of the three conditions (N_2, 0 hr, 2 hr).

Effects of Triglyceride Reduction on Expression of MH

A subgroup of MH swine tested in our laboratory was found to be deficient in skeletal muscle triglycerides.[4] These pigs, despite a previous episode of rigidity (barnyard challenge), did not exhibit muscle rigidity when challenged *in vivo* with halothane and succinylcholine. This challenge has previously been shown to cause marked rigidity and death in MH swine.[6] The swine did exhibit moderate changes in

Table 4. Phospholipid Fatty Esters in Longissimus Dorsi from Four Control and Eight MH-Susceptible Yorkshire × Doroc Swine

Group	Phospholipid Fatty Esters Percent of Total Distribution (mean ± SEM)									
	12:0	14:0	16:0	16:1	18:0	18:1	18:2	18:3	20:4	22:6
MH−	0.1 ±0.1	1.4 ±0.4	29 ±1	0.4 ±0.2	13 ±1	18 ±0	25 ±1	0.3 ±0.0	13 ±1	0.5 ±0.4
MH+	0.3 ±0.2	1.4 ±0.6	29 ±0	0.6 ±0.2	12 ±1	18 ±1	26 ±1	0.3 ±0.1	13 ±1	0.2 ±0.1

Abbreviations: see table 3; 12:0, lauric acid; 14:0, myristic acid; 22:6, docosahexaenoic acid.

Table 5. Metabolic Parameters Occurring in Pigs 48 Min after *in Vivo* Challenge with Halothane and Succinylcholine

	N	Lactate	HCO$_3$	Base Excess	[K$^+$]	Esoph. ΔT($^\circ$C)
Controls	4	3±1	29±1	−1±2	5.8±0.1	0.0±0.2
MH Susceptibles	9	13±2[b]	23±2[a]	−14±1[c]	7.2±0.3[b]	+0.7±0.1[b]

All values are mean ± SEM. Superscripts indicate that the values are significantly different from controls, as determined by a two-tailed t-test ([a]P < 0.05; [b]P < 0.01; [c]P < 0.001).

some of the parameters associated with MH (table 5). Additionally, we have recently observed in one MH-susceptible swine marked rigidity in the hind quarters to the extent that the leg could not be moved at the same time that the forelimbs were flaccid, suggesting that muscle rigidity need not be uniformly expressed throughout the musculature.

Effects of In Vivo Treatment With Phenytoin

A Quarter Horse with hyperkalemic periodic paralysis tested positive for MH before being placed on phenytoin treatment on two separate drug trials. After one week on phenytoin the horse tested negative for MH during both trials. The TCICR in HSR from this horse (2.4 ± 1.9 μmol Ca^{2+}/mg protein; mean ± SD; n = 3) was considerably lower than control equine values (*ca.* 10 μmol Ca^{2+}/mg protein)[4] before being placed on phenytoin treatment. After one week on phenytoin the TCICR increased toward control values. The skeletal muscle triglyceride values increased during phenytoin treatments in the first trial from 1.18 ± 0.07 to 1.97 ± 0.24 nmol/mg protein (P < 0.01), with a similar change occurring in the second trial. No changes were observed in phospholipid fatty esters or free fatty acids (data not shown).

DISCUSSION

The response to halothane in skeletal muscle is increased under conditions enhancing the production of free fatty acids. The addition of PLA$_2$ (to directly increase sarcolemmal and t-tubular free fatty acids) or isoproterenol (to activate HSL) to the tissue bath enhances not only the response to halothane, but also to caffeine, another agent used in the contracture test for diagnosis of MH. Similar conclusions with regard to the interaction between fatty acids and halothane were reported for hemolysis of red blood cells.[21] In this latter study, the susceptibility to halothane-induced hemolysis of red blood cells from five different species was inversely related to the amount of free saturated fatty acid in the membrane. Exogenously added unsaturated fatty acids greatly enhanced the hemolytic activity of halothane[21] and the TCICR in porcine and equine HSR fractions.[4] Preliminary studies suggest that the fatty acids also lower the threshold of halothane-induced Ca^{2+} release from HSR fractions (unpublished observations). These findings raise the possibility that any disorder raising the levels of free fatty acids could result in an MH-like reaction.

We previously suggested that the excessive fatty acids released during homogenization of MH muscle lowered the TCICR.[4] In agreement, the inclusion of a lipase inhibitor in the present study to decrease fatty acid production during preparation of the HSR fraction significantly increased the TCICR. We were previously unable to confirm an association between ryanodine binding to the Ca^{2+} release channel and MH[14] in swine depleted of triglycerides.[4] The relatively small change in K_d (about 20%) induced by fatty acids in the present study could not account for the much larger differences reported between normal and MH muscle.[13,25] Therefore, we conclude that the use of different strains of swine for the normal and MH susceptible groups by other investigators, and not MH susceptibility, is the only explanation for the reported differences in the K_d of ryanodine binding.

Addition of exogenous fatty acids modulates Ca^{2+} release,[4] halothane action[21] and even increases levels of IP_3.[27] This latter effect of fatty acids may account for the reports of reduced inositol 1,4,5-trisphosphate phosphatase activity in MH.[28] We have not established whether the fatty acids themselves and/or one or more of their metabolites (*e.g.*, acylCoAs) are involved. Since oleic acid is not metabolized by either the cyclooxygenase or lipoxygenase pathways, it is highly unlikely that the metabolites of these pathways are involved in modifying halothane action. It is of interest that Na^+ regulation is altered in primary skeletal muscle from MH susceptibles.[22] The Na^+ channel is an acylated protein that binds palmitic acid.[29] There is a difference in sensitivity of the Na^+ channel to activation by exogenously added fatty acids,[30] with the MH cells being less responsive.

The source of the elevated fatty acids in MH muscle has still not been conclusively determined. The original report of elevated fatty acid metabolism in MH muscle postulated PLA_2 as the defect in MH.[26] However, indirect evidence has since suggested that PLA_2 activity is normal in MH muscle.[4,17,20] Directly determining this activity on artificial substrates in the present study also revealed no difference between MH and normal PLA_2 activity. The levels of PLA_2 activity in porcine muscle are similar to those reported for human muscle by other investigators.[31] In agreement with the studies in human muscle,[31] we observed that PLA_2 activity was not Ca^{2+} dependent. Triglyceride metabolism is altered in human[20] and porcine[4] MH muscle, as well as in equine MH muscle and primary cultures of human MH skeletal muscle in the present study. We have suggested that a defect in the β-adrenergic-coupled HSL activity might account for MH in humans.[32] This protein is encoded on chromosome 19q13.1,[32] which is the proposed locus of the MH defect.[33] The present study demonstrates that triglyceride lipase activity (nonhormonally stimulated) is normal in MH tissue. The only remaining major lipase activity is HSL. The β-adrenergic system has been implicated by other investigators to be involved in porcine stress syndrome.[34] In this latter study the density of β-adrenergic receptors and activity of adenylate cyclase are greater in MH muscle. The antagonism by insulin of only the isoproterenol-enhanced component of the halothane contracture in porcine MH muscle in the present study does not support altered HSL activity or an increased β-adrenergic response as the MH defect in this strain of swine. Indeed, more recent studies in our laboratory suggest that *de novo* synthesis cannot be ignored as a possible source of the elevated fatty acids in MH muscle. Skeletal muscle is capable of significant fatty acid synthesis and produces about 20% of that synthesized by the total body.[35] It is possible that HSL plays a dual role in fatty acid regulation. In addition to hydrolyzing triglyceride fatty esters the HSL may be involved in esterification of triglycerides with newly synthesized fatty acids. Alternatively, one or more enzymes encoded at chromosome 19q13.1 may also be involved in fatty acid regulation.

The markedly reduced potential for expressing a full-blown MH syndrome (table 5) is consistent with the reduced triglyceride levels in these swine.[4] The reduced triglyceride levels coincide with decreased fatty acid levels and normal fatty acid generation in homogenates.[4] An equally important finding is that rigid and nonrigid MH can be accounted for with the same genetic defect. Reports of nonrigid MH may simply be variants of expression based on the physiological state of the subject at the time of challenge with triggering agents. However, it is still likely that more than one genetic defect could cause MH. For example, porcine (Yorkshire × Duroc) and human muscle differ in their responses to quinacrine (compare ref. 9 to ref. 8) and isoproterenol (compare ref. 4 to figure 1F), suggesting that different mechanisms controlled by different genes may account for the defect in these two species. Alternatively, the physiological mechanisms underlying the contractures could be very different, suggesting the pig is a poor model of human MH.

The only drug used clinically for MH is dantrolene. Other investigators have reported that acute administration of phenytoin was protective in a chick model of MH.[5] Our preliminary findings suggest that phenytoin can abolish any adverse effects of halothane on Ca^{2+} regulation in equine skeletal muscle.

In summary, free fatty acids markedly alter Ca^{2+} regulation and enhance the effects of halothane on Ca^{2+} regulation in skeletal muscle. The K_d of ryanodine binding to TC does not relate to MH or free fatty acids, but varies with the strain of pig. The source of elevated free fatty acids in MH muscle is linked to triglyceride metabolism and may be explained by an alteration in more than one mechanism, including β-adrenergic-dependent cAMP production, the HSL, β-oxidation and *de novo* synthesis of free fatty acids. Rigid and nonrigid MH can be accounted for by the same genetic defect and depend on the physiological state of the subject. Based on preliminary findings, phenytoin appears to completely antagonize the adverse effects of halothane in equine skeletal muscle.

ACKNOWLEDGEMENTS

The authors are deeply grateful to Drs. Steven J. Wieland, Pierre A. Conti, Gary M. Vita, Lauren L. Christian, Terry D. Heiman-Patterson, Roy C. Levitt, John Shutack, Gregory C. Allen, Philip Kistler, Eugene I. Tolpin, Janet Johnson and Robert Storella and to Kirsten Erwin, Linda Tripolitis, Florence Huggins, Leo Davidson, Ming-Shi Jiang, Susan Hanson, Michelle Yudkowsky, Scott Mayerberger, Gregory E. Conner, Merrill Hilf, Susan Lindborg and Qi-Hua Gong for their important contributions to the project (*sine qua non*). Supported by the Hahnemann Anesthesia Research Foundation.

REFERENCES

1. T. E. Nelson, Abnormality in calcium release from skeletal sarcoplasmic reticulum of pigs susceptible to malignant hyperthermia, *J Clin Invest* 72:862 (1983).
2. S. T. Ohnishi, S. Taylor, and G. A. Gronert, Calcium-induced Ca^{2+} release from sarcoplasmic reticulum of pigs susceptible to malignant hyperthermia: the effects of halothane and dantrolene, *FEBS Lett* 161:103 (1983).
3. B. A. Britt and W. Kalow, Malignant hyperthermia: a statistical review, *Can Anaesth Soc J* 17:293 (1970).
4. J. E. Fletcher, L. Tripolitis, K. Erwin, S. Hanson, H. Rosenberg, P. A. Conti, and J. Beech, Fatty acids modulate calcium-induced calcium release from skeletal muscle heavy sarcoplasmic reticulum fractions: implications for malignant hyperthermia, *Biochem Cell Biol* 68:1195 (1990).
5. A. D. Korczyn, S. Shavit, and I. Shlosberg, The chick as a model for malignant hyperpyrexia, *Eur J Pharmacol* 61:187 (1980).
6. G. C. Allen, J. E. Fletcher, F. J. Huggins, P. A. Conti, and H. Rosenberg, Caffeine and halothane contracture testing in swine using the recommendations of the North American Malignant Hyperthermia Group, *Anesthesiology* 72:71 (1990).
7. H. Rosenberg and S. Reed, In vitro contracture tests for susceptibility to malignant hyperthermia, *Anesth Analg* 62:415 (1983).

8. J. E. Fletcher, H. Rosenberg, and F. H. Lizzo, Effects of droperidol, haloperidol and ketamine on halothane, succinylcholine and caffeine contractures: implications for malignant hyperthermia, *Acta Anaesthesiol Scand* 33:187 (1989).

9. J. E. Fletcher, F. J. Huggins, and H. Rosenberg, The importance of calcium ions for in vitro malignant hyperthermia testing, *Can J Anaesth* 37:695 (1990).

10. J. E. Fletcher and H. Rosenberg, In vitro muscle contractures induced by halothane and suxamethonium: II. Human skeletal muscle from normal and malignant hyperthermia susceptible patients, *Br J Anaesth* 58:1433 (1986).

11. P. Palade, Drug-induced Ca^{2+} release from isolated sarcoplasmic reticulum. I. Use of pyrophosphate to study caffeine-induced Ca^{2+} release, *J Biol Chem* 262:6135 (1987).

12. A. Saito, S. Seiler, A. Chu, and S. Fleischer, Preparation and morphology of sarcoplasmic reticulum from rabbit skeletal muscle, *J Cell Biol* 99:875 (1984).

13. J. R. Mickelson, E. M. Gallant, L. A. Litterer, K. M. Johnson, W. E. Rempel, and C. F. Louis, Abnormal sarcoplasmic reticulum ryanodine receptor in malignant hyperthermia, *J Biol Chem* 263:9310 (1988).

14. G. M. Vita, J. E. Fletcher, L. Tripolitis, P. A. Conti and H. Rosenberg, Ryanodine binding in swine correlates with strain differences, not malignant hyperthermia (MH) susceptibility, *Anesthesiology* 73:A460 (1990).

15. J. E. Fletcher, K. Michaux, and M.-S. Jiang, Contribution of bee venom phospholipase A_2 contamination in melittin fractions to presumed activation of tissue phospholipase A_2, *Toxicon* 28:647 (1990).

16. J. E. Fletcher, K. Erwin, and L. J. Krueger, Fatty acid uptake and catecholamine-stimulated phospholipid metabolism in immortalized airway epithelial cells established from primary cultures, *Biochem Int* 21:733 (1990).

17. J. E. Fletcher, H. Rosenberg, K. Michaux, K. S. Cheah, and A. M. Cheah, Lipid analysis of skeletal muscle from pigs susceptible to malignant hyperthermia, *Biochem Cell Biol* 66:917 (1988).

18. J. E. Fletcher, P. Kistler, H. Rosenberg, and K. M. Michaux, Dantrolene and mepacrine antagonize the hemolysis of human red blood cells by halothane and bee venom phospholipase A_2, *Toxicol Appl Pharmacol* 90:410 (1987).

19. W. R. Morrison and L. M. Smith, Preparation of fatty acid methyl esters and dimethylacetals from lipids with boronfluoride-methanol, *J Lipid Res* 5:600 (1964).

20. J. E. Fletcher, H. Rosenberg, K. Michaux, L. Tripolitis, and F. H. Lizzo, Triglycerides, not phospholipids, are the source of elevated free fatty acids in muscle from patients susceptible to malignant hyperthermia, *Eur J Anaesth* 6:355 (1989).

21. J. E. Fletcher, M.-S. Jiang, L. Tripolitis, L. A. Smith, and J. Beech, Interactions in red blood cells between fatty acids and either snake venom cardiotoxin or halothane, *Toxicon* 28:657 (1990).

22. S. J. Wieland, J. E. Fletcher, H. Rosenberg, and Q. H. Gong, Malignant hyperthermia: slow sodium current in cultured human muscle cells, *Am J Physiol* 257:C759 (1989).

23. M. A. K. Markwell, S. M. Haas, L. I. Bieber, and N. E. Tolbert, A modification of the Lowry procedure to simplify protein determination in membrane and lipoprotein samples, *Anal Biochem* 87:206 (1978).

24. P. Stralfors, P. Bjorgell, and P. Belfrage, Hormonal regulation of hormone-sensitive lipase in intact adipocytes: identification of phosphorylated sites and effects on the phosphorylation by lipolytic hormones and insulin, *Proc Natl Acad Sci* 81:3317 (1984).

25. J. R. Mickelson, E. M. Gallant, W. E. Rempel, K. M. Johnson, L. A. Litterer, B. A. Jacobson, and C. F. Louis, Effects of the halothane-sensitivity gene on sarcoplasmic reticulum function, *Am J Physiol* 257:C781 (1989).

26. K. S. Cheah and A. M. Cheah, Skeletal muscle mitochondrial phospholipase A_2 and the interaction of mitochondria and sarcoplasmic reticulum in porcine malignant hyperthermia, *Biochim Biophys Acta* 638:40 (1981).

27. S. G. Laychock, Fatty acids and cyclooxygenase and lipoxygenase pathway inhibitors modulate inositol phosphate formation in pancreatic islets, *Mol Pharmacol* 37:928 (1990).

28. P. S. Foster, E. Gesini, C. Claudianos, K. C. Hopkinson, and M. A. Denborough, Inositol 1,4,5,-trisphosphate phosphatase deficiency and malignant hyperpyrexia in swine, *Lancet* 1:124 (1989).

29. J. W. Schmidt and W. A. Catterall, Palmitylation, sulfation, and glycosylation of the alpha subunit of the sodium channel. Role of post-translational modifications in channel assembly, *J Biol Chem* 262:13713 (1987).

30. S. J. Wieland, J. E. Fletcher, Q. H. Gong, and H. Rosenberg, Unsaturated fatty acids modulate sodium current in cultured skeletal muscle, *Soc Neurosci Abstr* in press (1990).

31. L. H. Schlisefeld and M. Barany, Hydrolysis of phosphatidylethanolamine, phosphatidylcholine and glycerophosphoylcholine in skeletal muscle, *Mol Physiol* 7:165 (1985).

32. R. C. Levitt, V. A. McKusick, J. E. Fletcher, and H. Rosenberg, Gene candidate (letter), *Nature* 345:297 (1990).

33. T. V. McCarthy, J. M. S. Healy, J. J. A. Heffron, M. Lehane, T. Deufel, F. Lehmann-Horn, M. Farrall, and K. Johnson, Localisation of the malignant hyperthermia susceptibility locus to human chromosome 19q12-13.2, *Nature* 343:562 (1990).

34. T. Stadler, W. Ebert, F. Kehlbach, E. Muller, and H. von Faber, Beta-adrenergic receptors and adenylate cyclase activity in heart, muscle and adipose tissue of German Landrace pigs selectively bred for differences in backfat deposition, *Horm Metabol Res* 22:145 (1990).

35. G. Gandemer, G. Durand, and G. Pascal, Relative contribution of the main tissues and organs to body fatty acid synthesis in the rat, *Lipids* 18:223 (1983).

Transitional Concepts: Skeletal - Cardiac Muscle

7

Why Does Halothane Relax Cardiac Muscle but Contract Malignant Hyperthermic Skeletal Muscle?

S. Tsuyoshi Ohnishi, Masayuki Katsuoka

SUMMARY

We have studied the question of the possible role of sarcoplasmic reticulum (SR) in the interaction of volatile anesthetics (such as halothane, enflurane and isoflurane) with muscle. We used two cardiac muscle models, *i.e.*, isolated rat myocytes and Langendorff perfused rat hearts. We compared the results with those for skeletal muscle SR from rabbits, rats and pigs susceptible to malignant hyperthermia (MH). In both skeletal and cardiac muscle SR, volatile anesthetics enhanced the calcium release from the SR. In cardiac muscle, these agents are known to decrease contracility (negative inotropism). We found that caffeine, a well-known agent which releases calcium from the SR, also had a negative inotropic effect in cardiac muscle, raising the possibility of an unexpected link between the potentiation of calcium release and mechanism underlying the observed negative inotropism. Current understanding of anesthetic mechanisms does not include this possibility. We further found that both volatile anesthetics and caffeine decrease the content of calcium in the SR, suggesting that the increase of calcium permeability results in the decrease of calcium ions in the SR which are available for excitation-contraction (E-C) coupling. In MH-susceptible skeletal muscle, a similar increase in calcium permeability does not cause a decrease of contractility, but rather may contribute to a fatal syndrome of temperature increase provoked by abnormal contracture. This difference may be because in skeletal myoplasm calcium ions recycle internally, while in the cardiac muscle cell they are in dynamic equilibrium with extracellular calcium ions.

INTRODUCTION

Skeletal Muscle and Cardiac Muscle

There are certain similarities and also dissimilarities between skeletal and cardiac muscle. A schematic presentation (figure 1) may help us to understand the basic mechanisms involved in the regulation of contraction of both muscle types. In

S. TSUYOSHI OHNISHI, Philadelphia Biomedical Research Institute, King of Prussia, Pennsylvania 19406. MASAYUKI KATSUOKA, Ihara Chemical Company, Shizuoka, Japan.

Mechanisms of Anesthetic Action in Skeletal, Cardiac, and Smooth Muscle
Edited by T.J.J. Blanck and D.M. Wheeler, Plenum Press, New York, 1991

73

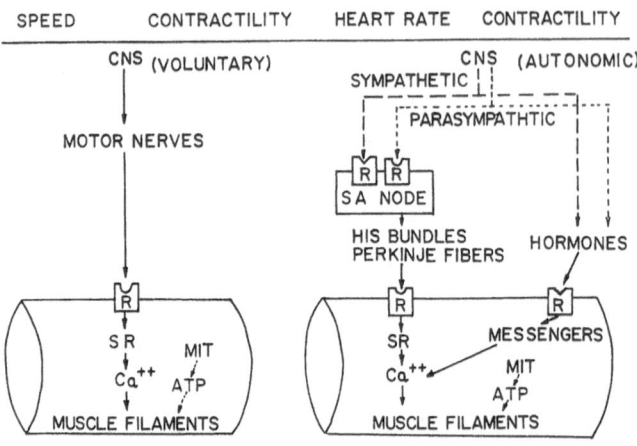

FIGURE 1. Schematic illustration of the regulation mechanism of skeletal and cardiac muscle.

skeletal muscle, the movement is controlled by motor nerve in accordance with the signal impulses dispatched from the central nervous system (CNS). In contrast, the movement of cardiac muscle is regulated by both the sympathetic and parasympathetic nervous system.

In skeletal muscle, we have shown that volatile anesthetics increased the calcium-induced calcium release from the SR.[1-7] Ethanol, which has an action similar to general anesthetics, increased the calcium permeability, thereby decreasing the calcium content of the SR.[5,6] We also demonstrated that these agents did not inhibit the calcium uptake activity, but increased the calcium permeability.[5,7] Halothane and caffeine triggered the release of calcium ions from the skeletal SR prepared from MH-susceptible pigs, while at the same time, they did not trigger release from the SR of normal pigs.[3,7] In the MH-susceptible SR, the calcium ion permeability was much higher than that of normal pigs, and the permeability was greatly increased in the presence of sub-anesthetic concentrations of halothane (less than 100 μM, as seen in figures 2 and 3).[4] Caffeine had a similar effect[4] (figure 4). Thus, as far as the ability of releasing calcium from the SR is concerned, both halothane and caffeine behaved similarly.

From the standpoint of E-C coupling mechanisms, the role of calcium-induced calcium release from the SR may play a more important role in cardiac muscle than in skeletal muscle.[8] Ohnishi found that the calcium-induced calcium release is very sensitive to drugs, such as caffeine and volatile anesthetics.[1-7] Therefore, it is to be expected that the effect of volatile anesthetics may be more pronounced in cardiac muscle than in skeletal muscle.

Although these agents have been known to decrease contractility of the heart (negative inotropism), they are also known to cause contracture in the skeletal muscle of MH-susceptible subjects (in humans and pigs). We will present a hypothesis that a "calcium recycling" movement is different between cardiac and skeletal muscle. This could explain why the same agents decrease contractility in one system but cause contracture in the other.

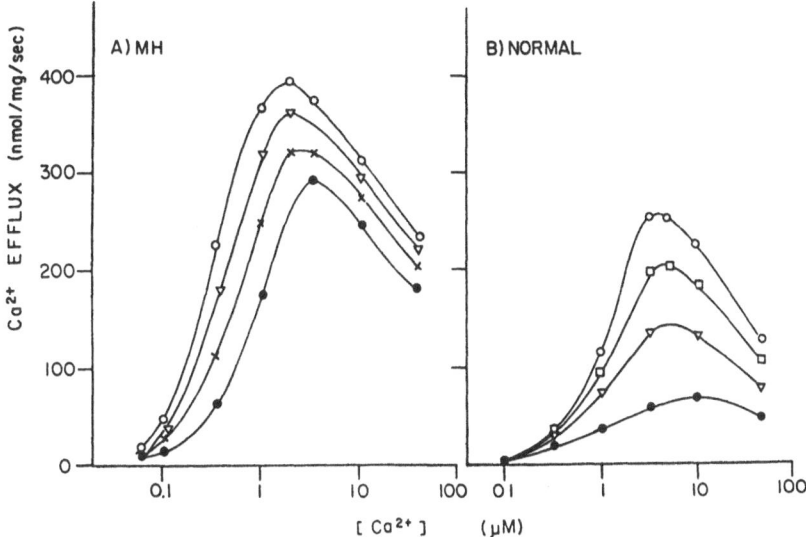

FIGURE 2. Effect of extravesicular calcium concentration at various halothane concentrations on the calcium permeability of SR from (A) MH pig and (B) normal pig. Experimental conditions: 120 mM KCl, 40 mM MES buffer (pH 6.8), 25°C. Halothane concentrations: 0 (•); 25 μM (×); 50 μM (▽); 100 μM (□) and 200 μM (○). Taken from S. T. Ohnishi, *Biochim Biophys Acta* 897:261-268 (1987).

Negative Inotropic Effect of Volatile Anesthetics

Volatile anesthetics have been used for decades, and they are known to cause a negative inotropic effect.[9-12] However, the exact mechanism underlying this effect is still unknown. Price *et al.* found that the negative inotropic effect was antagonized by an increase in extracellular calcium concentration,[13,14] and proposed that general anesthetics somehow interdict the movement of calcium ions involved in the

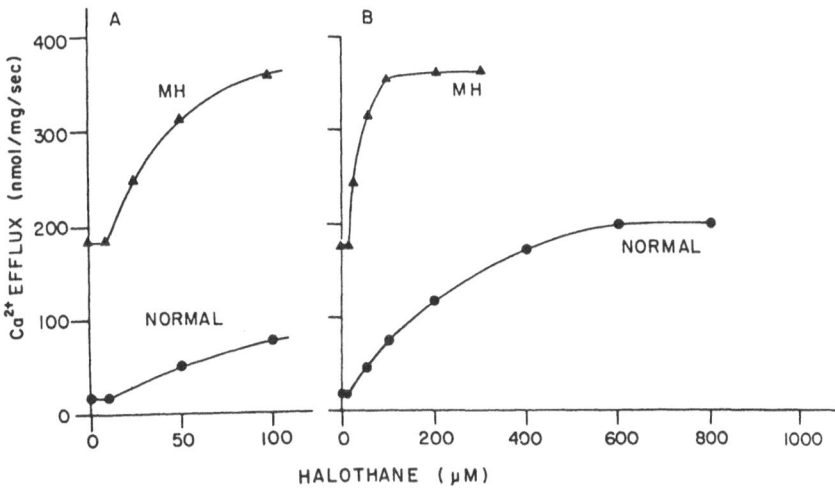

FIGURE 3. Effect of halothane on the calcium permeability of MH and normal SR. Panel (A) presents a low range of halothane concentrations and (B) a higher range. The free [Ca^{2+}] was 1 μM. Other conditions as in figure 2. Taken from S. T. Ohnishi, *Biochim Biophys Acta* 897:261-268 (1987).

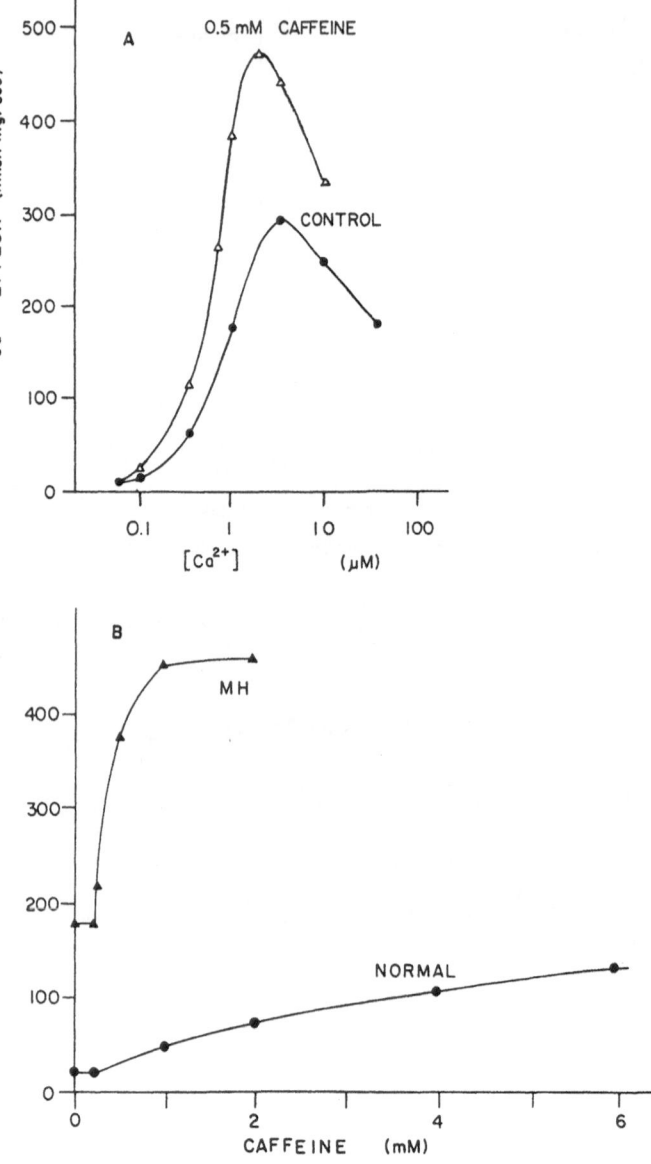

FIGURE 4. (A) Effects of caffeine (500 μM) on the calcium permeability of MH SR. (B) Effect of caffeine on the calcium permeability of MH and normal SR. Conditions were the same as shown in figure 2. Taken from S. T. Ohnishi, *Biochim Biophys Acta* 897:261-268 (1987).

mechanism of excitation-contraction (E-C) coupling. It is generally accepted that anesthetics interfere with the E-C coupling mechanism in the heart through several different mechanisms, namely on 1) the action potential,[15-18] 2) exchangeable Ca^{2+} in the plasma membrane,[19-20] 3) calcium channels,[25,26] 4) transverse tubule and 5) sarcoplasmic reticulum (SR),[21,22,29,30] and 6) contractile proteins.[22,30]

Several lines of evidence suggest that the interaction of calcium ions with the plasma membrane may play a role in this phenomenon.[19,20,23,24] Recent studies emphasize that the regulation of calcium entry through calcium channels by anesthetic agents may be important.[25,26] The effect of anesthetics on the transverse tubule have not been well studied.

In this paper, we focus our attention on the effect of volatile anesthetics on the SR. Caffeine has been known to increase calcium permeability in both skeletal and cardiac SR.[2-4,27,28] We have observed that in skeletal muscle SR, as well as in rat myocytes and rat Langendorff perfused hearts, we are proposing that the effect of volatile anesthetics is to increase calcium permeability of cardiac SR, thereby decreasing the calcium content of the SR to result in negative inotropic effects.[1,29-33]

MATERIALS AND METHODS

Myocyte Model

Myocytes were isolated from the ventricle of male Sprague-Dawley rats (300-400 g in body weight) by a modification of Kao's procedure.[34]

The free intracellular Ca concentration of these cells was measured using the Ca^{2+}-sensitive, fluorescent dye, fura-2. The fluorescence signal from fura-2 loaded cells was measured using a 2-channel (340 and 380 mm excitation, 490 nm emission), spinning wheel filter, fluorescence spectrophotometer, Model 2B (Biomedical Instrumentation Group, University of Pennsylvania). A myocyte suspension (2.5 mg dry weight/ml) was prepared with a calcium HEPES buffer which had been equilibrated with oxygen at 37°C, and was introduced to a 10 × 10 mm cuvette kept at 37 ± 0.1°C. The cuvette was sealed with a lid in order to reduce the loss of volatile anesthetics. The volume of the myocyte suspension was 2 ml, and the suspension was continuously stirred with a magnetic stirring device. Electric stimulation was with platinum electrodes connected to a Grass Stimulator (supra-threshold voltage pulses of 10 msec duration, 0.2 Hz). Data from 10 to 20 successful experiments at 340 nm and 380 nm were averaged using an IBM personal computer with a data translation board (DT 2801-first A/D Board, Marlboro, MA). Free calcium concentration was calculated using the equation described by Grynkiewicz et al.[35]

Volatile anesthetics were introduced into the cell suspensions in the following manner. Halothane was freshly distilled to remove thymol (preservative) and dissolved in dimethyl sulfoxide. This solution was directly added to the myocyte suspension in the cuvette. Enflurane and isoflurane were dissolved directly into dimethyl sulfoxide and used similarly. Gas phase anesthetic concentrations (the concentration which would be in equilibrium with the solution) were calculated from the known aqueous phase concentrations according to Renzi and Waud's partition coefficients[36] and shown in vol% in this paper.

The measurement of the Ca content of the SR was based on the property of caffeine to release SR Ca into the cytoplasm. After we recorded a train of intracellular calcium transients, electrical stimuli were stopped and EGTA (ethyleneglycol-bis-β-(aminoethyl ether) N,N'tetra-acetic acid) was added. When the fluorescence recording stabilized after 30 seconds, 4 mM caffeine was added and the fluorescence change, representing the amount of calcium release, was recorded. A similar experiment was performed in the presence of 4 mM halothane.

The total calcium content of the cells was measured using atomic absorption as follows: After calcium transient measurement with the fura-2 method, the myocyte suspension was wash-centrifuged in an ice-cold calcium-free HEPES buffer solution containing 5 mM EGTA. The myocytes were disrupted with 1 ml of 1% Brij-35 solution. From this solution, samples were taken out in triplicates and calcium content was assayed by a Zeeman effect atomic absorption spectrophotometer (Hitachi Model

Z-8000). The protein concentration of this solution was measured using the biuret method. The calcium content was expressed in nmoles calcium/mg protein.

Langendorff Perfusion Model

Male Sprague-Dawley rats weighting 300-400 g were anesthetized by intraperitoneal injection of 50 mg/kg sodium pentobarbital. The heart was quickly removed and immersed in ice cold Krebs-Henseleit bicarbonate solution (KHB solution; concentrations in mM: 118 NaCl, 4.7 KCl, 1.25 $CaCl_2$, 1.25 $MgSO_4$, 1.25 KH_2PO_4, 24 $NaHCO_3$, 15 glucose; pH adjusted to 7.4 when equilibrated with 95% O_2/5% CO_2 at 37°C). Then, extraneous tissue was removed and the aorta was cannulated on the perfusion apparatus. The left ventricular pressure was measured with a latex balloon connected to a Statham PB 23 transducer via a thin catheter. The baloon was inserted through the mitral valve and filled with water (volume, 0.1-0.2 ml).

The perfused heart was stimulated with electric pulses (10 V, 4-5 Hz, 10 msec duration). The KHB solution was switched to a KHB solution containing either caffeine or anesthetic agents (each anesthetic agent was vaporized by a Drager vaporizer and bubbled into the solution) and perfused for 10 minutes. The left ventricular pressure decreased and reached a new plateau level within the first 5 minutes. The perfusing solution was changed to a Ca^{2+}-free KHB solution, and perfused for 1 minute. The left ventricular muscle was trimmed into small pieces with scissors and homogenized with a Polytron homogenizer (volume 10 ml). The homogenate was extracted with 2 N HCl for 24 hours with constant shaking. The suspension was centrifuged at 2000 rpm for 10 minutes, and the calcium concentration of the supernatant was measured with a polarized Zeeman effect atomic absorption spectrophotometer (Hitachi Z-8000). The pellet was dissolved in 10 ml of 30% KOH and the protein concentration was determined by the biuret method. Measurements were done in triplicates.

The calcium content of the SR was estimated using the principle developed by Endo[27,37] and Heide et al.[38] for muscle fibers, with a slight modification. The heart was continuously stimulated (5 Hz). After equilibration with KHB solution for 15 minutes, the perfusing solution was switched to a Ca^{2+}-free KHB solution for 30 seconds, then changed to a Ca^{2+}-free KHB solution containing 40 mM caffeine. Endo found that the calcium release at this high concentration of caffeine represents the amount of calcium left in the SR.[27,37] Since several contractions were observed, the height of each contraction was measured. It was assumed that each contraction was related to calcium release from the SR. Thus, the summation of the height of each contraction was regarded as representative of the total calcium release from the SR.[33]

Concentrations of halothane and enflurane from respective vaporizers were determined by gas chromatography. The Student's t-test was used to compare differences in calcium transients, calcium content and time constants between the control and anesthetic-treated condition. The difference were considered significant when the P value was less than 0.05.

RESULTS

Myocyte Model

Signal-averaged calcium transients are shown in figure 5 illustrating the effects of caffeine and halothane. The traces on the left of each panel show the transient before the addition of an agent and the traces on the right show the effect of an agent.

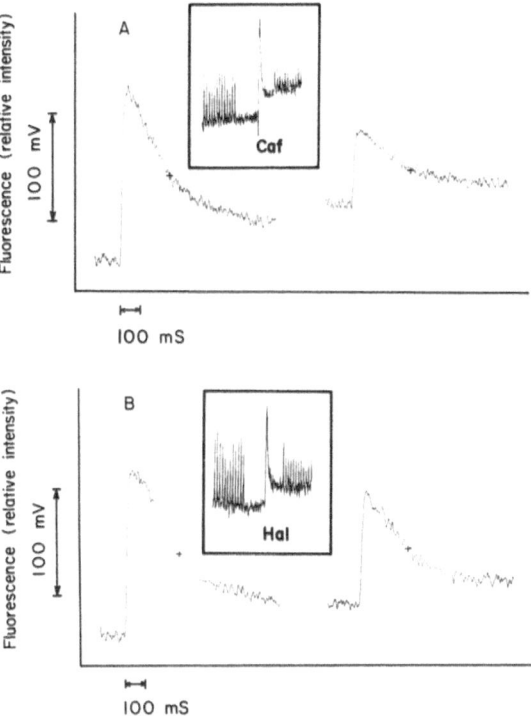

FIGURE 5. Effects of caffeine and halothane on the Ca transient. Myocytes were electrically stimulated for one-and-a-half minutes followed by a one minute rest. Then (A) 4 mM caffeine or (B) 1 mM halothane was added. After 1 minute, the myocytes were again stimulated as shown in the insets. Twenty transients were averaged. The left trace represents the Ca transient before the addition of agents, and the right trace represents effects of the agents. Taken from M. Katsuoka and S. T. Ohnishi, *Anesthesiology* 70:954-960 (1989).

The results demonstrate an increase in the resting calcium concentration and a decrease of the calcium transient.

The decay of calcium transient is mostly dependent upon the activity of calcium uptake of the SR, Na^+-Ca^{2+} exchange at the sarcolemmal membrane (SL), the activity of the calcium pump in the SL, and Ca^{2+} binding to troponin and parvalbumin. Therefore, the influence of an agent on the time constant of the decay of the calcium transient is determined by its net effect on all of these activities. We found that the value was not influenced significantly ($P > 0.2$) by these agents.[32]

Figure 6 shows 4 mM caffeine-stimulated calcium release from the SR (A), an index of the SR calcium content. This calcium release was decreased by pretreatment with 4 mM (13.6%) halothane (B). Halothane could also cause calcium release similar to caffeine (C). In the presence of caffeine, the calcium release caused by 4 mM halothane was reduced as shown in the trace marked D. Thus, caffeine and halothane had similar calcium releasing effects on the SR and their effects were interchangeable.

Figure 7 shows the relationship between calcium transients and the total calcium content of myocytes in the presence and absence of various agents. As shown in these figures, when the calcium transient decreased in the presence of an agent, the calcium content of the same myocyte preparation also decreased. It is interesting to note that all of these agents, caffeine, halothane, enflurane and isoflurane, had similar effects both on the calcium transient and on the calcium content.[32] A 50% depression of the calcium transient was caused by 1.3 mM halothane (4.4% or 5.9 MAC), 1.9 mM enflurane (6.5% or 3.9 MAC) or 3.2 mM isoflurane (14.8% or 12.9 MAC), where the values of MAC for these agents were taken from Eger.[39]

Whole Heart Model

When the heart was perfused with caffeine, the diastolic pressure increased. The systolic pressure showed an immediate transient increase, then decreased to reach a new plateau level which was lower than the control level (evidence of a negative inotropic effect). Similar changes were observed with both halothane and enflurane (figure 8). As shown in figure 9A, the pressure developed by the left ventricle (LVDP) was decreased by these agents in a dose-related manner. A 50% reduction was acheived with 1.8% halothane and 4.0% enflurane.

The effects of caffeine, halothane and enflurane on the Ca content of the left ventricular muscle are illustrated in figure 9B. All of these agents decreased the Ca content of the left ventricular muscle as measured by an atomic absorption spectrophotometer. When the heart was perfused with 2.8% halothane, 5.1%

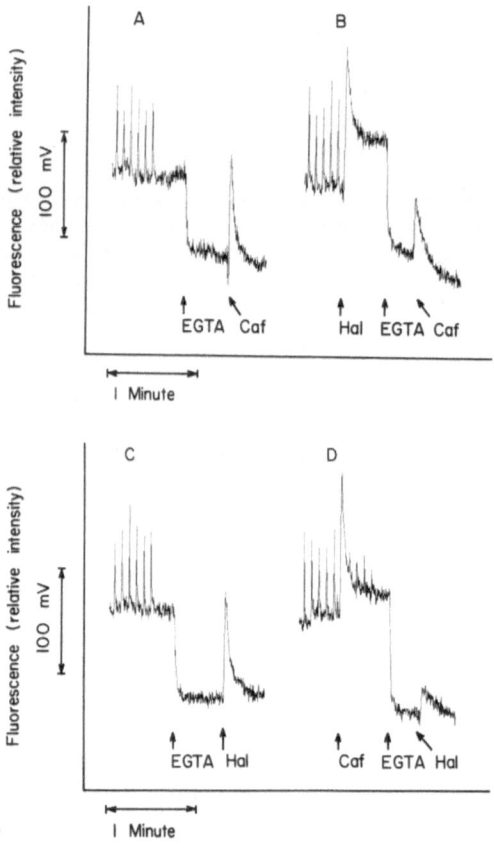

FIGURE 6. Effects of caffeine and halothane on Ca content of SR. (A) The Ca content of the SR was estimated by the addition of 4 mM caffeine in the absence of external calcium (4 mM EGTA). (B) Halothane (4 mM) was added before addition of EGTA and caffeine. The caffeine-induced calcium release was reduced in the presence of halothane. (C) Halothane (4 mM) caused a Ca release. (D) The halothane-induced Ca release was decreased in the presence of 4 mM caffeine. Taken from M. Katsuoka and S. T. Ohnishi, *Anesthesiology* 70:954-960 (1989).

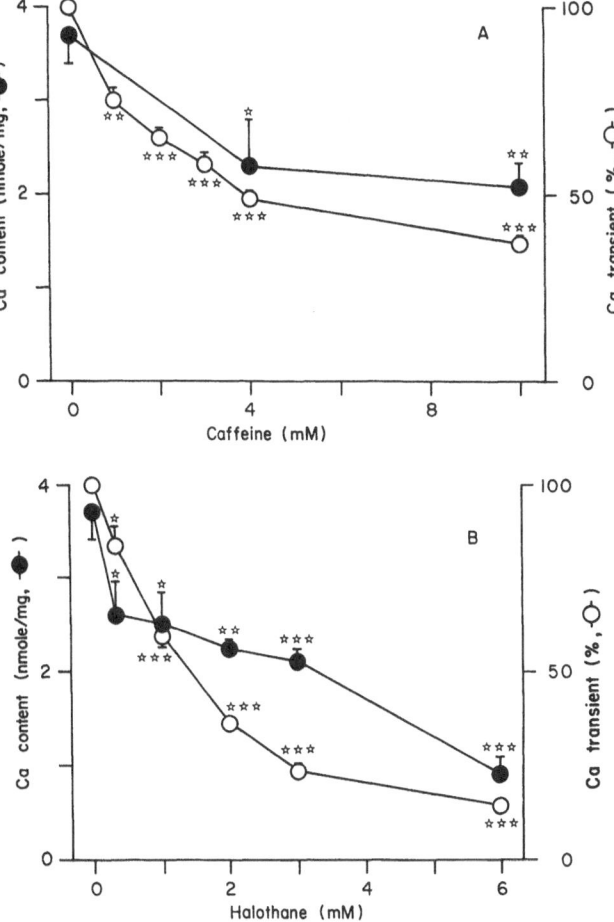

FIGURE 7. Dose-response relationship illustrating effects of (A) caffeine and (B) halothane on the Ca content and Ca transient of myocyte suspensiions. Closed circles represent Ca content measured by atomic absorption (left ordinates). Open circles represent Ca transcint (right ordinates). Each value indicates mean ± SEM. Asterisks show significant differences between control and agent (*P < 0.05, **P < 0.01, ***P < 0.001). Taken from M. Katsuoka and S. T. Ohnishi, *Anesthesiology* 70:954-960 (1989).

enflurane or 2 mM caffeine, the Ca content decreased to 70.3 ± 9.5% of control (P < 0.05), 69.4 ± 6.9% (P < 0.05), 56.1 ± 6.9% (P < 0.01), respectively.

In order to understand the relationship between the LVDP and the SR calcium content, the Ca content was estimated from the force of contractions related to caffeine-induced Ca release from the SR.[33] As shown in figure 10, there was a distinct correlation between the LVDP and the Ca content of the SR (r = 0.95, P < 0.001 and r = 0.91, P < 0.001 for halothane and for enflurane, respectively).

DISCUSSION

The results presented in this paper demonstrate that the effect of halothane on the cardiac SR may be similar to that on the skeletal SR. We observed that the time constant of decay of the calcium transient was not influenced by halothane.[32] This

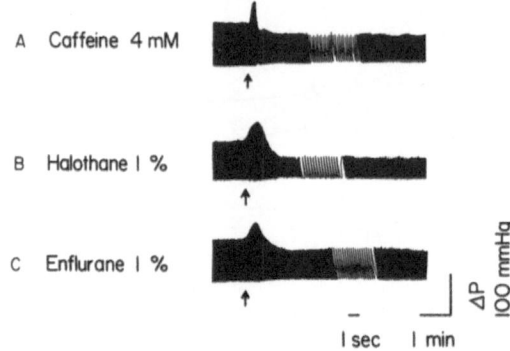

FIGURE 8. Effects of caffeine and inhalational anesthetics on the left ventricular developed pressure (LVDP) of the isolated heart. Representative recordings from isolated hearts perfused with (A) 4 mM caffeine, (B) 1% halothane and (C) 2% enflurane. Arrows indicate a point of changing perfusate. Taken from M. Katsuoka and S. T. Ohnishi, *Br J Anaesth* 62:669-673 (1989).

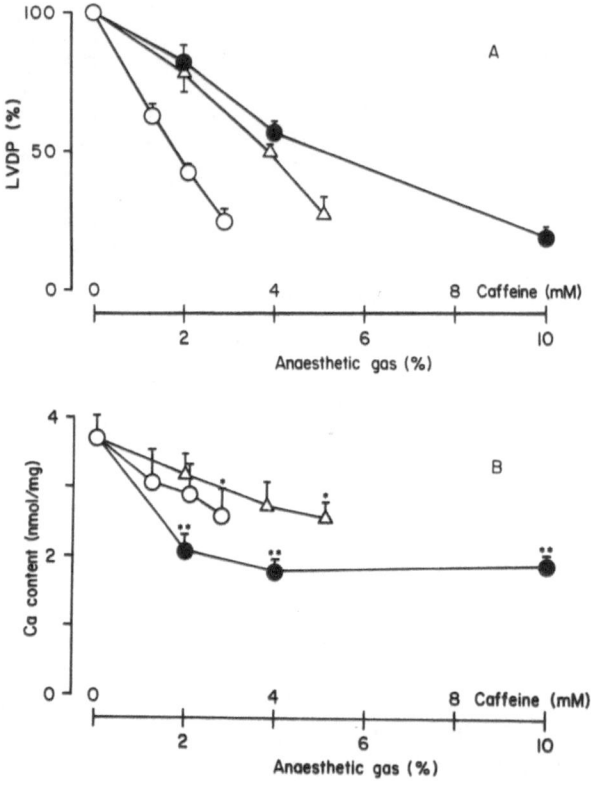

FIGURE 9. Effects of caffeine and inhalational anaesthetics on (A) the developed pressure (LVDP) and (B) calcium content. Symbols: Caffeine (filled circles), halothane (open circles) and enflurane (open triangles). Each point indicates mean ± S.E. (n = 6). Significance is indicated for difference between normal perfusate and perfusate with agents (*P < 0.05, **P < 0.01). Taken from M. Katsuoka and S. T. Ohnishi, *Br J Anaesth* 62:669-673 (1989).

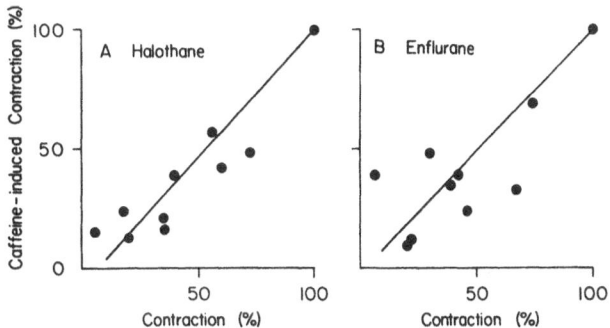

FIGURE 10. Relationship between the left ventricular developed pressure (LVDP) and Ca content in the SR. The LVDP was expressed as a percentage of control. The Ca content in the SR was measured by the caffeine-induced contraction method and expressed as a percentage of control. (A) Effect of halothane (r = 0.95, P < 0.001). (B) Effect of enflurane (r = 0.91, P < 0.001). Taken from M. Katsuoka and S. T. Ohnishi, *Br J Anaesth* 62:669-673 (1989).

suggests that activities of the calcium pumps in the SR and SL, Na^+-Ca^{2+} exchange in the SL, and Ca^{2+} binding to myofibrils may not be much influenced by halothane. Our results (from figure 5 and ref. 32) indicate that both caffeine and the volatile anesthetics caused 1) a calcium release, 2) a decrease in calcium transients, and 3) a sustained increase in the resting calcium concentration. These data strongly suggest that halothane, enflurane and isoflurane increased calcium release from the SR. Caffeine is, of course, a well-known calcium releasing agent in the cardiac SR.[27,28] The fact that caffeine, halothane, enflurane and isoflurane temporarily increased calcium release suggests that each of these agents increased the calcium permeability of the cardiac SR. Shortly after this temporary increase, the calcium transient gradually decreased with time. This seems to support our original hypothesis that anesthetics decrease the calcium content of SR by increasing the calcium permeability.[1,31] Our results are in good agreement with studies of Wheeler *et al.*[29] and Herland *et al.*,[30] who made similar observations using halothane, and with the study of Housmans and Murat[40] that isoflurane has less negative inotropic effect than halothane or enflurane.[32]

There may be a reason why halothane is a negative inotrope in the heart but is not in skeletal muscle. In the heart, it is well known that the calcium channels on the SR open upon depolarization and calcium influx takes place through these channels.[41-44] However, during the resting period, the calcium pump and Na^+-Ca^{2+} exchange, both located in the SL membrane, extrude calcium ions.[44-48] In contrast, in skeletal musacle there is no large influx and efflux of calcium ions across the SL membrane. Calcium released from the SR stays in the cell and is pumped back into the SR.[49-51] Since in cardiac muscle, the activities of the calcium pump and Na^+-Ca^{2+} exchange are very high, a portion of calcium ions which is released from the SR is transported out of the cell. The fact that the size of cardiac cells (diameter on the order of 20 microns) is smaller than that of skeletal muscle (diameter 60 microns or larger) would help this situation. The smaller the diameter, the greater the surface/volume ratio, and hence the greater would be the efficiency of exchanging the intracellular ions by the transport mechanisms across the SL membrane (figure 11).

Thus, in cardiac muscle calcium ions are in dynamic equilibrium with external calcium ions, and recycle with an extracellular link. This causes a decrease in the calcium content of the SR in the presence of anesthetics, and thus results in the observed negative inotropism in the heart. On the other hand, in skeletal muscle, calcium ions recycle "internally" (see figure 11). Thus, halothane could increase the

SKELETAL   CARDIAC

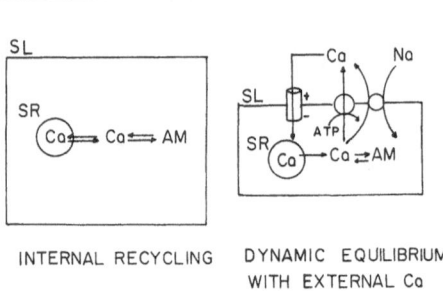

INTERNAL RECYCLING   DYNAMIC EQUILIBRIUM
                        WITH EXTERNAL Ca

FIGURE 11. Difference in calcium recycling between cardiac and skeletal muscle. AM stands for the actomyosin-troponin system and SL for the sarcolemma. On the SL of cardiac muscle, there are calcium channel, calcium pump and Na-Ca exchange transport (from left to right).

free calcium concentration in the muscle cells, especially in MH-susceptible individuals, where the calcium permeability of the SR is abnormally high[3,4], and could cause contracture.

In order to further verify this possibility, we examined the effects of caffeine and halothane. Endo *et al.* developed a method of measuring the calcium content of the SR by measuring caffeine-induced contraction of skinned fibers.[27,37] Using this method, Su and Kerrick demonstrated that the amount of calcium in cardiac SR decreased in the presence of halothane.[22] Applying this principle to the measurement of the calcium transient, we attempted to show that halothane and caffeine have similar properties. As shown in figure 6, we observed that the amount of calcium release by caffeine decreased in the presence of halothane and that the amount of calcium release by halothane decreased in the presence of caffeine. Their roles are interchangeable suggesting that the effect of both agents is to increase calcium permeability of cardiac SR.

Further evidence in support of our hypothesis is that the total calcium contents of both the myocytes and the whole heart as measured by an atomic absorption spectrophotometer decreased in the presence of caffeine, halothane, enflurane or isoflurane. Calcium binding to the plasma membrane was shown to increase in the presence of general anesthetics.[23,24] In brain mitochondria, 3% halothane was shown to decrease the calcium content. The effect of general anesthetics on the calcium content of cardiac mitochondria is not well known.

In this study, the calcium transient was significantly decreased at 0.5 mM (1.7% or 2.3 MAC) halothane, 0.7 mM (2.3 % or 1.4 MAC) enflurane and 1 mM (4.6% or 4 MAC) isoflurane. The fact that these concentrations were higher than 1 MAC may suggest the possibility that volatile anesthetics may also interfere with other components in muscle. It has been reported that halothane affects both action potentials[15-18] and calcium entry.[25,26] Further study is needed to clarify the contribution of these various other effects to the negative inotropism.

ACKNOWLEDGEMENTS

The authors acknowledge Dr. Henry Price, Department of Anesthesiology, Hahnemann University School of Medicine, who stimulated our interest in the anesthetic-induced negative inotropic effect, and gave us valuable suggestions. Gas chromatographic determinations of anesthetics were performed by Dr. Bryan Marshall of the Department of Anesthesiology, University of Pennsylvania School of Medicine. Thanks are also due to Dr. Andrew Thomas, Thomas Jefferson University, who allowed us to use his computer program to analyze fura-2 data, and to Dr. Makoto Endo, Tokyo University, for his valuable suggestions. The work was supported in part by NIH Grants GM 35681 and GM 33025.

REFERENCES

1. S. T. Ohnishi, Calcium-induced calcium release from fragmented sarcoplasmic reticulum, *J Biochem* 86:1147-1150 (1979).

2. S. T. Ohnishi, Calcium-Induced calcium release as a gated calcium transport, *in*: "Mechanism of Gated Calcium Transport Across Biological Membranes," S. T. Ohnishi, M. Endo, eds., Academic Press, New York (1981).

3. S. T. Ohnishi, A. J. Waring, S. G. Fang, K. Horiuchi, J. L. Flick, K. K. Sadanaga, T. Ohnishi, Abnormal membrane properties of the sarcoplasmic reticulum of pigs susceptible to malignant hyperthermia: Modes of action of halothane, caffeine, dantrolene and two other drugs, *Arch Biochem Biophys* 247:294-301, (1986).

4. S. T. Ohnishi, Effects of halothane, caffeine, dantrolene and tetracaine on the calcium permeability of skeletal sarcoplasmic reticulum of malignant hyperthermic pigs, *Biochem Biophys Acta* 897:261-268, (1987).

5. S. T. Ohnishi, J. L. Flick, F. Rubin, Ethanol increases calcium permeability of heavy sarcoplasmic reticulum of skeletal muscle, *Arch Biochem Biophys* 233:588-594, (1984).

6. S. T. Ohnishi, A. J. Waring, S. G. Fang, K. Horiuchi, T. Ohnishi, Sarcoplasmic reticulum membrane of rat skeletal muscle is disordered with chronic alcohol ingestion, *Membr Biochem* 6:49-63, (1984).

7. S. T. Ohnishi, S. Taylor, G. A. Gronert, Calcium-induced calcium-release from sarcoplasmic reticulum of pigs susceptible to malignant hyperthermia: The effect of halothane and dantrolene, *FEBS Lett* 161:103-107 (1983).

8. A. Fabiato, F. Fabiato, Contractions induced by a calcium triggered release of calcium from the sarcoplasmic reticulum of single skinned cardiac cells, *J Physiol (Lond)* 249:469-495 (1975).

9. W. R. Brewster, J. P. Isaacs, T. Waing-Anderson, Depressant effect of ether on myocardium of the dog and its modification by reflex release of epinephrine and norepinephrine, *J Pharmacol Exp Ther* 175:399-414 (1953).

10. H. L. Price, M. Helrich, The effect of cyclopropane, diethyl ether, nitrous oxide, thiopental and hydrogen ion concentration on the myocardial function of the dog heart-lung preparation, *J Pharmacol Exp Ther* 115:206-216 (1955).

11. B. R. Brown, J. R. Crout, A comparative study of the effect of five general anesthetics on myocardial contractility, *Anesthesiology* 34:236-245 (1971).

12. B. F. Rusy, H. Komai, Anesthetic depression of myocardial contractility: A review of possible mechanisms, *Anesthesiology* 67:745-766 (1987).

13. H. L. Price, Calcium reverses myocardial depression caused by halothane: Site of action, *Anesthesiology* 218:576-579 (1974).

14. H. L. Price, Myocardial depression by nitrous oxide and its reversal by Ca^{++}, *Anesthesiology* 44:211-215 (1976).

15. C. Lynch, S. Vogel, N. Sperelakis, Halothane depression of myocardial slow aciton potentials, *Anesthesiology* 55:360-368 (1981).

16. Z. J. Bosnjak, J. P. Kampine, Effects of halothane, enflurane, and isoflurane on the SA node, *Anesthesiology* 58:314-321 (1983).

17. Z. J. Bosnjak, J. P. Kampine, Effects of halothane on transmembrane potentials, Ca^{2+} transients, and papillary muscle tension in the cat, *Am J Physiol* 251:H374-H381 (1986).

18. C. Lynch III, Differential depression of myocardial contractility by halothane and isoflurane in vitro, *Anesthesiology* 64:620-631 (1986).

19. G. A. Langer, S. D. Serena, L. M. Nudd, Cation exchange in heart cell culture: Correlation with effects on contractile force, *J Mol Cell Cardiol* 6:149-161 (1974).

20. W. G. Nayler, J. Szeto, Effect of sodium pentobarbital on calcium in mammalian heart muscle, *Am J Physiol* 222:339-344 (1972).

21. E. S. Casella, N. D. A. Suite, Y. I. Fisher, T. J. J. Blanck, The effect of volatile anesthetics on the pH dependence of calcium uptake by cardiac sarcoplasmic reticulum, *Anesthesiology* 67:386-390 (1987).

22. J. Y. Su, W. G. L. Kerrick, Effects of halothane on caffeine-induced tension transients in functionally skinned myocardial fibers, *Pflugers Arch* 380:29-34 (1979).

23. S. T. Ohnishi, C. A. DiCamillo, M. Singer, H. L. Price, Correlation between halothane-induced myocardial depression and decreases in La^{3+}-displaceable Ca^{2+} in cardiac muscle cells, *J Cardiovasc Pharmacol* 2:67-75 (1980).

24. S. T. Ohnishi, D. M. Obzansky, H. L. Price, The increase in calcium binding of cardiac plasma membrane lipoprotein caused by general anesthetics and alcohol, *Can J Physiol Pharmacol* 58:525-530 (1980).

25. B. Drenger, T. J. J. Blanck, Volatile anesthetics depress the binding of calcium channel blocker to purified cardiac sarcolemma (abstract), *Anesthesiology* 69:A16 (1988).

26. Z. J. Bosnjak, F. D. Supan, N. J. Rusch, The effects of halothane, enflurane and isoflurane on calcium current in isolated canine ventricular cells, *Anesthesiology* 74:340-345 (1991).

27. M. Endo, Calcium release from the sarcoplasmic reticulum, *Physiol Rev* 57:71-108 (1977).

28. J. L. Sutko, L. J. Thompson, A. A. Kort, E. G. Lakatta, Comparison of effects of ryanodine and caffeine on rat ventricular myocardium, *Am J Physiol* 250:H786-H795 (1986).

29. D. M. Wheeler, R. T. Rice, R. C. Hansford, E. G. Lakatta, The effect of halothane on the free intracellular calcium concentration of isolated rat heart cells, *Anesthesiology* 69:578-583 (1988).

30. J. S. Herland, D. G. Stephenson, F. J. Julian, Halothane affects the contractile apparatus and sarcoplasmic reticulum of mechanically skinned rat ventricular fibers (abstract), *Biophys J* 53:335a (1988).

31. H. L. Price, S. T. Ohnishi, Effects of anesthetics on the heart, *Fed Proc* 39:575-1579 (1980).

32. M. Katsuoka, S. T. Ohnishi, Volatile anesthetics decrease calcium content of isolated myocytes, *Anesthesiology* 70:954-960 (1989).

33. M. Katsuoka, S. T. Ohnishi, Inhalation anaesthetics decrease calcium content of cardiac sarcoplasmic reticulum, *Br J Anaesth* 62:669-673 (1989).

34. R. L. Kao, E. W. Christman, S. L. Luh, J. M. Kraubs, G. F. Tylers, G. H. Williams, The effects of insulin and anoxia on the metabolism of isolated mature rat cardiac myocytes, *Arch Biochem Biophys* 203:587-599 (1980).

35. G. Grynkiewiez, M. Poenie, R. Y. Tsien, A new generation of Ca^{2+} indicators with greatly improved fluorescence properties, *J Biol Chem* 260:3440-3450 (1985).

36. F. Renzi, B. E. Waud, Partition coefficients of volatile anesthetics in Krebs'solution, *Anesthesiology* 47:62-63 (1977).

37. M. Endo, Mechanism of action of caffeine on the sarcoplasmic reticulum of skeletal muscle, *Proc Japan Acad* 51:479-484 (1975).

38. R. S. V. Heide, R. A. Altschuld, K. G. Lamka, C. E. Ganote, Modification of caffeine-induced injury in calcium-free perfused rat hearts, *Am J Pathol* 123:351-364 (1986).

39. E. I. Eger, Isoflurane: A review. *Anesthesiology* 55:559-576 (1981).

40. P. R. Housmans, I. Murat, Comparative effects of halothane, enflurane, and isoflurane at equipotent anesthetic concentrations on isolated ventricular myocardium of the ferret: I. Contractility, *Anesthesiology* 69:451-463 (1988).

41. M. Morad, Y. Goldman, Excitation-contraction coupling in the sarcoplasmic reticulum skinnet of tension, *Prog Biophys Mol Biol* 27:259-313 (1973).

42. G. B. McClellan, S. Winegrad, The regulation of the calcium sensitivity of the contractile system in mammalian cardiac muscle. *J Gen Physiol* 72:737-764 (1978).

43. H. Reuter, C. F. Stevens, R. W. Tsien, G. Yellin, Properties of single calcium channels in cardiac cell culture, *Nature* 297:501-504 (1982).

44. R. A. Chapman, Excitation-contraction coupling in heart muscle, *Prog Biophys Mol Biol* 35:1-52 (1979).

45. H. Reuter, Exchange of calcium ions in the mammalian myocardium: Mechanisms and physiological significance, *Circ Res* 34:599-606 (1974).

46. P. Caroni, E. Carafoli, An ATP-dependent Ca^{2+}-pumping system in dog heart sarcolemma, *Nature* 283:765-767 (1980).

47. E. Caraboeuf, P. Gautier, P. Guiraudou, Potential and tension changes induced by sodium removal in dog Purkinje fibers: Role of an electrogenic sodium-calcium exchange, *J Physiol (Lond)* 311:605-622 (1981).

48. E. Carafoli, The homeostasis of calcium in heart cells, *J Mol Cell Cardiol* 17:203-212 (1985).

49. F. F. Jobsis, M. J. O'Connor, Calcium release and reabsorption in the sartorius muscle of the toad, *Biochem Biophys Res Commun* 25:246-252 (1966).

50. S. Winegrad, Autoradiographic studies of intracellular calcium in frog skeletal muscle, *J Gen Physiol* 48:455-479 (1965).

51. S. Winegrad, Intracellular calcium movements of frog skeletal muscle during recovery from tetanus, *J Gen Physiol* 51:65-83 (1968).
52. A. J. Sweetman, A. F. Esmail, Evidence for the role of calcium ions and mitochondria in the maintenance of anesthesia in the rat, *Biochem Biophys Res Commun* 64:885-890 (1975).

Part II — Cardiac Muscle

8

Cardiac Effects of Anesthetics

Zeljko J. Bosnjak

The following chapters add to our knowledge of the possible mechanisms underlying the actions of anesthetics on the myocardium. These studies were designed to examine the effects of anesthetics on ionic fluxes across the sarcolemma, rapid changes in intracellular calcium, calcium transport functions of the sarcoplasmic reticulum (SR), isometric contractile force, and calcium sensitivity of the contractile apparatus.

Volatile anesthetics at clinically useful concentrations depress the contractile force of the heart, and these actions in part contribute to a significant decrement of cardiovascular homeostasis. Studies in isolated heart and papillary muscle preparations consistently demonstrate that these agents produce dose-dependent decreases of indices of contractility.[1-6] The mechanisms underlying the negative inotropic effects of the volatile anesthetics are not fully understood. Both past and present investigations of these actions have focused on cellular Ca^{2+} metabolism and/or the ultimate effect of Ca^{2+} on the myofibrillar apparatus. The reason behind such a focus is obvious: contractile force generated in the beating heart is associated with the rise and fall of intracellular calcium ion concentration. There are several mechanisms, then, by which agents may directly alter contractile performance of cardiac muscle. The first group of actions represents "upstream" mechanisms whereby the calcium transients themselves are altered, mostly by a variety of effectors at the surface membrane and sarcoplasmic reticulum. A second group, "downstream" mechanisms, involves changes in the sensitivity of troponin-C to calcium, or an altered response of the myofilaments to a given level of occupancy of the calcium binding sites on troponin-C. The sites of action of volatile anesthetics are difficult to separate because a change in the influx of sarcolemmal calcium alters the sequestration of calcium in the SR and ultimately the level of myoplasmic calcium available to activate the contractile proteins. Conversely, a change in the binding kinetics of Ca^{2+} to troponin-C may alter the magnitude or time course of the intracellular Ca^{2+} transient. Despite these difficulties, accumulating evidence suggests that the volatile anesthetics act in a number of specific ways, including 1) effects on the sarcolemmal flux of calcium; 2) alterations in SR function; and 3) modification of the responsiveness of the contractile proteins to activation by calcium.

The substantial attention to cellular Ca^{2+} in studies of the action of volatile anesthetics on the heart has been vindicated by an increasing body of evidence

ZELJKO J. BOSNJAK, Department of Anesthesiology, Medical College of Wisconsin, MFRC A-1000, Milwaukee, Wisconsin 53226.

Mechanisms of Anesthetic Action in Skeletal, Cardiac, and Smooth Muscle
Edited by T.J.J. Blanck and D.M. Wheeler, Plenum Press, New York, 1991

demonstrating that intracellular Ca transients are indeed altered by these agents (ref. 4 and Chapters 7, 9, 12, 18). Each of the three commonly used volatile anesthetics decreases the magnitude of the intracellular Ca^{2+} transient. The investigations contained in this volume, and the discussion which follows, attempt either to understand how the transient is reduced or to determine whether the entirety of the negative inotropic effect of the anesthetics can be attributed to a change in the Ca^{2+} transient.

The first and perhaps most obvious way to alter the intracellular Ca^{2+} transient would be through a change in Ca^{2+} influx across the sarcolemma. While not the sole sarcolemmal Ca^{2+} transporter, voltage-dependent Ca^{2+} channels are key in the beat-to-beat regulation of cardiac contractility. Although several types of voltage-dependent Ca^{2+} channels exist in various cell types,[7] the calcium channels of cardiac muscle include the low threshold, transient (T-type) channels and the high threshold, long-lasting (L-type) channels. In ventricular cells, the L-type channel is predominant, while current through T-type channels is small, decays quickly and contributes little to the total inward calcium current during the cardiac action potential.[8] Halothane has previously been shown to reduce a slow inward current in the isolated rat ventricular cells.[9] The influence on halothane on the binding of ^3H-nitrendipine and ^3H-D600 to voltage-dependent calcium channels was examined by Drenger et al. (Chapter 10). Halothane was found to inhibit the binding of two radiolabeled ligands which have distinct high affinity binding sites on the calcium channels. It is of interest that such decreases in binding were demonstrated in rat, rabbit and bovine membranes but not in the canine sarcolemmal membranes, but the functional significance of this species specificity has yet to be determined. Using patch-clamp methodology, Bosnjak et al. showed that halothane, enflurane and isoflurane, when tested in the same cardiac myocytes under identical conditions, produce equivalent depression of peak inward calcium current (I_{Ca}) at equianesthetic concentrations without shifting the current-voltage relationship for channel activation.[10] This observation is particularly important since equianesthetic concentrations of halothane and enflurane depress myocardial contractility more than isoflurane.[1-6] It is likely that the negative inotropic[4,11] and chronotropic[12] actions of halothane, enflurane and isoflurane on the myocardium are related, at least in part, to their inhibition of I_{Ca} at the sarcolemma.[10] Since all three agents depressed the I_{Ca} amplitude similarly at equianesthetic concentrations,[10] their quantitatively different effects on cardiac performance[1-3,6,13] are most likely due to differential actions at other cellular sites. However, differences between anesthetics at the sarcolemmal level cannot be completely excluded until potential differential effects on Na^+/Ca^{2+} exchange and membrane Ca^{2+} pump activity are examined.

Since halothane, enflurane and isoflurane decrease the Ca^{2+} current, it is expected that over many beats these effects will contribute to the decrease in SR loading, and therefore, the SR will release less calcium. In addition, it has been suggested that these agents also directly depress the SR Ca^{2+} content and thus additionally contribute to the negative inotropic effect.[14,15] Beside any effects on Ca^{2+} uptake, the possibility exists that inhalational anesthetics might also increase the rate of Ca^{2+} leak from the SR during rest and therefore contribute to depletion of SR Ca^{2+} content. In any event, since calcium transients are dominated by Ca^{2+} which is released from SR,[16] the most potent negative inotropic anesthetic is expected to have the greatest effect on the SR. Komai and Rusy have examined the effects of halothane, isoflurane and thiopental in several types of experiments designed to differentiate between SR and sarcolemmal effects in an intact muscle preparation (Chapter 11). They found that all three drugs depressed Ca^{2+} influx across the sarcolemma, but only halothane also exhibited a major effect at the SR. In another set of studies, Wilde *et al.* have shown that halothane is more effective than either isoflurane or enflurane in limiting availability of Ca^{2+} during the excitation-contraction cycle in single myocardial cells from rat ventricle (Chapter 12). Also using isolated rat myocytes, Wheeler *et al.* reported that halothane enhanced the SR Ca^{2+} release in beating cells and evoked spontaneous release from the SR of resting cells (Chapter 13). While enflurane generally acted like halothane in this preparation, it was less potent for induction of spontaneous Ca^{2+} release in resting cells. On the other hand, isoflurane did not appear to enhance SR Ca^{2+} release or or change SR Ca^{2+} content in quiescent cells. Ohnishi and Katsuoka similarly found that halothane evoked a release of SR Ca^{2+} from isolated rat heart cells and that both halothane and enflurane produced a transient increase in twitch amplitude which may be explained by an enhancement of SR Ca^{2+} release (Chapter 7). The results obtained from rat heart preparations need to be interpreted with caution because the rat cardiac SR accumulates Ca^{2+} during rest, and in some of the preparations the cells exhibit spontaneous contractile waves. The understanding of the above findings is also rendered somewhat more complex by the work of Lynch (Chapter 14). The local anesthetics, procaine and tetracaine, were shown to alter the control of some component of SR Ca^{2+} stores *via* a mechanism which is pharmacologically different from that of ryanodine. This selective depression of tension development was also observed in the presence of alcohols, octanol, heptanol and benzyl alcohol. It is thus possible that an anesthetic could act at the SR but not have effects on ryanodine-sensitive endpoints. It is interesting to note also that Komai and Rusy reported that isoflurane altered Ca^{2+} release from the SR during rapid cooling contractures, while in many of the other experiments cited above, isoflurane had minimal SR-attributable effects.

The major determinant of the decline of the Ca^{2+} transient in heart muscle is likely to be re-uptake of Ca^{2+} by the SR.[16] In addition to the SR Ca^{2+} pumps, sarcolemmal Ca^{2+} pumps and Na^+/Ca^{2+} exchange also contribute to the decline of the Ca^{2+} transient.[17] However, the SR Ca^{2+} pumping ATPase appears to be primarily responsible for the rate of fall of the Ca^{2+} transients once release of Ca^{2+} is over.[18,19] Relative to that of the sarcolemmal Ca^{2+} ATPase, the greater role of the SR for Ca^{2+} re-uptake is due to the SR Ca^{2+} ATPase's greater affinity for Ca^{2+} and higher V_{max}. Therefore, if anesthetic agents decrease the rate of Ca^{2+} uptake by the SR, it is expected that they would have a corresponding effect on the rate of decline of the Ca^{2+} transient, as measured by the aequorin signal. As seen from the current results, halothane was most effective in lengthening the time-to-peak duration of the aequorin signal as well as the time of the calcium transient measured at half of peak amplitude (Chapter 9). The amplitudes of the aequorin signal obtained in the presence of anesthetics were also adjusted electronically for comparisons of their time courses. In addition, since the upper part of the aequorin signal due to this adjustment might be disproportionately prolonged,[16] the time course of aequorin signals that have been increased to the equal magnitude by increases in extracellular calcium, were also compared with control. During all measurements, in the presence of higher concentrations of halothane, the aequorin signal was longer than that recorded during control measurements, while isoflurane at similar concentrations decreased the duration of the aequorin signal. This finding might imply that halothane delays the rate of Ca^{2+} removal by the SR and/or Na^+/Ca^{2+} exchange. In the presence of isoflurane, calcium removal is well maintained as indicated by the decrease in the duration of the calcium transient.

Many of the effects described here could be attributed to various second messengers within the cardiac cell. However, Vulliemoz has found that the negative inotropic effects of halothane and isoflurane are not mediated by alterations in arachidonate metabolites or a pertussis toxin-sensitive G-protein (Chapter 15). It was further shown that halothane inhibits the stimulatory effect of norepinephrine on phosphoinositide hydrolysis in the heart. While such an effect would not bear on the results of any of the studies in this section, it may be relevant in the whole animal.

Several studies of anesthetic effects on skinned muscle preparation suggest that volatile anesthetics may depress myofibrillar calcium sensitivity (refs. 20 and 21 and Chapter 16). Although valuable information was obtained regarding the relationship between intracellular Ca^{2+} and contractile force in the mammalian cardiac tissue using the skinned muscle preparations, extrapolation of the findings to the intact myocytes might be difficult.[22] Specifically, it appears that the myofilament sensitivity to Ca^{2+} may be greater in intact muscle than in skinned preparations. It was reported that the Ca^{2+} required for maximal activation of intact muscle fiber is less than or equal to 1 μM.[17] On the other hand, the concentrations of Ca^{2+} producing maximal constrictions in skinned muscles are approximately 10 μM.[23] Using a rabbit papillary muscle with intact membranes, Berman et al. have examined the effects of halothane on the generation of contractile force during tetanus (Chapter 17). They concluded that a decrease in interaction of myosin with actin during clinically relevant concentrations of halothane contributes only to a small degree to the overall negative inotropic effect of halothane. In intact ferret papillary muscle, Housmans has measured force generation and intracellular Ca^{2+} transients in the presence and absence of volatile anesthetics (Chapter 18). He concludes that while each agent has an inhibitory effect on myofibrillar Ca^{2+} responsiveness (isoflurane's being the largest effect), the relative magnitude of the effect is small relative to the ability of the agents to decrease Ca^{2+} availability. The results of another study in intact papillary muscle suggest that while halothane and enflurane do not alter myofibrillar calcium sensitivities, isoflurane depresses Ca^{2+} sensitivity despite relatively less influence on the calcium transient and overall contractile force development (Chapter 9). This study could not confirm the depressant effects of halothane and enflurane on calcium sensitivity obtained form either mechanically[20] or chemically[21] treated cardiac fibers. In work which compares Ca^{2+} transient amplitude and twitch tension, inference is made about the calcium/tension relationship, but one should be cautious about comparing these and other findings because maximal contractile force was not achieved and the relationship between the Ca^{2+} and tension was not obtained at steady-state. Overall, however, it appears that depression of Ca^{2+} sensitivity probably does not play a major role in the negative inotropic effects of halothane and enflurane. On the other hand, a decrease of Ca^{2+} sensitivity by isoflurane may play a minor role in depression of contractile force, although this appears to be compensated for by less depression of the Ca^{2+} transient. This lesser depression of the Ca^{2+} transient by isoflurane (Chapter 9) is in agreement with results obtained using different methodology[24,25] showing that halothane and enflurane are more potent in depressing cellular accumulation and release of intracellular Ca^{2+} than isoflurane. These effects of inhalational agents could lead to a decrease in calcium content of cardiac cells[26,27] and most likely contribute to the cardiac protection following ischemia and calcium paradox.[28]

ACKNOWLEDGEMENTS

Supported in part by National Institutes of Health grants HL 34708, HL 39776 and HL 01901, and Anesthesiology Research Training Grant GM 08377.

REFERENCES

1. B. F. Rusy and H. Komai, Anesthetic depression of myocardial contractility: A review of possible mechanisms, *Anesthesiology* 67:745-766 (1987).
2. P. R. Housmans and I. Murat, Comparative effects of halothane, enflurane, and isoflurane at equipotent anesthetic concentrations on isolated ventricular myocardium of the ferret: I. Contractility, *Anesthesiology* 69:451-463 (1988).
3. H. Komai and B. F. Rusy, Negative inotropic effects of isoflurane and halothane in rabbit papillary muscles, *Anesth Analg* 66:29-33 (1987).
4. Z. J. Bosnjak and J. P. Kampine, Effects of halothane on transmembrane potentials, Ca^{2+} transients, and papillary muscle tension in the cat, *Am J Physiol* 251:H374-H381 (1986).
5. C. Lynch III, S. Vogel, M. G. Pratila and N. Sperelakis, Enflurane depression of myocardial slow action potentials, *J Pharmacol Exp Ther* 222:405-409 (1982).
6. C. Lynch III, Differential depression of myocardial contractility by halothane and isoflurane in vitro, *Anesthesiology* 64:620-631 (1986).
7. E. W. McCleskey, A.P. Fox, D. Feldman and R.W. Tsien, Different types of calcium channels, *J Exp Biol* 124:177-190 (1986).
8. B. Nilius, P. Hess, J. B. Lansman and R. W. Tsien, A novel type of cardiac channel in ventricular cells, *Nature* 316:443-446 (1985).
9. Y. Ikemoto, A. Yatani, H. Arimura and J. Yoshitake, Reduction of the slow inward current of isolated rat ventricular cells by thiamylal and halothane, *Acta Anaesthesiol Scand* 29:583-586 (1985).
10. Z. J. Bosnjak, F. D. Supan and N. J. Rusch, The effects of halothane, enflurane and isoflurane on calcium current in isolated canine ventricular cells, *Anesthesiology* 74:340-345 (1991).
11. J. L. Seagard, Z. J. Bosnjak, F. A. Hopp, K. J. Kotrly, T. J. Ebert and J. P. Kampine, Cardiovascular effects of general anesthesia, in: "Effects of Anesthesia," B. G. Covino, H. A. Fozzard, K. Rehder and G. Strichartz, eds., Baltimore, Williams & Wilkins (1985) pp 149-177.
12. Z. J. Bosnjak and J. P. Kampine, Effects of halothane, enflurane and isoflurane on the SA node, *Anesthesiology* 58:314-321 (1983).
13. J. L. Atlee and Z. J. Bosnjak, Mechanisms for cardiac dysrhythmias during anesthesia, *Anesthesiology* 72:347-374 (1990).
14. J. Y. Su and W. G. L. Kerrick, Effects of halothane on caffeine-induced tension transients in functionally skinned myocardial fibers, *Pflugers Arch* 380:29-34 (1979).
15. J. Y. Su and W. G. L. Kerrick, Effects of enflurane on functionally skinned myocardial fibers from rabbits, *Anesthesiology* 52:385-389 (1980).
16. J. R. Blinks, W. G. Wier, P. Hess and F. G. Prendergast, Measurements of Ca^{2+} concentration in living cells. *Prog Biophys Mol Biol* 40:1-114 (1982).
17. W. G. Wier, Cytoplasmic $[Ca^{2+}]$ in mammalian ventricle: Dynamic control by cellular processes, *Annu Rev Physiol* 52:467-485 (1990).
18. E. Carafoli, Membrane transport of calcium: An overview, *Methods Enzymol* 157:3-11 (1988).
19. G. Inesi, Mechanism of calcium transport, *Annu Rev Physiol* 47:573-601 (1985).
20. J. Y. Su and W. G. L. Kerrick, Effects of halothane on Ca^{2+}-activated tension development in mechanically disrupted rabbit myocardial fibers, *Pflugers Arch* 375:111-117, (1978).
21. I. Murat, R. Ventura-Clapier and G. Vassort, Halothane, enflurane and isoflurane decrease calcium sensitivity and maximum force in detergent-treated rat cardiac fibers, *Anesthesiology* 69:892-899 (1988).
22. D. T. Yue, E. Marban and W. G. Wier, Relationship between force and intracellular $[Ca^{2+}]$ in tetanized mammalian heart muscle, *J Gen Physiol* 87:223-242 (1986).
23. E. J. Krane and J. Y. Su, Comparison of the effects of halothane on skinned myocardial fibers from newborn and adult rabbit: I. Effects on contractile proteins, *Anesthesiology* 70:76-81 (1989).
24. M. C. DeTraglia, H. Komai and B. F. Rusy, Differential effects of inhalation anesthetics on myocardial potentiated-state contractions in vitro, *Anesthesiology* 68:534-540 (1988).
25. C. Lynch III, Differential depression of myocardial contractility by volatile anesthetics in vitro: Comparison with uncouplers of excitation-contraction coupling, *J Cardiovasc Pharmacol* 15:655-665 (1990).
26. M. Katsuoka, K. Kobayashi and S. T. Ohnishi, Volatile anesthetics decrease calcium content of isolated myocytes, *Anesthesiology* 70:954-960 (1989).

27. D. M. Wheeler, R. T. Rice and E. G. Lakatta, The action of halothane on spontaneous contractile waves and stimulated contractions in isolated rat and dog heart cells, *Anesthesiology* 72:911-920 (1990).

28. Z. J. Bosnjak, S. Hoka, L. A. Turner and J. P. Kampine, Cardiac protection by halothane following ischemia and calcium paradox, *in:* "Cell Calcium Metabolism," G. Fiskum, ed., Plenum Press, New York (1989) pp 593-601.

9

Effects of Volatile Anesthetics on the Intracellular Calcium Transient and Calcium Current in Cardiac Muscle Cells

Zeljko J. Bosnjak

INTRODUCTION

Volatile anesthetics at clinically useful concentrations depress the contractile force of the heart, and this effect in part contributes to a significant decrement of cardiovascular homeostasis. The mechanism of action underlying the negative inotropic effect of volatile anesthetics is not fully understood. There are several mechanisms by which agents may directly alter contractile performance of cardiac muscle. The first group represents the "upstream" mechanisms whereby intracellular calcium transients are mostly influenced by a variety of effectors at the surface membrane and sarcoplasmic reticulum. In addition, changes in sensitivity of troponin-C to calcium, and an altered response of the myofilaments to a given level of occupancy of the calcium binding sites on troponin C ("downstream" mechanisms) need to be considered as well. Accumulating evidence suggests that volatile anesthetics have multiple actions relevant to cardiac contractility, including a decrease in the sarcolemmal flux of calcium, a change in SR function, and a decrease in the level of intracellular ionized calcium during systole as well as a modification in the responsiveness of the contractile proteins to activation by calcium. A variety of the newest technology and methods, including the measurements of inward calcium current (I_{Ca}), measurement of calcium transients, and myofibrillar responsiveness to calcium, were utilized in the studies reported here in order to improve our knowledge of the effects of volatile anesthetics on cardiac muscle.

Studies in the isolated heart and papillary muscle demonstrate that halothane, enflurane and isoflurane depress myocardial contractility in a dose-dependent matter with equianesthetic doses of halothane and enflurane depressing cardiac function more than isoflurane.[1-6] The mechanism(s) responsible for these differences are controversial. Some investigators have concluded that the major effect of isoflurane occurs *via* inhibition of calcium influx[6] while others have attributed the difference to a greater effect on the SR.[3] Determining which cellular sites are targets for the action of volatile anesthetics is difficult in intact cardiac preparations since changes in contractile force reflect interaction between Ca^{2+} influx through the sarcolemma,

ZELJKO J. BOSNJAK, Departments of Anesthesiology and Physiology, Medical College of Wisconsin, MFRC A-1000, Milwaukee, Wisconsin 53226.

Mechanisms of Anesthetic Action in Skeletal, Cardiac, and Smooth Muscle
Edited by T.J.J. Blanck and D.M. Wheeler, Plenum Press, New York, 1991

release and sequestration of Ca^{2+} by the sarcoplasmic reticulum, activity of membrane Ca^{2+} pumps and ionic exchanges, and the Ca^{2+} sensitivity of the contractile proteins. Likewise, predictions of Ca^{2+} influx drawn from the cardiac action potential configuration are complicated by interaction between different ionic channel types.[1,7] Despite these drawbacks, increasing evidence suggests important quantitative differences between the depressant action of volatile agents on myocardial function, with equianesthetic doses of halothane and enflurane depressing cardiac function more than isoflurane. Although the direct cardiac depression seen *in vitro* may be reduced in intact animals and humans,[8] isoflurane also appears to be a less potent cardiac depressant than either halothane or enflurane in patients.

The cellular basis for the pharmacological differences between general anesthetics agents is unclear. In the first part of this report, direct comparisons of these agents were made on the voltage dependent L-channel Ca^{2+} current in isolated canine ventricular fibers using the whole-cell voltage-clamp technique. This was done in order to separate the effects of these agents on calcium influx across the sarcolemma from their other cellular actions. In the second part of this report, the effects of halothane, enflurane, and isoflurane on rapid changes in intracellular calcium (calcium transient) in papillary muscles from the guinea pig were determined and the effects of these agents on mechanical performance were measured in order to examine their potencies. In addition, calcium sensitivity curves were obtained in the presence of inhalational anesthetics for comparison of calcium sensitivity in the intact cardiac muscle.

METHODS

Calcium Current

Adult mongrel dogs were anesthetized with halothane, the chest was opened, and the hearts were excised rapidly and rinsed in cold oxygenated Krebs solution. Thin strips of ventricular tissue were obtained using a biopsy needle and cut into 5 mm lengths and washed in cold cardioplegic solution. The washed ventricular segments were then placed in calcium-free Tyrode's solution containing collagenase and bovine serum albumin. The solution containing the ventricular tissue was incubated at 37°C for 1-1.5 hr in a slow shaker. After this incubation, single ventricular cells were washed 3 times in cold potassium glutamate solution. This procedure produced myocytes which were calcium tolerant as evidenced by the lack of contraction when external calcium concentration was increased to millimolar levels. Dispersed cells were placed in a perfusion chamber (22°C) on the stage of an inverted microscope. A hydraulic micromanipulator was used to position heat polished borosilicate patch pipettes with tip resistance of 4-6 MΩ on the membranes of single cardiac cells. High-resistance seals were formed, after which the pipette patch was removed by negative pressure to give electrical access to the whole cell, as previously described.[9] Whole cell Ca^{2+} currents were elicited every 5-10 seconds by 200 ms depolarizing pulses generated by a computerized system. The currents were amplified by a List EPC 7 patch-clamp amplifier. Substitution of tetraethylammonium chloride for sodium and potassium in the external solution and cesium for potassium in the pipette solution eliminated the sodium and potassium currents and permitted isolated measurement of the calcium current. For recording of Ca^{2+} channel current, barium (10 mM) rather than calcium usually has been used as a charge carrier, since it augments current through L-type Ca^{2+} channels and slows its decay. Leak and capacitive currents were subtracted from each record by linearly summating scaled currents obtained during 10-mV hyperpolarizing pulses. Prior to the onset of the experiments, current-voltage (I-V curves) were obtained 5-10 min after rupturing the cell membrane and again 5 min later to monitor time-dependent changes in calcium current amplitude. To determine the effect of the anesthetic agents on I_{Ca}, the inflow superfusate was changed to one containing a given concentration of anesthetic agent. To compare halothane, enflurane and isoflurane at equianesthetic concentrations, agents were prepared in final bath concentrations of 0.21, 0.40 and 0.88 mM for halothane; 0.39, 0.80 and 1.41 mM for enflurane; and 0.29, 0.61 and 1.12 mM for isoflurane. The concentration of these agents at 22°C in the Krebs' solution are equivalent to the following percentages in the gas phase: halothane (0.36, 0.68 and 1.50%), isoflurane (0.5, 1.0 and 1.9%) and enflurane (0.66, 1.36 and 2.39%). At the same temperature the partition coefficient of these anesthetics is approximately 1.5. The above concentrations compare well with the anesthetic potency ratio, based on MAC values in the dog, of 1:1.5:2.3 for halothane, isoflurane and enflurane, respectively.

Calcium Transient

After i.p. ketamine injection, guinea pigs were decapitated and the hearts quickly removed and perfused briefly with cold, oxygenated Krebs' solution. Right ventricular papillary muscles having a width of less than 1 mm (mean O.D. = 0.7 mm) were excised from 13 guinea pigs and mounted horizontally in an organ bath. Oxygenated Krebs' solution (97% O_2-3% CO_2 mixture) was circulated at 30°C as reported earlier.[10] Highly purified aequorin used in this study was prepared in the laboratory of J. R. Blinks, Rochester, MN. Lyophilized aequorin was reconstituted to a concentration of 2 mg/ml with calcium-free distilled water to give an aqueous solution containing 150 mM KCl and 5 mM HEPES buffer at pH 7.5. The aequorin solution was filtered and placed into fine microelectrodes. Platinum wire was positioned inside the micropipettes to permit simultaneous recording of cellular potentials while applying N_2 pressure of up to 100 p.s.i. to the pipette. It was necessary to inject at least 50 cells in order to obtain satisfactory light signals. The intracellular aequorin light signals provide a good qualitative indication of the overall magnitude and time course of the intracellular myoplasmic Ca^{2+} concentrations.[11,12] The light emitted by the aequorin was recorded in a light-proof setting using a photomultiplier cathode. Successive light signals were averaged (100 consecutive beats) to obtain satisfactory signal-to-noise ratios. Light signals were expressed in terms of anode current in nanoamps while tension was expressed as mN per square mm (tension being normalized for the cross sectional area of the muscle). Resting light emission was low and so close to the threshold for detection that it was not feasible to investigate the effects of different anesthetics on the resting emission. The papillary muscle was stimulated at frequency of 0.5 or 1 Hz by field stimulation throughout the experiment with 2 ms pulses at slightly above the threshold strength. The corda tendinae of the muscle was connected with a fine 10-0 thread to the arm of the isometric force transducer. At the beginning of each experiment, the muscle length was adjusted to a point where the tension developed was maximal. These experiments were conducted at 30°C since it was observed that the muscles maintained a stable level of developed tension and light signal for a longer period of time as compared to that obtained at 37°C. Anesthetic concentrations in the tissue bath were measured during the control periods and during anesthetic exposure using a gas chromatograph with a flame ionization detector. These concentrations at 30°C were converted to their equivalent volume percentages in the gas phase. The average percentages for halothane were 0.65 and 1.15%, for enflurane 1.0 and 2.2%, and for isoflurane 0.77 and 1.6%. The potency ratios for these agents were reasonably close to the estimated potency ratios for the guinea pig (1 : 2.15 : 1.14 for halothane, enflurane and isoflurane, respectively).[13] Statistical differences were determined using two-way ANOVA and least significant difference means comparison test. In order to directly examine the effects of halothane, enflurane and isoflurane on the relationship between intracellular Ca^{2+} and tension development in the papillary muscle of guinea pig, simultaneous measurements of Ca^{2+} transient with bioluminescent protein aequorin and isometric contractile force were performed in order to determine myofibrillar responsiveness to calcium by increasing the extracellular calcium from 1 to 15 mM.

RESULTS

Calcium Currents

Superfusion with halothane, enflurane and isoflurane similarly depressed I_{Ca} amplitude in dog ventricular cells. This was particularly evident in experiments where I_{Ca} amplitude was measured in a single cell exposed consecutively to three different anesthetic agents (full washout in between). Figure 1 shows the effects of halothane, enflurane and isoflurane on I_{Ca} in the same cell. I_{Ca} elicited by voltage steps from -80 mV to $+10$ mV was reduced by approximately the same amount at equianesthetic concentrations (medium dose) of the three anesthetic agents. The depression of peak I_{Ca} was remarkably similar among all three agents. At lower doses of halothane, enflurane and isoflurane, the peak reduction was 84 ± 2%, 81 ± 3% and 86 ± 3% of control, respectively. Depression of peak I_{Ca} during medium doses of these anesthetic agents was 69 ± 3%, 68 ± 5% and 62 ± 4% of control. The greatest reduction was measured at high concentrations at which peak I_{Ca} was reduced to 43 ± 7%, 38 ± 5% and 42 ± 5% of control by halothane, enflurane and isoflurane, respectively.

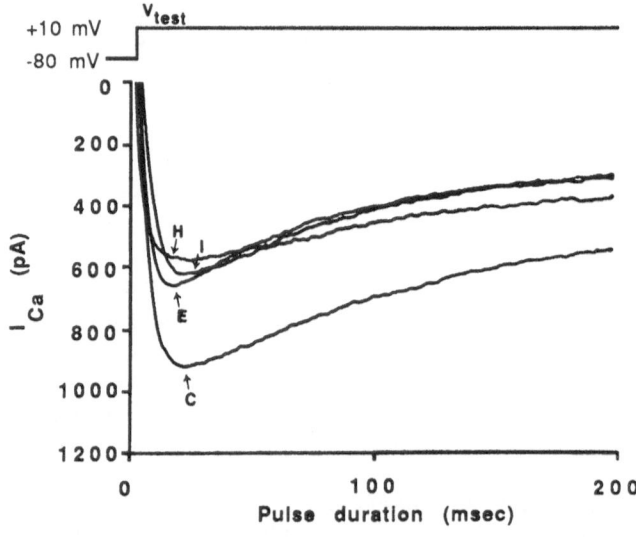

FIGURE 1. Calcium channel currents (I_{Ca}) elicited by test pulses from -80 to $+10$ mV in a single dog ventricular cell. Recordings were obtained in control solution (C) and in the presence of 0.45% halothane (H), 0.7% isoflurane (I) and 0.9% enflurane (E).

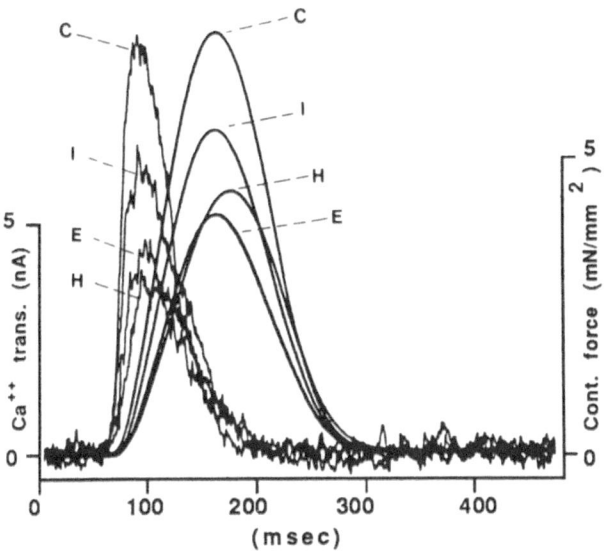

FIGURE 2. Effects of higher doses of halothane (H, 1.1%), enflurane (E, 2.2%) and isoflurane (I, 1.6%) on aequorin signal and isometric contractions of a single isolated guinea pig papillary muscle. C = control; pacing rate 1 Hz; 30°C.

From the findings described, it can be concluded that significant depression of I_{Ca} occurred at all 3 concentrations of these agents but there were no quantitatively different effects among halothane, enflurane and isoflurane at equianesthetic concentrations. In the control solution and after the application of anesthetic agents, families of Ca^{2+} currents were generated by step-wise depolarizing pulses from negative holding potentials (-80 and -40 mV) to more positive command potentials. Peak I_{Ca} was plotted as a function of membrane potential in order to analyze the effects of anesthetic agents on the current-voltage relationship for I_{Ca} activation. This analysis showed that exposure to low, medium and high concentration of these agents dose-dependently depressed I_{Ca} amplitude over the entire voltage range studied with preservation of the initial activation range and reversal potential.

Calcium Transient

The direct effects of halothane, enflurane and isoflurane on aequorin luminescence were tested *in vitro* by injecting aliquots of aequorin (10^{-8} M) into a buffer solution containing several concentrations of the anesthetic agents (halothane: 0.8 and 1.6%, isoflurane 1.1 and 2% and enflurane 1.3 and 2.5%). At these concentrations anesthetics had no significant effect on aequorin light signals.

The effects of inhalational anesthetic agents at higher concentrations on Ca^{2+} transients and contractile force in a typical preparation from a papillary muscle of the guinea pig are seen in figure 2. Depression of intracellular calcium concentration was accompanied by depression of contractile force. As seen, the depression of calcium transient in the presence of the isoflurane was less than that produced in the presence of halothane or enflurane. The rapid increase in intracellular Ca^{2+} concentration following the initiation of the action potential is thought to be primarily due to Ca^{2+} release from the SR, and this rapid release is generally believed to involve a Ca^{2+}-induced release mechanism.[14,15] Therefore, the Ca^{2+} which enters the cytoplasm *via* calcium channels contributes to the Ca^{2+} transient by acting both as a trigger for release of calcium and as the primary source of Ca^{2+} for the SR. Although it has been shown that Ca^{2+} influx *via* L-type Ca^{2+} channels is required for SR Ca^{2+} release in heart muscle,[16] the role of Ca^{2+} influx *via* calcium channels as a prerequisite for SR Ca^{2+} release has been challenged.[17] The authors have suggested that the triggered Ca^{2+} enters *via* sodium/calcium exchange instead.

The effects of halothane, enflurane and isoflurane on calcium transient and isometric force are summarized in figure 3. Values are means ± SEM as percent of control. As shown, the negative inotropic effects of halothane and enflurane were dose-dependent and closely related to the decrease in intracellular calcium. Isoflurane also reduced contractile force in a dose-dependent manner, but the decrease was significantly less than that of halothane and enflurane. A more striking feature observed with isoflurane was a dissociation between intracellular calcium availability and contractile force. Although the magnitude of the Ca^{2+} transient did not change when the concentration of isoflurane was increased from medium to high concentration, the contractile force decreased.

Figure 4 represents the effect of halothane, enflurane and isoflurane on time-to-peak amplitude of the aequorin signal and on the isometric contractile force. Halothane was most effective in increasing the time-to-peak calcium transient and time-to-peak tension. These effects are most likely due to a slower Ca^{2+} release from the SR secondary to a lesser Ca^{2+} gradient between the SR and the cytoplasm following the attenuated Ca^{2+} uptake by the SR. On the other hand, the falling phase of the calcium transient and contractile force contributed to a change in the duration of the force and light measured at half of peak amplitude (figure 5). Generally, the slower falling phase of the aequorin signal may suggest that there is a slower removal of calcium from the myoplasm by the SR or a slower sodium/calcium exchange. These results indicate that halothane was most potent in increasing the duration of calcium transient as measured at half of the peak amplitude and that isoflurane shortened the calcium transient. This abbreviation of the light signal suggests a faster removal of calcium from cytoplasm by the SR in the presence of isoflurane as compared to halothane and enflurane. During studies on alteration of calcium sensitivity at the

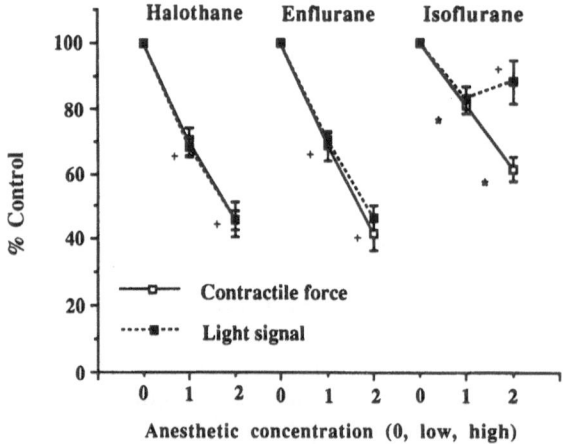

FIGURE 3. Effects of halothane, enflurane and isoflurane at lower and higher concentrations on the peak tension development and peak aequorin signal as a percent of control in the isolated guinea pig papillary muscle. [+] $P < 0.05$ *vs.* 0, [*] $P < 0.05$ *vs.* 0 and other anesthetics.

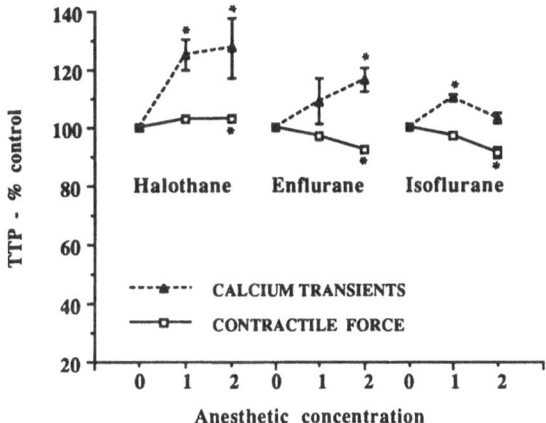

FIGURE 4. Effects of two doses of halothane, enflurane and isoflurane on time to peak duration (TTP) of the aequorin signal and isometric contractile force as a percent of control in the isolated guinea pig papillary muscle. Anesthetic concentration: 0, low (1), high (2). * $P < 0.05$ *vs.* 0.

myofilaments, it was found that isoflurane shifts the Ca^{2+} isometric tension curve toward higher intracellular calcium concentrations (figure 6) while no differences were observed between calcium sensitivity curves obtained in the absence and the presence of equianesthetic concentrations of halothane and enflurane.

DISCUSSION

Although different types of voltage-dependent Ca^{2+} channels exist in various cell types,[18] calcium channel types in cardiac muscle include the low threshold, transient (T-type) channels and the high threshold, long-lasting (L-type channels). In ventricular cells, the L-type channel is predominant, while current through T-type channels is

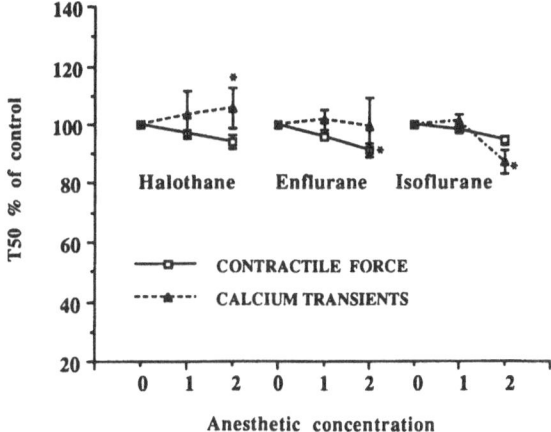

FIGURE 5. Effects of two doses of halothane, enflurane and isoflurane on the duration of calcium transients and contractile force measured at half of peak amplitude (T50) in the guinea pig papillary muscle. Anesthetic concentration: 0, low (1), high (2). * $P < 0.05$ *vs.* 0.

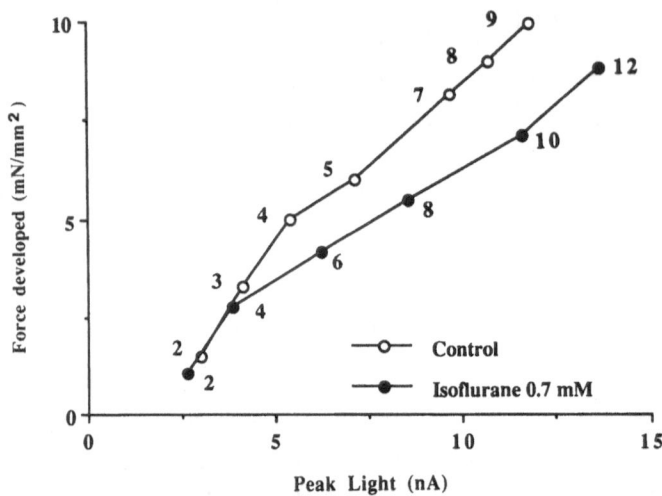

FIGURE 6. Comparison of the relationship between peak isometric contractile force and peak aequorin light measured in the same muscle at various extracellular calcium concentrations before and after isoflurane (2%). The numbers beside the points indicate extracellular Ca^{2+} in mM.

small, decays quickly and contributes little to the total inward calcium current during the cardiac action potential.[19] Halothane has been shown to reduce a slow inward current in the isolated rat ventricular cells.[20] This study shows that halothane, enflurane and isoflurane, when tested in the same cardiac myocytes under identical conditions and at equianesthetic concentrations, produce a concentration-dependent depression of peak I_{Ca} without shifting the current-voltage relationship for channel activation. Therefore, these findings suggest that the negative inotropic and chronotropic actions of halothane, enflurane and isoflurane on the ventricular myocardium are related, at least in part, to their inhibition of I_{Ca} at the sarcolemma. However, since all three anesthetic agents depress I_{Ca} amplitude similarly, their quantitatively different effects on cardiac performance are most likely due to differential actions at other cellular sites.

Measurements of changes in free-intracellular calcium using the bioluminescent protein aequorin and contractile force, as performed in a comparative study of inhalational anesthetics in the guinea pig papillary muscle, indicate that the weaker negative inotropic effect of isoflurane as compared to halothane and enflurane is associated with less depression of systolic intracellular calcium concentration. The decrease in calcium transient by these agents is likely related to the inhibition of I_{Ca} at the sarcolemma which in turn could affect the quantity of calcium released by the sarcoplasmic reticulum. Since there was no quantitative difference between the effects of isoflurane and the other two agents on I_{Ca} amplitude in the present study, the sarcolemma is an unlikely site for their differential cellular effects.

Since halothane, enflurane and isoflurane decrease the Ca^{2+} current, it is expected that over many beats, these fluxes will contribute to the decrease in SR loading and therefore the SR will release less calcium. In addition, it was suggested that these agents also depress the net SR Ca^{2+} uptake and contribute to the negative inotropic effect.[21,22] Beside the effects on Ca^{2+} uptake, the possibility exists that inhalational anesthetics might also increase the rate of Ca^{2+} leak from the SR during rest and therefore contribute to depletion of SR Ca^{2+} content. In any event, since calcium transients are dominated by the Ca^{2+} which is released from SR, the most potent negative inotropic anesthetic is expected to have the greatest effect on the SR.

The major determinant of the decline of the Ca^{2+} transient in heart muscle, appears to be re-uptake of Ca^{2+} by SR.[23] The SR Ca^{2+}-pumping ATPase appears to be responsible for the rate of fall of the Ca^{2+} transient once release of Ca^{2+} is over.[24,25] The greater role of the SR in Ca^{2+} re-uptake is due to a greater Ca^{2+} ATPase affinity for Ca^{2+} and a higher V_{max} relative to that of sarcolemmal Ca^{2+} ATPase. In addition to these pumps, the sodium/calcium exchange also contributes to the decline of the Ca^{2+} transient.[26] Therefore, if anesthetic agents decrease the rate of Ca^{2+} uptake by the SR, it is expected that they would have a corresponding effect on the rate of decline of the aequorin signal. As seen from these results, halothane was most effective in lengthening the time-to-peak duration of the aequorin signal as well as the duration of the calcium transient measured at half of peak amplitude. The amplitudes of aequorin signal obtained in the presence of anesthetics were also adjusted electronically for comparisons of their time courses. In addition, since the upper part of the aequorin signal due to this adjustment might be disproportionately prolonged,[23] the time course of aequorin signals that have been increased to equal magnitude by increases in extracellular calcium, were also compared with control. During all measurements, in the presence of halothane, the aequorin signal was longer than that recorded during control measurements.

Although valuable information was obtained regarding the relationship between intracellular Ca^{2+} and contractile force in the mammalian cardiac tissue using the skinned muscle preparations, extrapolation of the findings to the intact myocytes might be difficult.[27] Specifically, it appears that the myofilament sensitivity to Ca^{2+} may be greater in intact muscle than in skinned preparations. It was reported that the Ca^{2+} required for maximal activation of the fiber is less than or equal to 1 μM.[27] On the other hand, the concentrations of Ca^{2+} producing maximal constrictions in skinned muscles are approximately 10 μM.[28] Several laboratories have used skinned muscle preparations[29,30] and have shown that inhalational anesthetics decrease the Ca^{2+} sensitivity of the contractile apparatus. In this study using intact muscle, inference is made about the calcium-tension relationship, but one should be cautious about comparing these findings with those obtained using skinned muscle, since in these experiments, maximal force was not achieved and the relationship between the Ca^{2+} and tension was not obtained at steady-state. Nevertheless, the same muscles were exposed to identical protocols with and without inhalational anesthetics and these results suggest that only isoflurane exhibits a depressant effect on calcium sensitivity or other downstream mechanisms.

In summary, these results support previous findings[20,31,32] showing the reduction of inward current in the presence of inhalational anesthetics and further indicate that all three agents tested depress I_{Ca} amplitude similarly when used at equianesthetic concentrations.[33] Utilizing intact papillary muscle preparations, these studies could not confirm the depressant effects of halothane and enflurane on calcium sensitivity.[29,30] Furthermore, results obtained from the calcium transients are in agreement with results obtained using different methodology[34,35] showing that halothane and enflurane are more potent in depressing cellular accumulation and release of intracellular Ca^{2+}. These effects of inhalational agents could lead to a decrease in calcium content of cardiac cells[36,37] and most likely contribute to the cardiac protection following ischemia and calcium paradox.[38]

ACKNOWLEDGEMENTS

Supported in part by NIH grants HL 39776, 34708 and 01901. Enflurane and isoflurane were kindly provided by Mr. Rick Deutsch of Anaquest.

REFERENCES

1. B. F. Rusy and H. Komai, Anesthetic depression of myocardial contractility: A review of possible mechanisms, *Anesthesiology* 67:745-766 (1987).
2. P. R. Housmans and I. Murat, Comparative effects of halothane, enflurane, and isoflurane at equipotent anesthetic concentrations on isolated ventricular myocardium of the ferret. I. Contractility, *Anesthesiology* 69:451-463 (1988).
3. H. Komai and B. F. Rusy, Negative inotropic effects of isoflurane and halothane in rabbit papillary muscles, *Anesth Analg* 66:29-33 (1987).
4. Z. J. Bosnjak and J. P. Kampine, Effects of halothane on transmembrane potentials, Ca^{2+} transients, and papillary muscle tension in the cat, *Am J Physiol* 251:H374-H381 (1986).
5. C. Lynch III, S. Vogel, M. G. Pratila and N. Sperelakis, Enflurane depression of myocardial slow action potentials, *J Pharmacol Exp Therap* 222:405-409 (1982).
6. C. Lynch III, Differential depression of myocardial contractility by halothane and isoflurane *in vitro*, *Anesthesiology* 64:620-631 (1986).
7. J. L. Atlee and Z. J. Bosnjak, Mechanisms for cardiac dysrhythmias during anesthesia, *Anesthesiology* 72:347-374 (1990).
8. J. L. Seagard, Z. J. Bosnjak, F. A. Hopp, K. J. Kotrly, T. J. Ebert and J. P. Kampine, Cardiovascular effects of general anesthesia, *in:* "Effects of Anesthesia," B. G. Covino, H. A. Fozzard, K. Rehder and G. Strichartz, eds., Williams & Wilkins, Baltimore (1985) pp. 149-177.
9. O. P. Hamill, A. Marty, E. Neher, B. Sakmann and F. J. Sigworth, Improved patch-clamp techniques for high-resolution current recording from cells and cell-free membrane patches, *Pflugers Arch* 391:85-100 (1981).
10. Z. J. Bosnjak, A. Aggarwal, L. A. Turner, J. Marijic and J. P. Kampine, Differential effects of inhalational anesthetics on calcium sensitivity (abstract), *Biophys J* 57:338a (1990).
11. D. G. Allen and C. H. Orchard, The effects of changes of pH on intracellular calcium transients in mammalian cardiac muscle, *J Physiol (Lond)* 335:555-567 (1983).
12. J. R. Blinks, W. G. Wier, P. Hess and F. G. Prendergast, Measurements of Ca^{2+} concentration in living cells, *Prog Biophys Mol Biol* 40:1-114 (1982).
13. A. B. Seifen, R. H. Kennedy, J. P. Bray and E. Seifen, Estimation of minimum alveolar concentration (MAC) for halothane, enflurane and isoflurane in spontaneously breathing guinea pigs, *Lab Anim Science* 39:579-581 (1989).
14. A. Fabiato, Calcium-induced release of calcium from the cardiac sarcoplasmic reticulum, *Am J Physiol* 245:C1-C14 (1983).
15. M. Morad and Y. Goldman, Excitation-contraction coupling in heart muscle: Membrane control of development of tension. *Prog Biophys Mol Biol* 27:257-313 (1973).
16. M. Nabauer, G. Callewaert, L. Cleemann and M. Morad, Regulation of calcium release is gated by calcium current not gating charge, in cardiac myocytes, *Science* 244:800-803 (1989).
17. N. Leblanc and J. R. Hume, Sodium current-induced release of calcium from cardiac sarcoplasmic reticulum, *Science* 248:372-376 (1990).
18. E. W. McCleskey, A. P. Fox, D. Feldman and R. W. Tsien, Different types of calcium channels, *J Exp Biol* 124:177-190 (1986).
19. B. Nilius, P. Hess, J. B. Lansman and R. W. Tsien, A novel type of cardiac channel in ventricular cells, *Nature* 316:443-446 (1985).
20. Y. Ikemoto, A. Yatani, H. Arimura and J. Yoshitake, Reduction of the slow inward current of isolated rat ventricular cells by thiamylal and halothane, *Acta Anaesthesiol Scand* 29:583-586 (1985).
21. J. Y. Su and W. G. L. Kerrick, Effects of halothane on caffeine-induced tension transients in functionally skinned myocardial fibers, *Pflugers Arch* 380:29-34 (1979).
22. J. Y. Su and W. G. L. Kerrick, Effects of enflurane on functionally skinned myocardial fibers from rabbits, *Anesthesiology* 52:385-389 (1980).
23. M. Endo and J. R. Blinks, Actions of sympathomimetic amines on the Ca^{2+} transients and contractions of rabbit myocardium: Reciprocal changes in myofibrillar responsiveness to Ca^{2+} mediated through α- and β-adrenoceptors, *Circ Res* 62:247-265 (1988).
24. E. Carafoli, Membrane transport of calcium: An overview, *Methods Enzymol* 157:3-11 (1988).
25. G. Inesi, Mechanism of calcium transport, *Annu Rev Physiol* 47:573-601 (1985).
26. W. G. Wier, Cytoplasmic $[Ca^{2+}]$ in mammalian ventricle: Dynamic control by cellular processes, *Annu Rev Physiol* 52:467-485 (1990).
27. D. T. Yue, E. Marban and W. G. Wier, Relationship between force and intracellular $[Ca^{2+}]$ in tetanized mammalian heart muscle, *J Gen Physiol* 87:223-242 (1986).

28. E. J. Krane and J. Y. Su, Comparison of the effects of halothane on skinned myocardial fibers from newborn and adult rabbit. I. Effects on contractile proteins, *Anesthesiology* 70:76-81 (1989).

29. J. Y. Su and W. G. L. Kerrick, Effects of halothane on Ca^{2+}-activated tension development in mechanically disrupted rabbit myocardial fibers, *Pflugers Arch* 375:111-117 (1978).

30. I. Murat, R. Ventura-Clapier and G. Vassort, Halothane, enflurane and isoflurane decrease calcium sensitivity and maximum force in detergent-treated rat cardiac fibers, *Anesthesiology* 69:892-899 (1988).

31. D. A. Terrar and J. G. G. Victory, Isoflurane depresses membrane currents associated with contraction in myocytes isolated from guinea-pig ventricle, *Anesthesiology* 69:742-749 (1988).

32. D. A. Terrar and J. G. G. Victory, Effects of halothane on membrane currents associated with contraction in single myocytes isolated from guinea pig ventricle, *Br J Pharmacol* 94:500-508 (1988).

33. Z. J. Bosnjak, F. D. Supan and N. J. Rusch, The effects of halothane, enflurane, and isoflurane on calcium current in isolated canine ventricular cells, *Anesthesiology* 74:340-345 (1991).

34. M. C. DeTraglia, H. Komai and B. F. Rusy, Differential effects of inhalation anesthetics on myocardial potentiated-state contractions *in vitro*, *Anesthesiology* 68:534-540 (1988).

35. C. Lynch III, Differential depression of myocardial contractility by volatile anesthetics *in vitro*: Comparison with uncouplers of excitation-contraction coupling, *J Cardiovasc Pharmacol* 15:655-665 (1990).

36. M. Katsuoka, K. Kobayashi and S. T. Ohnishi, Volatile anesthetics decrease calcium content of isolated myocytes, *Anesthesiology* 70:954-960 (1989).

37. D. M. Wheeler, R. T. Rice and E. G. Lakatta, The action of halothane on spontaneous contractile waves and stimulated contractions in isolated rat and dog heart cells, *Anesthesiology* 72:911-920 (1990).

38. Z. J. Bosnjak, S. Hoka, L. A. Turner and J. P. Kampine, Cardiac protection by halothane following ischemia and calcium paradox, *in:* "Cell Calcium Metabolism," G. Fiskum, ed., Plenum Press, New York (1989) pp. 593-601.

10

Halothane Inhibits Binding of Calcium Channel Blockers to Cardiac Sarcolemma

Benjamin Drenger, Susan Riggs Runge, Paul Hoehner, Mary Quigg, Thomas J.J. Blanck

INTRODUCTION

Several pieces of evidence exist which suggest that the volatile anesthetics exert their negative inotropic effect by interfering with the calcium homeostasis of the myocardial cell.[1] Our laboratory has examined three important sites in the myocardial cell that are known to be involved with the phasic changes in intracellular calcium which occur during contraction. These sites are the sarcolemma, the sarcoplasmic reticulum, and the myofibrils. This paper is directed towards our investigations of one constituent of the sarcolemma, the voltage-dependent calcium channel (VDCC), and its probable alteration by the volatile anesthetic, halothane.

Other investigators have shown that the calcium current resulting from depolarization of the myocardium is inhibited by halothane.[2-4] This calcium current is due to the flow of calcium through the VDCC. We postulated that this alteration in calcium entry might be due to a change either in the conformation or environment of the VDCC. We examined the binding of two radiolabeled ligands to the VDCC as a measure of the integrity of the channel in the isolated sarcolemmal membrane. We have previously shown that halothane inhibited the binding of the dihydropyridine, [³H]nitrendipine, to the VDCC in crude membrane preparations from both rabbit and rat myocardium.[5] The inhibition of binding was also found to be a reversible process, as is necessary for any relevant anesthetic mechanism. The purpose of the present investigation was to quantitate the binding of [³H]nitrendipine to VDCCs from purified bovine and canine sarcolemma VDCCs and to compare these properties to the effect of halothane on the binding of the phenylalkylamine, [³H]D600, to bovine VDCCs. Since these two classes of drugs have separate and distinct binding sites on the VDCC, such a comparison might yield a clue to the nature of halothane's interaction with the VDCC.

BENJAMIN DRENGER, Department of Anesthesiology, Cardiothoracic Anesthesia Unit, Hadassah Hospital, Jerusalem, Israel. SUSAN RIGGS RUNGE, PAUL HOEHNER, MARY QUIGG, THOMAS J. J. BLANCK, Division of Cardiac Anesthesia, Department of Anesthesiology and Critical Care Medicine, Johns Hopkins University, Baltimore, Maryland 21205.

Mechanisms of Anesthetic Action in Skeletal, Cardiac, and Smooth Muscle
Edited by T.J.J. Blanck and D.M. Wheeler, Plenum Press, New York, 1991

MATERIALS AND METHODS

The methods used are those described in a previous publication.[6] Isolation of the bovine and canine heart sarcolemmal membranes was performed using a modification of the method of Jones et al.[7] and Caroni et al.[8] Ventricular muscle was homogenized and fractionated in a series of centrifugations and extractions. This series of homogenizations gradually removed nuclei, cell debris, and mitochondria, followed by elimination of the contractile proteins and sarcoplasmic reticulum. In order to evaluate the purity of the sarcolemmal preparation, the final four stages of the sarcolemma isolation procedure were tested by $^{45}Ca^{2+}$ uptake studies in the presence of oxalate. As sarcolemmal $^{45}Ca^{2+}$ uptake is not amplified by oxalate, we used the diminution of oxalate-supported $^{45}Ca^{2+}$ uptake as a marker for the decrease in sarcoplasmic reticulum contamination of our sarcolemmal preparations. In parallel, we performed [^{3}H]nitrendipine binding studies in each of the final four stages of the preparation. Binding assays were carried out in 31 ml glass vials, sealed with Teflon liners to maintain constant anesthetic concentration.[9] Anesthetic was added to the solution in liquid form with a Hamilton microliter syringe. Addition of 3 μl halothane, 5 μl isoflurane and 8 μl enflurane produced anesthetic partial pressures equivalent to 1.9, 2.3 and 4.8 volume% respectively. These concentrations are close to those published in the anesthetic literature for human equipotent ratios.

The membranes were incubated during the experiment at a constant temperature of 25°C, using a water shaker bath. This temperature was used in order to slow membrane degradation. It has been shown[10] that radioligand experiments at 25°C correlate well with patch clamp experiments on the effect of nitrendipine on inhibition of cardiac calcium currents.

In the equilibrium binding studies the sarcolemmal membranes (40 - 90 μg) were added to 0.01 - 1 nM of [^{3}H]nitrendipine in 50mM Tris buffer (pH 7.5) in a final volume of 1 ml. Parallel vials incubated in the presence or absence of 500 nM unlabelled nitrendipine were used to define non-specific and total binding, respectively. All measurements at all concentrations were made in triplicate. In each experiment the control [^{3}H]nitrendipine binding values were compared to those achieved when 1.9% halothane, 2.3% isoflurane, or 4.8% enflurane were added. In another set of six experiments, the effect of increasing concentration of halothane (0.78%, 1.33%, 1.90%, 2.57%) on the binding of 1 nM [^{3}H]nitrendipine was evaluated. After 60 min of incubation, during which the vials were protected from light to minimize the photochemical decomposition of nitrendipine, the reaction was terminated when an aliquot of 800 μl was filtered under vacuum through 2.5 cm Whatman GF/C glass fiber filters and washed three times with 10 ml of cold 20mM Tris buffer. Samples were counted in a Beckman LS2800 scintillation counter. Counting efficiency was approximately 55%.

Experiments with D600 were performed in a similar manner under similar conditions. [^{3}H]D600 concentration varied from 5 to 100 nM, and 1 mM unlabeled D600 was used as displacer. Non-specific binding is defined as that binding of [^{3}H]nitrendipine or [^{3}H]D600 which occurs in the presence of a high concentration of unlabelled ligand (500 nM nitrendipine or 1 mM D600 in this study). Specific binding, which indicates binding only to the voltage dependent Ca^{2+} channels, is calculated from the difference between total binding and non-specific binding.

Equilibrium binding data were analyzed using the Enzfitter program (Robin J. Leatherbarrow, Elsevier Science Publishers, Amsterdam) which yields a nonlinear, least-squares fit of the data to the following equation:

$$NTP\ Bound = \frac{B_{max} \cdot [NTP]_{free}}{K_d + [NTP]_{free}}$$

where B_{max} is the total number of binding sites, K_d is the dissociation constant, and $[NTP]_{free}$ is the concentration of nitrendipine in the reaction mixture. D600 binding was analyzed in a similar fashion. The program yields estimates of B_{max} and K_d for nitrendipine and D600 binding to the VDCC.

Statistical analysis of the binding studies was performed using paired t-test, in which the effect of halothane was compared to the control binding values for each concentration. All data are given as the mean ± SEM of independent experiments.

RESULTS AND DISCUSSION

The VDCC in isolated sarcolemma are in the inactivated state, which exhibits high affinity binding characteristics for dihydropyridines. Figure 1 demonstrates the marked effect of 1.9% halothane on the binding of [³H]nitrendipine to the VDCC. There is a marked decrease in binding over the entire range of [³H]nitrendipine concentration. In this specific experiment, the result of exposure of the membranes to halothane was a decrease in the maximal number of specific dihydropyridine binding sites, which is in agreement with the more general data previously reported.[6,12]

Figure 2 demonstrates the dose-dependent decrease in the specific binding of [³H]nitrendipine to purified sarcolemmal vesicles as a function of increasing halothane concentration. This figure, demonstrating data obtained with bovine cardiac sarcolemma, is similar to our results with crude membranes from rabbit and rat hearts,[5] and gives additional support that the alteration in the VDCC by halothane

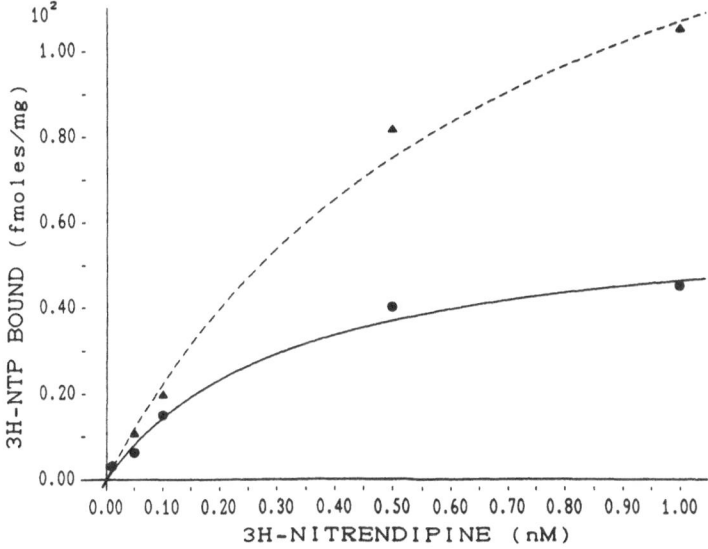

FIGURE 1. The specific binding of [³H]nitrendipine as a function of the concentration of [³H]nitrendipine. Data are the mean of 3 experimental measurements in the presence and absence of 1.9% halothane. The equilibrium binding was done for 60 min at 25°, pH 7.5. Equivalent samples were incubated with 500 nM of unlabelled nitrendipine in order to determine specific binding.

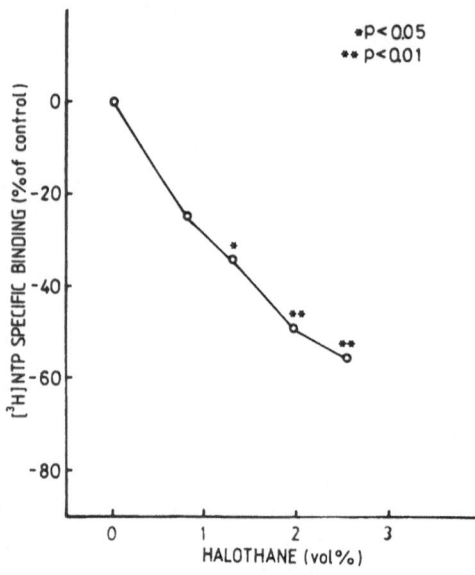

FIGURE 2. Dose-dependent depression of specific [³H]nitrendipine binding by increasing concentrations of halothane. The data are expressed as percentage of control. Each point represents a triplicate measurement of six experiments.

might be a common theme in terms of anesthetic action. We have observed that approximately 30% depression of [³H]nitrendipine binding occurs at only 0.78% halothane in purified bovine sarcolemma. This concentration, even at 25°C, represents a clinically relevant concentration, and is associated with an effect that is within the range of cardiac depression that is observed clinically. Furthermore, both our data and that of Nakao *et al.* indicate that there is no plateau to the inhibitory effect of halothane on the binding of [³H]nitrendipine, suggesting that all the VDCC sites in the sarcolemmal are equivalent and can be totally obscured by higher halothane concentration.

Table 1 indicates similar experiments which were performed with D600, a verapamil analog, which is known to bind to VDCC but at a separate locus from the

Table 1. Effect of Halothane on [³H]D600 Binding to Bovine Cardiac Sarcolemma

Halothane Concentration (vol%)	K_d	B_{max}
0	35.2 ± 6.1	263 ± 18
0.7	35.6 ± 10.4	174 ± 20
1.3	24.4 ± 9.1	160 ± 20
2.5	31.0 ± 4.8	102 ± 6
2.5 reversed*	27.6 ± 9.7	236 ± 29

The K_d and B_{max} are reported as the mean ± SEM. * These samples were exposed to anesthetic for 1 hour, then the halothane was allowed to evaporate and incubation with labelled [³H]D600 was initiated (see ref. 11).

dihydropyridine binding site. Halothane decreased the binding of D600 to the VDCC in bovine cardiac sarcolemma as indicated by the marked decrease in B_{max}. This effect was also reversible. The fact that two separate binding sites, the dihydropyridine and the phenylalkalamine sites, on the α_1 subunit of the VDCC could be equally affected by halothane suggests a rather significant alteration in the conformation or exposure of the channel. It further suggests that the effect of halothane on the VDCC is most likely due to an alteration of the sarcolemmal lipid environment surrounding the VDCC rather than a direct effect on the VDCC itself. We believe this to be the case since halothane is a small molecule and yet can interfere with the binding of two distinct ligands that have distinct and separate binding sites on the VDCC.

The questions that these data prompt are why the number of binding sites decrease and whether the decrease in VDCC binding sites actually occurs *in vivo* during anesthetic exposure. It is probable that the binding sites for both the dihydropyridine and the phenylalkylamine are intimately related to the surrounding lipid environment which is likely to be disrupted by exposure to halothane which is highly hydrophobic. The problem with proving whether the decrease in VDCC upon anesthetic exposure is an actual *in vivo* mechanism is compounded by the volatility of halothane and the reversibility of this process.

Although we have observed decreases in [³H]nitrendipine binding in the cardiac membranes from three species, rat, rabbit and cattle, we have noted a significant exception in the case of the VDCC from canine cardiac sarcolemmal membranes. The data in figure 3 demonstrates the lack of effect of halothane on the total and non-specific binding of [³H]nitrendipine to canine cardiac sarcolemma. At this time we have no experimental evidence to explain this difference. We suspect that the insensitivity of canine cardiac membranes might be related to a difference in the lipid environment surrounding the canine VDCC or to a modification of the canine VDCC during sarcolemmal purification that renders it insensitive to volatile anesthetics.

In summary, we have described the inhibition by halothane of the binding of two radiolabelled ligands that have high affinity binding sites on the VDCC. The inhibition is reversed when anesthetic is removed and occurs at clinically relevant anesthetic concentration. We suggest that these observations point to the VDCC as an important negative inotropic site of halothane in the heart.

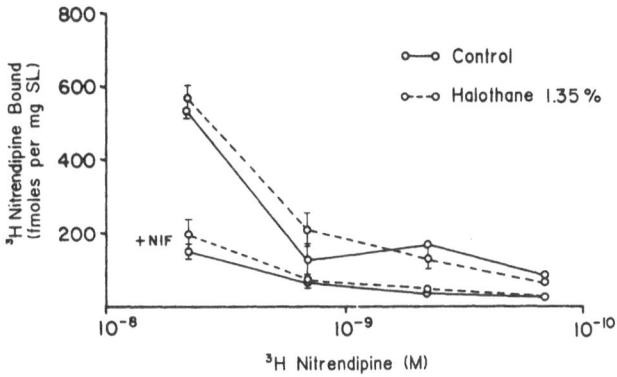

FIGURE 3. Total and nonspecific binding of [³H]nitrendipine in the presence and absence of 1.35% halothane to canine cardiac sarcolemma. The conditions of the experiment are as described in figure 1. [³H]nitrendipine binding to canine sarcolemmal vesicles in the presence of halothane. This graph illustrates the effect of halothane (1.35%) on [³H]nitrendipine binding to sarcolemmal vesicles in the presence and absence of nifedipine (NIF) 10^{-5} M.

REFERENCES

1. B. F. Rusy and H. Komai, Anesthetic depression of myocardial contractility: A review of possible mechanisms, *Anesthesiology* 67:745 (1987).
2. C. Lynch III, Differential depression of myocardial contractility by halothane and isoflurane in vitro, *Anesthesiology* 64:620 (1986).
3. Z. J. Bosnjak and J.P. Kampine, Effects of halothane on transmembrane potentials, Ca^{2+} transients, and papillary muscle tension in the cat, *Am J Physiol.* 251:H374 (1986).
4. D. A. Terrar and J. G. G. Victory, Isoflurane depresses membrane currents associated with contraction in myocytes isolated from guinea-pig ventricle, *Anesthesiology* 69:742 (1988).
5. T. J. J. Blanck, S. Runge and R. L. Stevenson, Halothane decreases calcium channel antagonist binding to cardiac membranes, *Anesth Analg* 67:1032 (1988).
6. B. Drenger, M. Quigg and T. J. J. Blanck, Volatile anesthetics depress calcium channel blocker binding to bovine cardiac sarcolemma, *Anesthesiology* 74:155-165 (1991).
7. L. R. Jones, H. R. Besch, J. W. Fleming, M. M. McConnaughey and A. M. Watanabe, Separation of vesicles of cardiac sarcolemma from vesicles of cardiac sarcoplasmic reticulum, *J Biol Chem* 254:530 (1979).
8. P. Caroni, M. Zurini, A. Clark and E. Carafoli, Further characterization and reconstitution of the purified Ca^{2+} pumping ATPase of heart sarcolemma, *J Biol Chem* 258:7305 (1983).
9. T. J. J. Blanck, A simple closed system for performing biochemical experiments at clinical concentrations of volatile anesthetics, *Anesth Analg* 60:435 (1981).
10. B. F. Bean, Nitrendipine block of cardiac calcium channels: High-affinity binding to the inactivated state. *Proc Natl Acad Sci* (USA) 81:6388 (1984).
11. P. J. Hochner, M. Quigg and T. J. J. Blanck, Halothane depresses gallopamil (D600) binding to bovine heart sarcolemma. *Anesthesiology* 73:A370 (1990).
12. S. Nakao, H. Hirata and Y. Kagawa, Effects of volatile anesthetics on cardiac calcium channels. *Acta Anaesthesiol Scand* 33:326-330, 1989.

11

Contribution of the Known Subcellular Effects of Anesthetics to Their Negative Inotropic Effect in Intact Myocardium

Hirochika Komai and Ben F. Rusy

INTRODUCTION

Various anesthetics have been shown to affect virtually every step involved in myocardial excitation-contraction coupling. What is not known is the relative importance of these multiple effects to the overall negative inotropic response in intact myocardium. We have used isolated rabbit papillary muscles and left atrial muscles to evaluate the relative contribution of anesthetic effects on the transsarcolemmal Ca^{2+} influx, on the function of the sarcoplasmic reticulum, and on the response of the myofibril.

Figure 1 schematically illustrates the flux of Ca^{2+} in reference to cardiac excitation-contraction coupling. Ca^{2+} entering the cell through the slow Ca^{2+} channel or by means of Na/Ca exchange 1) directly activates the myofibril, 2) fills the sarcoplasmic reticulum, and 3) triggers the release of Ca^{2+} stored in the sarcoplasmic reticulum. Thus, when an anesthetic reduces the transsarcolemmal Ca^{2+} influx, even if the anesthetic has no direct effect on the function of the sarcoplasmic reticulum, the amount of Ca^{2+} stored in the sarcoplasmic reticulum may be reduced and the amount of triggering Ca^{2+} may be reduced. Likewise, the reduction in Ca^{2+} influx most likely reduces the total amount of activator Ca^{2+} that reacts with the myofibril, and reduces the magnitude of myofibrillar response to the excitation even if the anesthetic has no direct effect on the myofibril.

RESULTS

Direct Effect of Anesthetics on Sarcoplasmic Reticulum Function

To evaluate the direct effect of an anesthetic on the function of the sarcoplasmic reticulum in intact myocardium, it is necessary to subtract the force reduction due to a decrease in the availability of sarcoplasmic reticular Ca^{2+} secondary to the anesthetic-induced reduction in the transsarcolemmal Ca^{2+} influx. In a

HIROCHIKA KOMAI, BEN F. RUSY, Department of Anesthesiology, University of Wisconsin, Madison, Wisconsin 53792.

Mechanisms of Anesthetic Action in Skeletal, Cardiac, and Smooth Muscle
Edited by T.J.J. Blanck and D.M. Wheeler, Plenum Press, New York, 1991

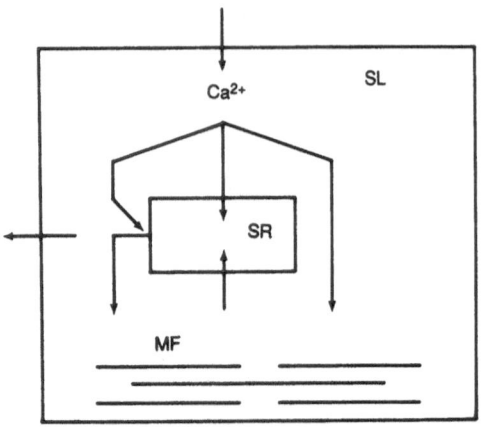

FIGURE 1. Calcium flux in myocardial excitation-contraction coupling. Abbreviations: SL, sarcolemma; SR, sarcoplasmic reticulum; MF, myofibril.

previously published study,[1] we evaluated the direct effect of halothane and isoflurane on the function of the sarcoplasmic reticulum by comparing their negative inotropic effects in a medium containing normal (2.5 mM) Ca^{2+} with the force depression in a medium containing no anesthetic but a lower concentration of Ca^{2+}. As the force of contraction in the presence of ryanodine is considered to be activated solely by transsarcolemmal Ca^{2+} influx, we used depression of developed force by an anesthetic or by low extracellular Ca^{2+} measured in the presence of ryanodine as an index of the reduction in transsarcolemmal Ca^{2+} influx. We determined, in the presence of 1 μM ryanodine, the concentrations of anesthetic and extracellular Ca^{2+} which caused equal depression of force and, thus, equal depression of transsarcolemmal Ca^{2+} influx. If, in the absence of ryanodine, where sarcoplasmic reticular function is intact, depressant effects of these same concentrations of anesthetic and the decreased concentration of extracellular Ca^{2+} are again equal, it is likely that the anesthetic effect on the transsarcolemmal Ca^{2+} influx and the secondary reduction in the amount of Ca^{2+} released from the sarcoplasmic reticulum account for the negative inotropic effect of the anesthetic. If, however, the depressant effect of an anesthetic in the absence of ryanodine exceeds that of low Ca^{2+}, it follows that the anesthetic has a direct depressant effect on the sarcoplasmic reticulum in addition to its effect to reduce transsarcolemmal Ca^{2+} influx. Since we were interested in the effect on the sarcoplasmic reticulum, we used rabbit atria, a tissue whose contractile activity is strongly dependent on Ca^{2+} released from the sarcoplasmic reticulum, and further, we measured the force of (postrest) potentiated-state contractions, which are highly dependent on Ca^{2+} released from the sarcoplasmic reticulum and strongly inhibited by ryanodine.

The force of contraction measured in the presence of ryanodine showed a steep dependence on the extracellular Ca^{2+} concentration (figure 2), suggesting that if Ca^{2+} is bound to a saturable site, the dissociation constant is high. In contrast, the force measured in the absence of ryanodine appeared to saturate at a much lower concentration of extracellular Ca^{2+} (figure 3). Note that this difference in extracellular Ca^{2+} dependence of the force measured in the presence and absence of ryanodine for rabbit atria is similar to that of the extracellular Ca^{2+} dependence of the contractile activity of rabbit and rat myocytes reported by Capogrossi *et al.*[2] It is very likely that the direct activation of the myofibril by the influx of extracellular Ca^{2+} shows a steep

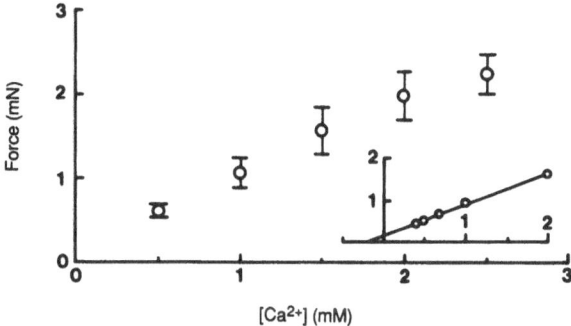

FIGURE 2. Extracellular Ca^{2+} dependence of the force of contraction in the presence of ryanodine. Postrest contractions of isolated rabbit atria were elicited 2 s after steady-state contraction at 3 Hz and measured in the presence of 1 μM ryanodine. Points are mean ± SEM (n = 10). The inset is a double reciprocal plot of the mean values. The inset abscissa label and units are $[Ca^{2+}]_o{}^{-1}$ (mM^{-1}); ordinate, $Force^{-1}$ (mN^{-1}). The estimated $[Ca^{2+}]_o$ for half-maximal force of contraction was 4.6 mM.

dependence on extracellular Ca^{2+} concentration, whereas filling of the sarcoplasmic reticulum and triggering of Ca^{2+} release from the sarcoplasmic reticulum are saturable at relatively low concentrations of extracellular Ca^{2+}. Figure 4 shows the effects of halothane, isoflurane, and low extracellular Ca^{2+} concentration on the force of contraction measured in the presence of ryanodine. The control medium contained 2.5 mM Ca^{2+} and no anesthetic. The negative inotropic effect of 0.6% halothane in a medium containing 2.5 mM Ca^{2+} was not significantly different from the force decrease in a medium containing 1.5 mM Ca^{2+} without the anesthetic. Similarly, there was no significant difference between the relative force decrease in 1.0% halothane or 1.5% isoflurane and that in a medium containing 1.0 mM Ca^{2+}, as well as between 2.4% isoflurane and that in 0.5 mM Ca^{2+}. In the absence of ryanodine, the negative inotropic effect of 0.6% halothane was significantly larger than the force decrease in

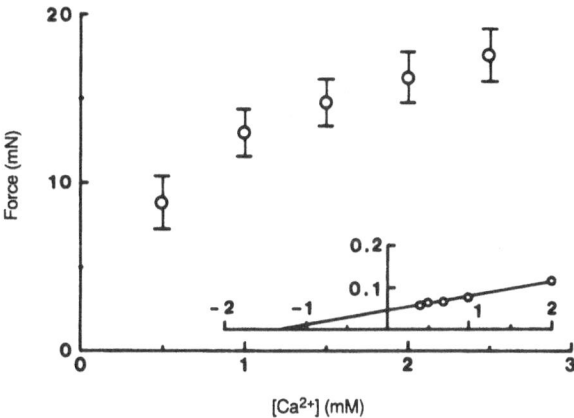

FIGURE 3. Extracellular Ca^{2+} dependence of the force of contraction in the absence of ryanodine. Postrest contractions of isolated rabbit atria elicited 2 s after steady-state contraction at 3 Hz. Points are mean ± SEM (n = 10). The inset is a double reciprocal plot of the mean values, with axis labels and units as follows: abscissa, $[Ca^{2+}]_o{}^{-1}$ (mM^{-1}); ordinate, $Force^{-1}$ (mN^{-1}). Estimated $[Ca^{2+}]_o$ for half-maximal force of contraction was 0.8 mM.

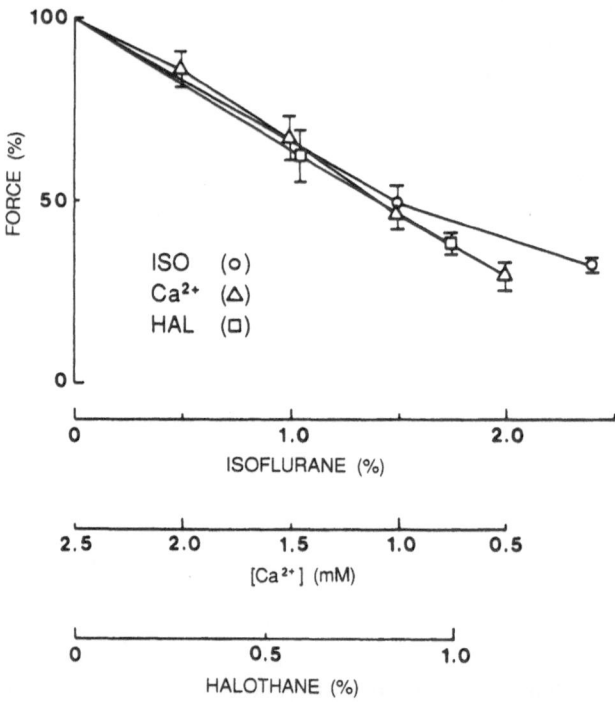

FIGURE 4. Negative inotropic effect of isoflurane, halothane and low Ca^{2+}, measured in the presence of ryanodine (1 μM). Force was measured as the developed tension during postrest potentiated-state contractions in isolated rabbit atria at 30°C. All points are expressed as a percent of control (either no anesthetic or normal Ca^{2+} medium) and given as mean ± SEM. ISO (o): effect of isoflurane (1.5% and 2.4%) in normal Ca^{2+} (2.5 mM) medium. Control force was 1.7 ± 0.3 mN. Ca^{2+} (△): effect of Ca^{2+} concentration in the absence of an anesthetic. Control force was 2.2 ± 0.2 mN. HAL (□): effect of halothane (0.6% and 1.0%) in normal Ca^{2+} (2.5 mM) medium. Control force was 1.7 ± 0.2 mN. The number of muscles was 8 for the points on the isoflurane curve, 10 for the points on the varying Ca^{2+} curve and 9 for the points on the halothane curve. The scales of abscissas were chosen so that the three curves approximately overlap. Reprinted from H. Komai and B. F. Rusy, *Anesthesiology* 72:694-698 (1990) with permission.

1.5 mM Ca^{2+} medium, and the effect of 1.0% halothane was larger than that in 1.0 mM Ca^{2+} medium (figure 5). These results suggest that halothane does have a direct effect on the function of the sarcoplasmic reticulum in intact myocardium in addition to the effect secondary to the anesthetic-induced reduction in the transsarcolemmal Ca^{2+} influx. From the known effect of halothane to induce a transient increase in intracellular Ca^{2+} and to induce a transient increase in the force of contraction,[3-6] it is very likely that halothane, by rendering the sarcoplasmic reticulum leaky, reduces the amount of Ca^{2+} stored in this organelle rather than inhibiting calcium-induced release of calcium. The effect of isoflurane, on the other hand, was less than that found in the low extracellular Ca^{2+} medium (figure 5). It is possible that isoflurane inhibits a Ca^{2+} leak form the sarcoplasmic reticulum, although we cannot exclude the possibility that isoflurane reduced the myofibrillar response to the low concentrations of activator Ca^{2+} in the presence of ryanodine and not to the high concentrations of activator Ca^{2+} in the absence of ryanodine.

Rapid-Cooling-Induced Contractures

In an attempt to determine whether anesthetics alter sarcoplasmic reticular Ca^{2+} loading in the intact myocardium, we examined the effects of anesthetics on rapid cooling contractures using rabbit papillary muscles. Rapid cooling contractures were elicited by changing the 30°C bathing medium to a 3°C medium after steady state beating at 2 Hz had been achieved. These contractures are activated by calcium released from the sarcoplasmic reticulum[7,8] by a mechanism different from that involved in contractile activity induced by depolarization of the myocardium. Rapid cooling contractures can be considered to be a measure of the amount of Ca^{2+} stored in the sarcoplasmic reticulum. If an anesthetic has no effect on the magnitude of the rapid cooling contracture, it is very likely that the anesthetic has no effect on the amount of Ca^{2+} stored in the sarcoplasmic reticulum. Such a situation was observed for thiopental (figure 6). Note that Blanck and Stevenson have shown that thiopental does not alter Ca^{2+} uptake by isolated sarcoplasmic reticulum.[9] When an anesthetic reduces the magnitude of the rapid cooling contracture, it may mean that the amount of Ca^{2+} stored in the sarcoplasmic reticulum is reduced or, alternatively, that Ca^{2+} release by rapid cooling is inhibited. As we have reported at the last meeting and subsequently published,[10] halothane reduced the magnitude of the contracture without slowing the rate of contracture development, whereas isoflurane markedly reduced the

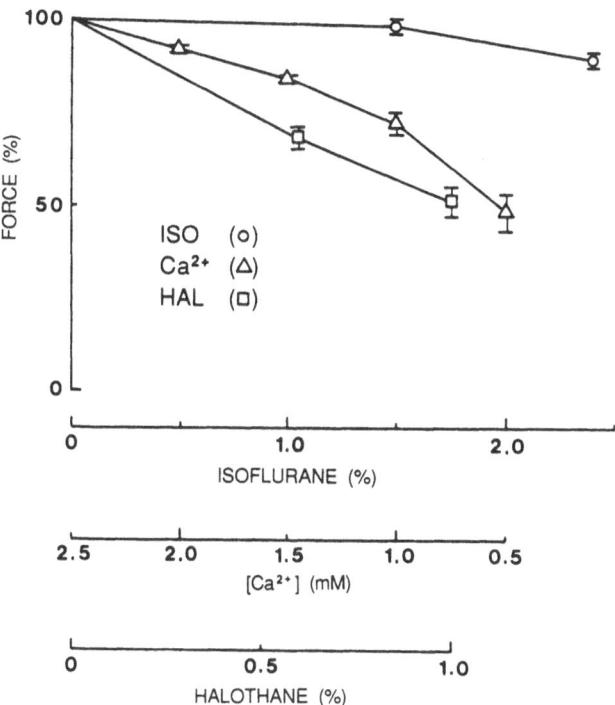

FIGURE 5. Depression of contractile force by isoflurane, halothane and low Ca^{2+} medium, measured in the absence of ryanodine. The experimental preparation and design are the same as in figure 4. Data presentation and scales of abscissas are also as in figure 4. ISO (o): effect of isoflurane (1.5% and 2.4%) in normal Ca^{2+} (2.5 mM) medium. Control force was 17.3 ± 1.4 mN. Ca^{2+} (▲): effect of Ca^{2+} concentration in the absence of an anesthetic. Control force was 17.6 ± 1.6 mN. HAL (□): effect of halothane (0.6% and 1.0%) in normal Ca^{2+} (2.5 mM) medium. Control force was 16.2 ± 1.4 mN. Reprinted from H. Komai and B. F. Rusy, *Anesthesiology* 72:694-698 (1990) with permission.

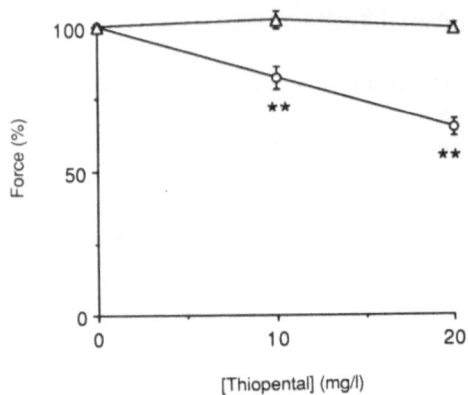

FIGURE 6. Effect of thiopental on steady-state contractions at 2 Hz and on the rapid cooling contracture. Open circles (○) represent steady-state contractions at 2 Hz; control value 29 ± 5 mN/mm^2. Open triangles (△) represent rapid cooling contractures; control value 20 ± 3 mN/mm^2. Each point is mean ± SEM (n = 6).

rate of contracture development in addition to decreasing the magnitude of the contracture. Figure 7 illustrates the effects of halothane (0.6%), enflurane (1.7%), and isoflurane (1.5%) on the force of steady-state contraction at 2.0 Hz and on the force of the rapid cooling contracture. These results suggest that halothane reduces the amount of Ca^{2+} stored in the sarcoplasmic reticulum whereas isoflurane selectively inhibits the Ca^{2+} release induced by rapid cooling. The effect of enflurane on the time course of rapid cooling contracture was somewhere in between that of halothane and that of isoflurane.

Anesthetic Effects on the Force-Length Relationship

The increase in the force of contraction accompanying an increase in muscle length is known to reflect the myofibrillar response to activator Ca^{2+},[11] and it is possible that this relationship may be altered if an anesthetic changes the response of

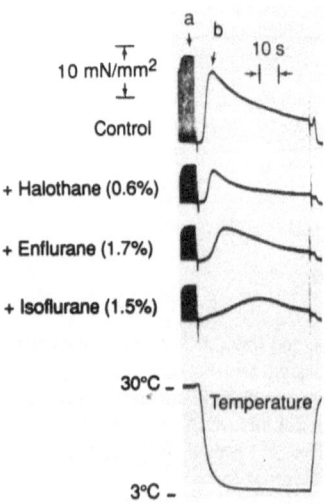

FIGURE 7. Effects of volatile anesthetics on steady-state contractions at 2 Hz and on the rapid cooling contracture. Steady state at 2 Hz is indicated by "a"; the rapid cooling contracture by "b."

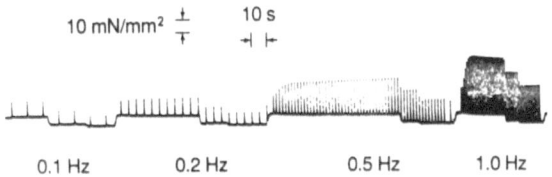

FIGURE 8. Effects of muscle length on the force of contraction at different stimulation frequencies.

the myofibril to Ca^{2+}. It is important to note that the magnitude of the force change accompanying a change in muscle length is dependent on the amount of activator Ca^{2+} available. For this reason, the anesthetic effect needs to be evaluated at the same level of activator Ca^{2+}. What we did was to vary the frequency of stimulation to match the force of contraction of rabbit papillary muscles at the L_{max} (the muscle length at which the developed force is maximal) in the absence and in the presence of an anesthetic, and then evaluate the anesthetic effect on the force decrease accompanying a decrease in the muscle length from the L_{max} to 95% L_{max} or 90% L_{max}. Figure 8 shows that when the muscle length was shortened from L_{max} to 95% L_{max} and then to 90% L_{max} at different stimulation frequencies, the developed force decreased. When the decrease in the force of contraction accompanying a shortening of the muscle was plotted against the force at L_{max} obtained at different stimulation frequencies, a linear relationship was obtained (figure 9). The line was shifted in the presence of 0.6% halothane, indicating that the force decrease accompanying shortening of the muscle was larger in the presence of the anesthetic than in its absence (figure 9). The results with 14 muscles are summarized in table 1. The slope of the regression line was significantly increased by 0.6% halothane when the muscle was shortened from the L_{max} to 90% L_{max}. These results suggest that halothane increases the magnitude of the

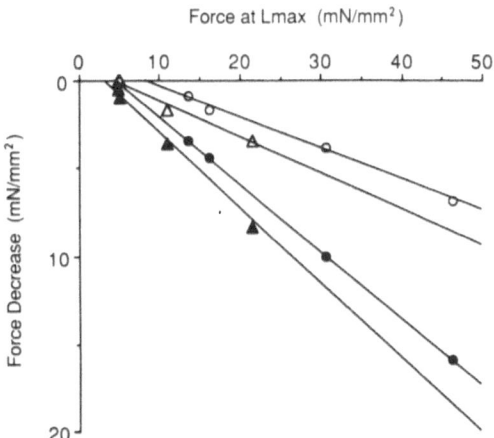

FIGURE 9. Effect of halothane (0.6%) on the relationships between the force at L_{max} and the force decrease accompanying a decrease in muscle length from L_{max} to 95% L_{max} and to 90% L_{max}. The force at L_{max} and the force decrease accompanying a length change from L_{max} to 95% L_{max} or to 90% L_{max} was measured at stimulation frequencies of 0.1, 0.2, 0.5 and 1.0 Hz, and the force decrease was plotted against the force at L_{max}. Open symbols represent points generated by a resting length change from L_{max} to 95% L_{max}, and filled symbols, L_{max} to 90% L_{max}. The circles represent controls (no anesthetic), and the triangles are points obtained in the presence of 0.6% halothane.

121

Table 1. The Effect of Halothane (0.6%) on the Coefficients Which Characterize
the Relationship between Resting Muscle Length and Developed Force

Change in Muscle Length	Halothane Present?	A (mN/mm^2)	B
L_{max} to 95% L_{max}	No	-0.4 ± 0.3	0.26 ± 0.02
	Yes	-0.4 ± 0.2	0.29 ± 0.03
L_{max} to 90% L_{max}	No	-0.6 ± 0.2	0.54 ± 0.03
	Yes	-0.6 ± 0.2	$0.58 \pm 0.03^*$

Coefficients defined as follows: (force decrease) = A + B × (force at L_{max}). Similar curves are shown in figure 9. The values are means ± SEM and N = 14. The asterisk (*) indicates a significant effect of halothane (P < 0.05).

decrease in myofibrillar response to a given level of activator Ca^{2+} accompanying shortening of the muscle.

SUMMARY

The results of these studies suggest that halothane, in addition to its effect of reducing the transsarcolemmal Ca^{2+} influx, has a direct effect on the function of the sarcoplasmic reticulum by making the organelle leaky to Ca^{2+} and reducing the amount of Ca^{2+} stored. Isoflurane, on the other hand, does not appear to make the sarcoplasmic reticulum leaky to Ca^{2+}. In fact, our observations suggest that isoflurane makes the sarcoplasmic reticulum less leaky to Ca^{2+}. Halothane may enhance the reduction in the myofibrillar response to activator Ca^{2+} when the muscle fiber length is shortened. Thiopental appears to reduce the transsarcolemmal Ca^{2+} influx without reducing the amount of Ca^{2+} stored in the sarcoplasmic reticulum.

ACKNOWLEDGEMENTS

Supported in part by NIH grant GM29527 and by a grant-in-aid from the University of Wisconsin, Department of Anesthesiology Research and Development Fund.

REFERENCES

1. H. Komai and B. F. Rusy, Direct effect of halothane and isoflurane on the function of the sarcoplasmic reticulum in intact rabbit atria, *Anesthesiology* 72:694-698 (1990).
2. M. C. Capogrossi, M. M. Kort, H. A. Spurgeon and E. G. Lakatta, Single adult rabbit and rat cardiac myocytes retain the Ca^{2+}- and species-dependent systolic and diastolic contractile properties of intact muscle, *J Gen Physiol* 88:589-613 (1986).
3. D. M. Wheeler, R. T. Rice, R. G. Hansford and E. G. Lakatta, The effect of halothane on the free intracellular calcium concentration of isolated rat heart cells, *Anesthesiology* 69:578-583 (1988).
4. C. Lynch III, S. Vogel and N. Sperelakis, Halothane depression of myocardial slow action potentials, *Anesthesiology* 55:360-368 (1981).
5. H.-N. Luk, C.-I. Lin, C.-L. Chang and A.-R. Lee, Differential inotropic effects of halothane and isoflurane in dog ventricular tissues, *Eur J Pharmacol* 136:409-413 (1987).
6. D. M. Wheeler, R. T. Rice and E. G. Lakatta, The action of halothane on spontaneous contractile waves and stimulated contractions in isolated rat and dog heart cells, *Anesthesiology* 72:911-920 (1990).

7. S. Kurihara and T. Sakai, Effects of rapid cooling on mechanical and electrical responses in ventricular muscle of guinea-pig, *J Physiol* (Lond) 361:361-378 (1985).

8. J. H. B. Bridge, Relationships between the sarcoplasmic reticulum and sarcolemmal calcium transport revealed by rapidly cooling rabbit ventricular muscle, *J Gen Physiol* 88: 437-473 (1986).

9. T. J. J. Blanck and R. L. Stevenson, Thiopental does not alter Ca^{2+} uptake by cardiac sarcoplasmic reticulum, *Anesth Analg* 67:346-348 (1988).

10. H. Komai, D. Redon and B. F. Rusy, Effects of isoflurane and halothane on rapid cooling contractures in myocardial tissue, *Am J Physiol* 257:H1804-H1811 (1989).

11. D. G. Allen and J. C. Kentish, The cellular basis of the length-tension relation in cardiac muscle. *J Mol Cell Cardiol* 17:821-840 (1985).

12

Effects of Volatile Anesthetics on the Intracellular Ca^{2+} Concentration in Cardiac Muscle Cells

Dixon W. Wilde, Ravi Gutta, Michael F. Haney, Paul R. Knight

INTRODUCTION

The myocardial depressant effects of the three halogenated volatile anesthetics, halothane, enflurane and isoflurane, have been well documented clinically and experimentally.[1-4] In recent years multiple mechanisms of action of these agents have been demonstrated. The results of a number of studies have pointed to anesthetic effects on transmembrane signalling in a variety of cell types as a central facet of anesthetic action (for brief review see Maze[5]). Many efforts have been made to elucidate the mechanisms by which these agents alter excitation-contraction coupling in the heart to produce negative inotropy. Halothane, for example, has been observed to depress myocardial contractility[4] through depletion of the sarcoplasmic reticulum (SR) stores of calcium (Ca^{2+}).[6-9] The evidence suggests that the negative inotropic effect of halothane is mediated by enhancement of Ca^{2+} efflux through SR Ca^{2+} release channels.[10] Halothane's effects to limit availability of Ca^{2+} for contraction do not appear restricted to actions on SR Ca^{2+} release, however. Direct effects of halothane on myocardial Ca^{2+} channels have also been proposed,[11] although it remains unclear whether this is a direct effect on the L-type Ca^{2+} channels, or instead the indirect effect of a change in some intracellular mediator. Halothane may also alter Ca^{2+} uptake by the myocardial SR[12,13] and have effects on mitochondrial Ca^{2+} metabolism and storage. These various putative mechanisms may act alone or in combination to limit Ca^{2+} availability for contraction in the heart.

Although studies of volatile anesthetic effects on the heart have focused on the actions of halothane, comparative studies of halothane, enflurane and isoflurane have been reported. Results of experiments with equipotent concentrations of halothane, enflurane and isoflurane have clearly indicated that halothane is more potent than isoflurane and enflurane in depressing contractility of isolated papillary muscles of various species.[14-16] These studies, though, do not identify the cellular or molecular mechanisms by which the volatile anesthetics exert their effects. Multicellular

DIXON W. WILDE, RAVI GUTTA, MICHAEL F. HANEY, PAUL R. KNIGHT, Department of Anesthesiology, University of Michigan Medical Center, Ann Arbor, Michigan 48109-0572.

Mechanisms of Anesthetic Action in Skeletal, Cardiac, and Smooth Muscle
Edited by T.J.J. Blanck and D.M. Wheeler, Plenum Press, New York, 1991

preparations may afford individual cells a degree of protection in studies of anesthetic action. Single, isolated cell techniques or isolated myocardial fiber preparations such as those of Wheeler et al.[7] or Murat et al.[17] provide effective models for studies of a more fundamental mechanism of volatile anesthetic actions. Experiments using single cells may lead to specific theories of the molecular mechanisms governing anesthetic-induced negative inotropy in the heart.

In this report, we directly compare the effects of halothane to those of enflurane and isoflurane in single myocardial cells isolated from rat ventricle. The advent of intracellular, Ca^{2+}-sensitive fluorescent indicators (for review see Blinks[18]) has provided non-invasive techniques for examining the changes in the free Ca^{2+} concentration in the cytosol ($[Ca^{2+}]_i$) during anesthetic exposure. The isolated rat ventricular myocyte is a desirable model for examination of anesthetic effects on Ca^{2+} release from the SR since these cells depend primarily on SR sources of Ca^{2+} during contraction.[19] Clinically relevant concentrations of halothane, enflurane and isoflurane were examined for their effects on caffeine-stimulated Ca^{2+} transients and on depolarization-induced elevations of $[Ca^{2+}]_i$ caused by high extracellular potassium and direct electrical stimulation of the myocytes. The results suggest that halothane is more effective than either isoflurane or enflurane in limiting availability of Ca^{2+} during the excitation-contraction cycle. In addition, enflurane and isoflurane appear to induce spontaneous activity in these cells whereas halothane does not. The mechanisms of action of these agents is considered.

METHODS

Single quiescent myocytes were isolated *via* enzymatic dispersion from adult female Sprague-Dawley rats according to a variation of the method of Mitra and Morad.[20] Briefly, animals were heparinized (200 U/kg Na-heparin, i.p.) and anesthetized with halothane. The heart was removed under halothane anesthesia, cannulated and perfused on a modified Langendorff apparatus with a solution of collagenase (150 U/ml, Sigma, St.Louis), 0.1% bovine serum albumin (BSA, Sigma), 15 mM taurine (Sigma) and 50 mM Ca^{2+} in a Tyrode's solution buffered by 5 mM N-2-hydroxyethylpiperazine-N'-2-ethanesulfonic acid (HEPES, Sigma). The standard Tyrode's solution used for this procedure and for all experimentation contained, in addition to HEPES: NaCl 150.0 mM, KCl 5.4 mM, $MgCl_2$ 1.2 mM and glucose 5.0 mM. Solution pH was adjusted to 7.29 with NaOH. Calcium was added from a 1 M $CaCl_2$ stock. During the heart dispersion, preparation temperature was maintained at 37°C. Following the Langendorff perfusion, the ventricles were removed, minced and treated to a second incubation in enzyme supplemented by 1% BSA. Isolated cells were recovered from this medium by centrifugation and were washed with HEPES-Tyrode's containing 1.8 mM Ca^{2+} and 1% bovine serum albumin (BSA). This procedure yielded many quiescent, relaxed myocytes possessing clear striations. These cells contracted vigorously when stimulated. Once isolated, the cell suspensions were allowed to cool to room temperature.

Suspensions of myocytes were loaded with the intracellular, fluorescent Ca^{2+} indicator fura-2 by incubation with the membrane-permeant, acetoxy-methylester form, fura-2 AM (Molecular Probes, Eugene, OR), at a concentration of 4 μM for 10 minutes. Excess fura-2 was removed from the cell suspension by washing the cells with HEPES-Tyrode's containing 1.8 mM Ca^{2+} and 1% BSA. Loaded cell suspensions were kept in light-proof tubes to minimize photolysis of the fluorescent probe. Loaded cells were allowed to sit for 30 minutes to allow for complete cytosolic de-esterification of the fura-2 AM to its Ca^{2+}-sensitive form, fura-2. Small aliquots of fura-2 loaded cells were placed on cover slips in a controlled-atmosphere, controlled-temperature chamber. This device was then positioned on the stage of a Leitz Diavert inverted microscope equipped with quartz optics and a 75 W Xe lamp. A bath superfusion system delivered the HEPES-Tyrode's with 1.8 mM Ca^{2+} at a rate of 3 ml/minute. Given a bath volume of 3 ml, complete solution exchange was attained in 5 minutes. Preparation temperature was maintained at 30°C. The low temperature slowed the Ca^{2+} release events to a point at which capture of very fast Ca^{2+} transients was possible.

Fura-2 yields maximum fluorescence emissions at 505-512 nm resulting from excitation at either of two wavelengths: when bound to Ca^{2+}, fura-2 fluoresces at 340 nm; and it fluoresces at 380 nm when free of Ca^{2+}.[23] From the ratio of fluorescence emission at each wavelength, one may calculate an estimated $[Ca^{2+}]_i$. For these experiments, fluorescence excitation was provided by a computer-controlled filter wheel which alternated between the 340 nm and 380 nm filters. The fluorescence emissions were detected by a Leitz MPV photomulitplier and were digitized and stored using either of two computer systems. For continuous, long term monitoring of fura-2 signals, data were digitized and stored on a Hewlett-Packard computer using software for wheel control and data storage modified from a program obtained from Leitz. This program allowed sampling at rates up to 4 pts/sec, useful for recording relatively slow changes in $[Ca^{2+}]_i$ such as occur during caffeine or potassium stimulation. For recording of fast Ca^{2+} transients such as those elicited by electrical pacing of single myocytes, new software was developed using the Labview® programming package (version 2.0) which allows photomultiplier output sampling at up to 40 μsec/pt. We estimated the maximal fura-2 binding time constant for our cells to be approximately 160 μsec for each Ca^{2+}-fura association,[21,22] or approximately four times slower than the computer sampling. Therefore, this system was more than adequate for recording fast, electrically stimulated transients.

Single cells were selected for use on the basis of their morphology and their quiescence following observation. For each experimental protocol, single cells were initially superfused with 1.8 mM Ca^{2+} Tyrode's equilibrated with 100% O_2 while the 340 nm and 380 nm fluorescence emissions were recorded. Cell autofluorescence was subtracted electronically. Each cell was then exposed to one of three stimuli: 15 mM caffeine (Sigma) to elicit Ca^{2+} release from SR, 50 mM K^+ solution to depolarize the sarcolemma to approximately -25 mV (based on the Nernst potential for that K^+ gradient and assuming a resting potential of -70 mV), or electrical stimulation by extracellular suction pipette electrode at 1 Hz (square wave depolarization, constant current with stimulus isolation). Glass suction pipettes were pulled from 2.0 mm o.d. borosilicate glass tubing (WPI, New Haven, CT) and were filled with Ca^{2+}-free Tyrode's solution. In the caffeine and high K^+_o experiments, each cell received a control exposure to the stimulant in 100% O_2. After washout, the cells were exposed to one of the three anesthetics by equilibration of the superfusates with anesthetic supplied by vaporizer using 100% O_2 as the carrier gas. Anesthetic concentration was monitored in the gas phase by an anesthesia circuit evaluator (Traverse Medical Monitors, Saline, MI) and in solution by N-heptane extraction of anesthetic and measurement on a Gow-Mac gas chromatograph. Control experiments revealed that anesthetic equilibration in both the superfusate reservoirs and the bath fluid was acheived within 5 minutes. Following 10 minutes of anesthetic exposure, the stimulation used during the control period was repeated. The anesthetic was then washed out for 10 minutes and a washout response obtained.

Cytosolic $[Ca^{2+}]$ was measured by alternately exciting fura-2 at 340 nm and 380 nm *via* the filter wheel, which switched position in 100 msec. The computer sampled at 4 pts/sec at each wheel position. A cutoff filter of 460 nm was used. The ratio of fluorescence emission at each wavelength was used to calculate $[Ca^{2+}]_i$ according to the equation of Grynkiewicz et al.,[23]

$$[Ca^{2+}]_i = K_d \frac{F - F_{min}}{F_{max} - F} \frac{Sf_2}{Sb_2}$$

where F_{max} is the fluorescence ratio in a 1.8 mM Ca^{2+} solution containing the calcium ionophore, ionomycin (Calbiochem, San Diego, CA), F_{min} is the ratio in a Ca^{2+}-free solution containing ethylene glycol bis(b-aminoethyl ether)-N,N,N′,N′ tetraacetic acid (EGTA, Sigma), and the expression Sf_2/Sb_2 represents the ratio of 380 nm signals in EGTA and ionomycin solutions. The K_d for fura-2 pentapotassium salt was determined *in vitro* in this optical system to be 391 nM. Prior to Ca^{2+} permeabilization with ionomycin, each cell was treated with a glucose-free, 1.8 mM Ca^{2+} Tyrode's solution containing 2 mM carbonyl cyanide p-(trifluoromethoxy)-phenylhydrazone (FCCP, Sigma) to induce a rigor state in the cell, thereby eliminating ATP-dependent Ca^{2+} pumping at the sarcolemma as a source of interference with the calibration process.[24,25]

Since each cell acted as its own control, simple statistical tests using paired analyses were possible. The paired Student's t-test was utilized with differences considered significant at $P < 0.05$. All results are presented as mean ± SEM.

RESULTS

Effects of Volatile Anesthetics on Response of $[Ca^{2+}]_i$ to Caffeine

The Ca^{2+}-tolerant, single cardiac myocytes used in these studies exhibited a mean resting $[Ca^{2+}]_i$ of 79 ± 5 nM. Superfusion of single cardiac myocytes with Tyrode's solutions containing 15 mM caffeine produced a transient elevation in $[Ca^{2+}]_i$. This response was characterized by an initial rapid rise in $[Ca^{2+}]_i$ which rapidly fell below pre-stimulus levels (undershoot) and then gradually recovered. Observations of cells exposed to caffeine in this manner revealed a transient contraction of the cell corresponding to the rise in $[Ca^{2+}]_i$. The "undershoot" probably results from stimulation of Ca^{2+} extrusion and/or reuptake mechanisms, *i.e.*, Na^+/Ca^{2+} exchange.[26] Figure 1A depicts a typical experiment in which halothane, at a concentration of 0.6% (0.8 MAC), reduced the net caffeine-stimulated Ca^{2+} transient from 68 nM to 32 nM following 10 minutes of exposure to the anesthetic. The effect was reversible upon removal (washout) of the anesthetic. Halothane caused a concentration-dependent decrease in the caffeine-stimulated Ca^{2+} transient (figure 2A). Halothane at 0.6% (0.8 MAC) produced a mean reduction in transient amplitude to 67 ± 12% of control ($P < 0.05$, n = 5). Halothane's attenuation was most pronounced at 1.3% (1.7 MAC), to 35 ± 14 percent of control ($P < 0.05$, n = 6). Halothane caused no significant change in the resting $[Ca^{2+}]_i$ at concentrations between 0.6 and 1.3%. However, there was some indication that a concentration of 2.2% (2.8 MAC) halothane did elevate basal $[Ca^{2+}]_i$ to 103 ± 14 nM ($P > 0.05$, n = 7). Spontaneous Ca^{2+} transients were not observed in myocytes following exposure and removal of halothane from the bathing medium in any of these experiments.

Isoflurane also reversibly reduced the caffeine-stimulated Ca^{2+} transient in single ventricular myocytes. However, the concentration-dependence was less uniform. Figure 1B is a representative ratio trace from one of these experiments. In this

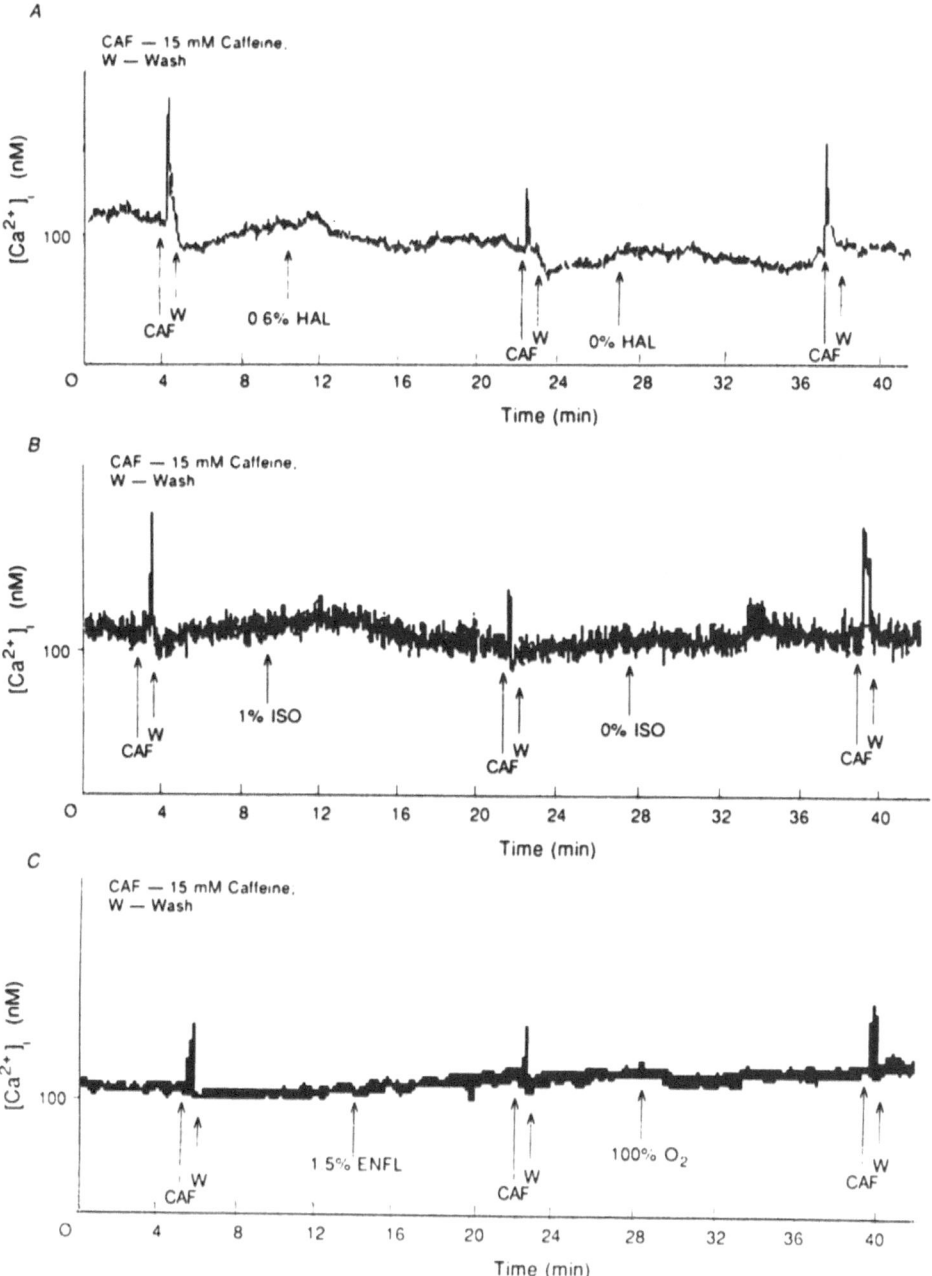

FIGURE 1. Representative experimental traces depicting the changes in $[Ca^{2+}]_i$ in response to 15 mM caffeine during exposure to halothane (A), isoflurane (B) and enflurane (C). Records are ratio plots of fura-2 fluorescence intensity after subtraction of background fluorescence. Cells were superfused with caffeine-containing Tyrode's solution until maximum response was attained. The cells were then washed with 1.8 mM Ca^{2+} Tyrode's solution (at "W"). Note the "undershoot" in $[Ca^{2+}]_i$ as the caffeine effect reverses.

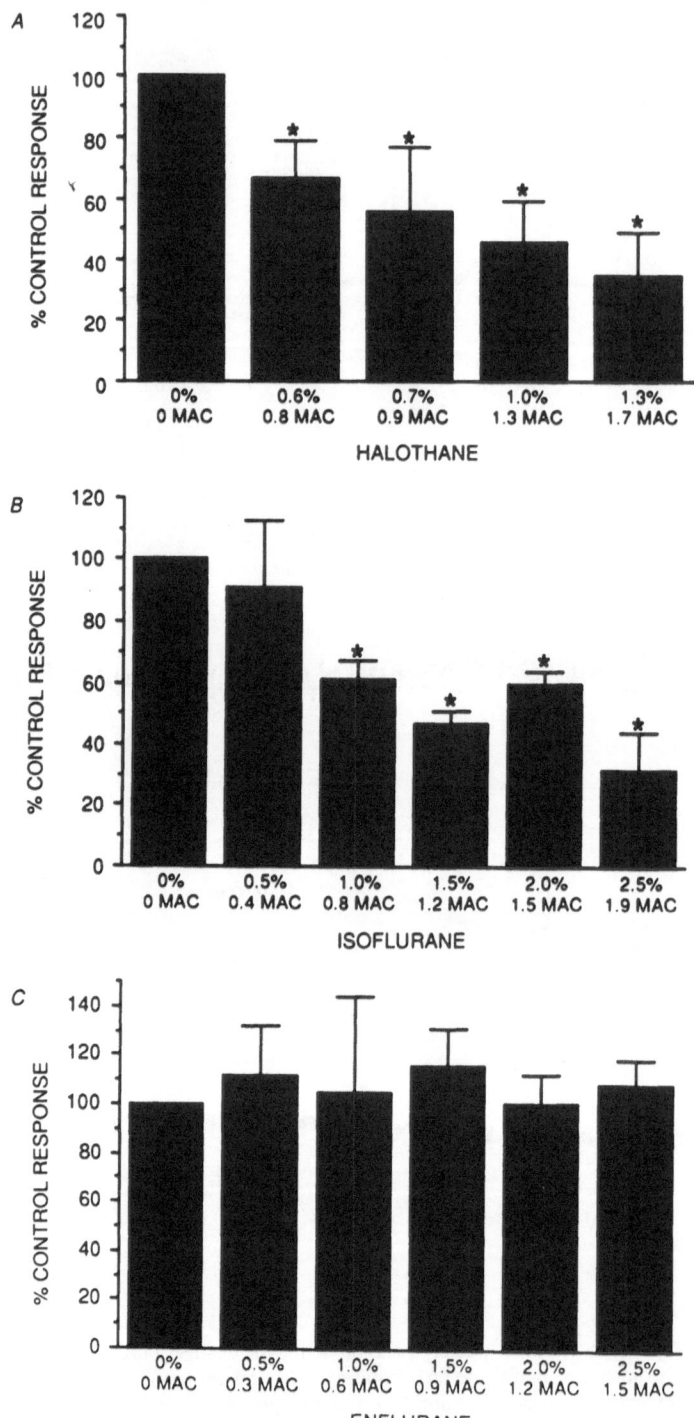

FIGURE 2. Graphic comparison of the effects of increasing concentrations of halothane (A), isoflurane (B) and enflurane (C) on the net Ca^{2+} transient evoked by 15 mM caffeine superfusion of single cardiac myocytes. Net response is taken as the maximum response minus baseline. Values are presented as mean ± SEM. Significant differences ($P < 0.05$) from control are denoted by asterisks (*).

example, 1% isoflurane reduced the net caffeine-stimulated Ca^{2+} transient (peak response minus baseline in nM) from a control value of 170 nM to 33 nM. The inhibition was reversed following washout of the anesthetic by 100% O_2. Isoflurane caused no significant or consistent change in the resting $[Ca^{2+}]_i$. Spontaneous transient activity was observed in 60.5% of the cells (n = 82) either during the initial isoflurane exposure or, more commonly, during and after anesthetic washout. This spontaneous activity; characterized by large, often tonic increases in $[Ca^{2+}]_i$; coincided with contraction of the myocyte. The effect of increasing concentrations of isoflurane on the net Ca^{2+} transient caused by caffeine application is shown in figure 2B. At roughly equivalent MAC values, isoflurane (1 MAC = 1.2%) was approximately equipotent with halothane (1 MAC = 0.78%) in reducing the caffeine-stimulated Ca^{2+} transient.

Enflurane had no effect on the caffeine-induced Ca^{2+} transient at any concentration (0.5-2.5%, which corresponds to 0.3-1.5 MAC; 1 MAC = 1.68%; see figures 1C and 2C). Mean values for net transient amplitudes were slightly, but not significantly, elevated over control for some of the concentrations. However, enflurane, like isoflurane, did induce spontaneous Ca^{2+} transients and contractions during initial anesthetic exposure or during anesthetic washout in 69.7% of the cells studied (n = 43). The time course and magnitude of the "undershoot" in $[Ca^{2+}]_i$ following development of the caffeine response did not appear altered by any of the three anesthetics. Although this component of the transient was not examined thoroughly, this suggests that sarcolemmal Ca^{2+} extrusion mechanisms, such as Na^+/Ca^{2+} exchange,[26] may not be affected by these agents.

$[Ca^{2+}]_i$ during Membrane Depolarization with High $[K^+]_o$

Superfusion of myocytes with 50 mM K^+ Tyrode's solution produced a tonic rise in $[Ca^{2+}]_i$. Assuming a resting potential in the quiescent myocyte of -70 mV, 50 mM K^+ should depolarize the cell to approximately -25 mV (according the Nernst equation), or to a point where L-channel current is approximately 30% of maximum. The characteristic response to 50 mM K^+_o was an initial rapid rise in $[Ca^{2+}]_i$ followed by a slow phase until a plateau was attained. Control experiments revealed that the myocytes exhibited consistent responses over time (repeated high K^+ exposures at 10-15 minute intervals).

Halothane reversibly limited the elevation of $[Ca^{2+}]_i$ in response to extracellular 50 mM K^+. Figure 3A is a typical example of the effect of halothane on the high K^+ response. In this example, high K^+ elevates $[Ca^{2+}]_i$ from an initial resting level of approximately 100 nM to 242 nM. Following 10 minutes of exposure to 1.5% halothane, a repeated exposure to 50 mM K^+ leads to a reduced response which recovers to approximately 80-85% of control following washout of the anesthetic. Halothane did not cause any change in resting $[Ca^{2+}]_i$ at any concentration tested in this series of experiments. The halothane inhibition exhibited a concentration dependence (figure 4A) which was quite steep. Halothane did not induce any spontaneous Ca^{2+} transients in the myocytes used in this series.

Isoflurane was also a potent inhibitor of the K^+-induced elevation in $[Ca^{2+}]_i$ in these cells. A representative experimental record is depicted in figure 3B. However, there was minimal concentration dependence over the range of concentrations used (0.5-2.5% or 0.6-1.9 MAC; figure 4B). The reduction of the response to 50 mM K^+ was obvious at concentrations as low as 0.5% with the net change in $[Ca^{2+}]_i$ reduced to 64 ± 10 percent of control. The effect of isoflurane on the K^+-stimulated Ca^{2+} transients was significantly different from control at all concentrations (P < 0.05). More recent data have shown that 0.25% isoflurane reduces the response to 50 mM K^+ to 83 ± 9 percent of control. Although there was a slight increase in the degree

131

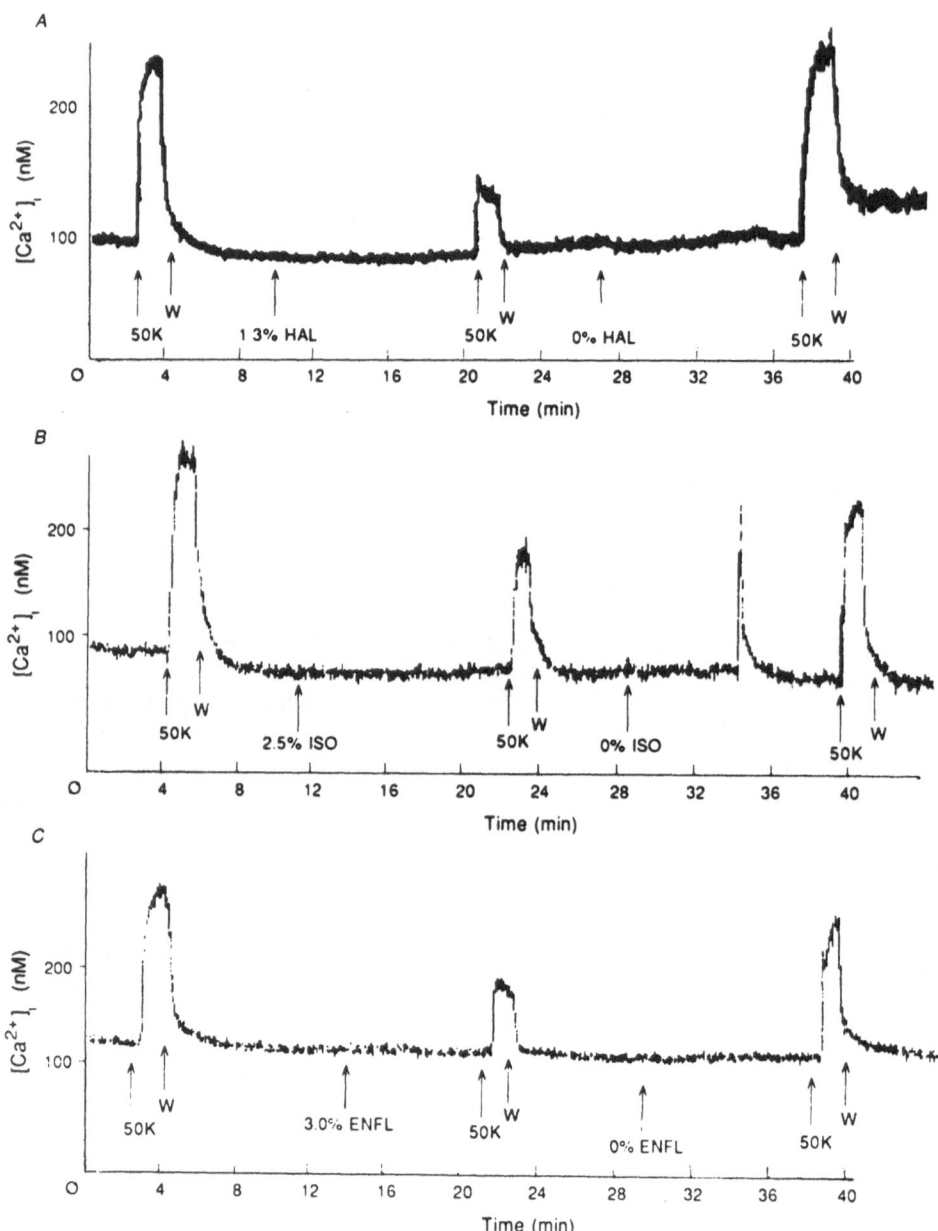

FIGURE 3. Representative experimental traces of the effects of halothane (A), isoflurane (B) and enflurane (C) on the rise in $[Ca^{2+}]_i$ evoked by superfusion of single cardiac myocytes with 1.8 mM Ca^{2+} HEPES-Tyrode's containing 50 mM K^+. Assuming a resting potential of -70 mV for a quiescent, healthy appearing myocyte; depolarization by 50 mM K^+_o should proceed to -25 mV. This is the point at which L-channel current is approximately 30% of maximum in these cells. Traces are ratio plots of fura-2 fluorescence at 340 nm and 380 nm excitation wavelengths. Washout of 50 mM K^+ is denoted by W.

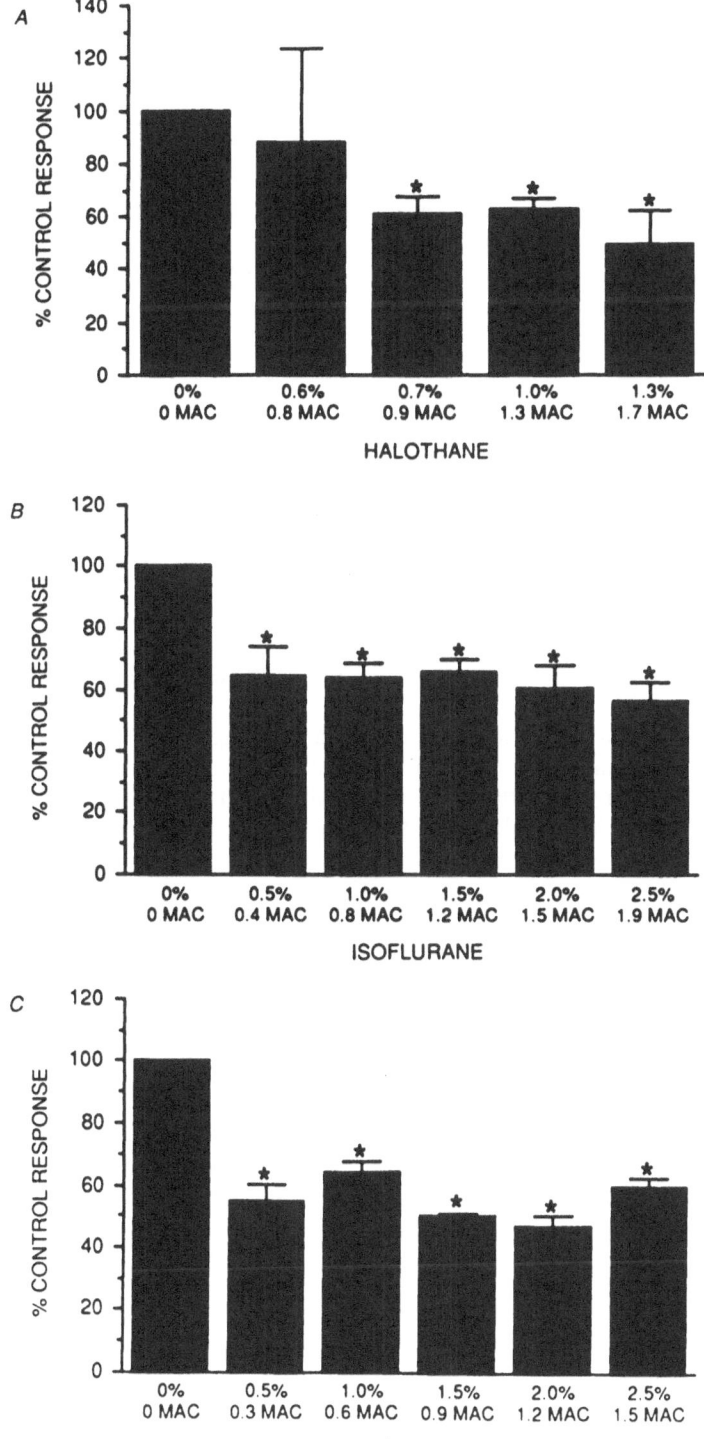

FIGURE 4. Effect of increasing concentrations of halothane (A), isoflurane (B) and enflurane (C) on tonic elevation of $[Ca^{2+}]_i$ in myocytes stimulated by 50 mM K^+_o. Results are expressed as percent of control net response; mean ± SEM. Values significantly different from control ($P < 0.05$) are denoted by asterisks (*). In panels B and C, recent data have revealed that 0.25% isoflurane and enflurane reduce the transients to 83 ± 9 and 79 ± 6 percent of control, respectively.

of inhibition with increasing concentration of isoflurane, the steep concentration effect observed with halothane was not apparent. The inhibition was minimally reversible and ranged from 39 to 67 percent of the control net transient amplitude. The response following anesthetic washout was occasionally obscured by spontaneous Ca^{2+} transients. Isoflurane-induced spontaneous Ca^{2+} transients were observed in 67 percent of the cells examined (n = 33) and were coincident with cell shortening. Spontaneous activity commenced during either initial anesthetic exposure or during anesthetic washout with 100% O_2.

Enflurane, in concentrations ranging from 0.5 to 2.5% (0.3-1.5 MAC), was effective in significantly reducing the K^+-stimulated elevation in $[Ca^{2+}]_i$ (P < 0.05). The effect of enflurane on the response to K^+ is shown in figure 3C. However, in a manner similar to the effects of isoflurane, enflurane caused minimal concentration-dependent inhibition of the $[Ca^{2+}]_i$ change (figure 4C). Recent data have shown that 0.25% enflurane reduces the K^+-stimulated response to 79 ± 6 percent of control. At 0.5% concentration, the K^+-induced $[Ca^{2+}]_i$ change was reduced to 55 ± 6 percent of control but at 2% enflurane, the inhibition only caused a mean inhibition to 47 ± 3 percent of the control response. During washout, spontaneous activity usually commenced. Recovery of the response to K^+ following anesthetic washout was partial and ranged from 77 to 90 percent of the net control response.

Spontaneous Activity Induced by Enflurane and Isoflurane

The spontaneous contractile activity and changes in $[Ca^{2+}]_i$ observed during application and washout of isoflurane and enflurane, but not halothane, prompted an investigation into this mechanism underlying this spontaneous activity. We tested the efficacy of nitrendipine to inhibit the spontaneous Ca^{2+} transients following exposure of myocytes to enflurane and isoflurane. Nitrendipine (1 μM) rapidly and reversibly blocked the spontaneous transients occurring after either isoflurane or enflurane exposure. Figures 5A and 5B depict representative recordings from such experiments. In these cases, the cells had been stimulated before and during anesthetic application with 50 mM K^+ as part of the normal protocol. Upon initiation of a spontaneous Ca^{2+} transient, the superfusate was rapidly changed to one containing 1 μM nitrendipine. The transient was immediately abolished. Washout of nitrendipine by 1.8 mM Ca^{2+} Tyrode's resulted in the redevelopment of the spontaneous Ca^{2+} transients. Similar results were obtained with enflurane-induced spontaneity. Exposure of cells to isoflurane or enflurane alone and without any prior stimulation did not induce spontaneous activity. The spontaneous transients were dependent on L-channel activation. The spontaneous activity could be repeatedly abolished by nitrendipine until the cells underwent hypercontracture.

Effects of Volatile Anesthetics on Electrically Stimulated Ca^{2+} Transients

All three anesthetic agents depressed the fast Ca^{2+} transients initiated by direct extracellular membrane stimulation *via* suction pipette. Square-wave depolarization of the cardiac cell membrane by 3 msec pulses at 1 Hz resulted in a rapidly rising Ca^{2+} transient (figure 6A) which had a duration of 150-250 msec and was coincident with cell shortening. Intracellular Ca^{2+} transient amplitudes, under conditions in which the cell was superfused only with oxygen-equilibrated HEPES-Tyrode's solution with 1.8 mM Ca^{2+}, had a mean amplitude for the net change in $[Ca^{2+}]_i$ of 210 ± 23 nM (n = 51). Halothane, at concentrations of 0.4-1.5% (0.5-1.9 MAC), caused a concentration-dependent inhibition of the transient. For example, following treatment

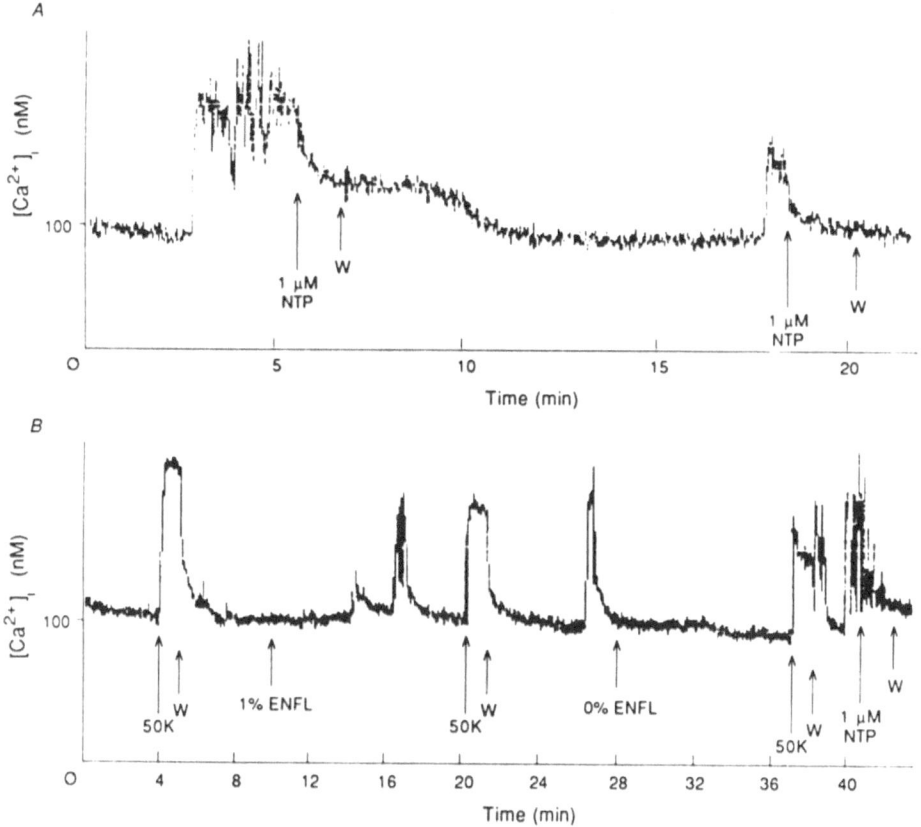

FIGURE 5. Effect of 1 mM nitrendipine on spontaneous $[Ca^{2+}]_i$ transients in single cardiac myocytes induced during washout of isoflurane (A) and enflurane (B). Upon commencement of the spontaneous transient, the superfusate was rapidly switched to one containing nitrendipine (1 mM). Traces are ratio plots of fura-2 fluorescence at 340 nm and 380 nm excitation wavelengths.

of cells with 1.3% (1.7 MAC) halothane, the net transient amplitude was depressed from a mean control value of 163 nM (mean for this group) to 44 nM, a reduction of 73 percent. The inhibition was completely reversible. Figure 6B depicts a typical experiment in which halothane is applied to a cell under constant pacing. Upon return to 100% oxygen, the halothane effect is reversed. Although apparent in figure 6B, halothane did not significantly shift diastolic $[Ca^{2+}]_i$ in the cells used for this group of experiments. Figure 6C shows the effect of 1% halothane on transients recorded at the 340 nm excitation wavelength only, using a fast computer sampling protocol. These traces, recorded at a sampling rate of 40 msec/pt (5000 point acquisition), clearly show that the amplitude of the Ca^{2+} transient is reduced. The rate of rise of the Ca^{2+} transient also appears affected by halothane. Detailed examination of the time course of the transient development before and during halothane exposure is currently underway and may reveal halothane-induced changes in rate of Ca^{2+} mobilization from either SR stores or from sarcolemmal Ca^{2+} channel activation.

Enflurane and isoflurane produced similar reductions in electrically stimulated Ca^{2+} transient amplitude. These experiments have revealed that enflurane, at concentrations as low as 0.25%, may reduce transient amplitude to 58 ± 17 percent of control transient amplitude ($P < 0.05$). Isoflurane is less potent, reducing transients

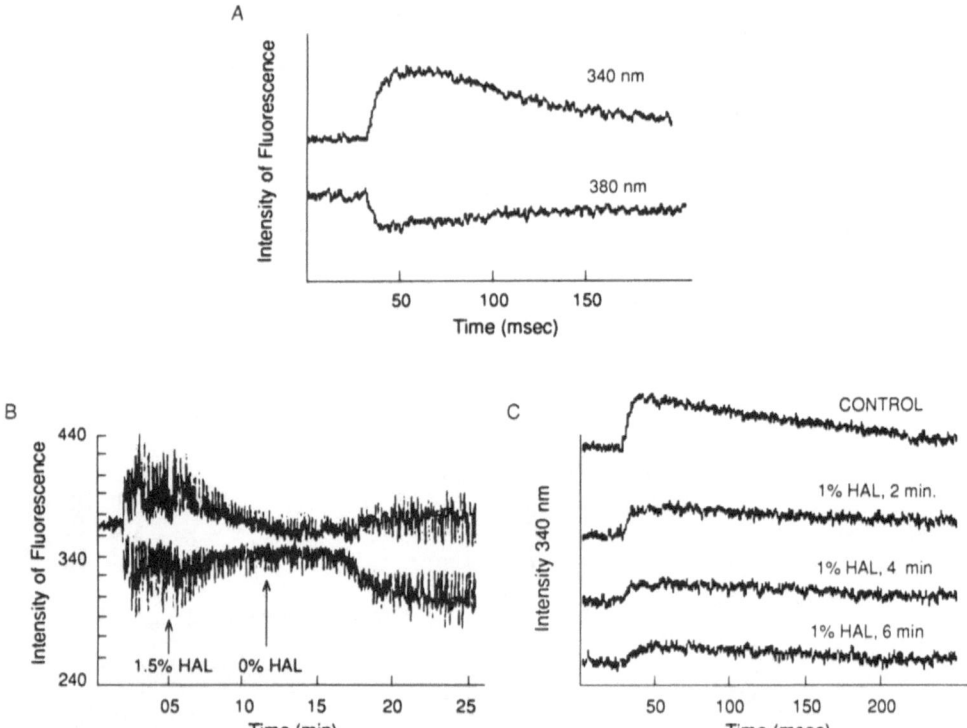

FIGURE 6. A: Single Ca^{2+} transients evoked by electrical pacing of a single cardiac myocyte via an extracellular suction pipette electrode. Traces were recorded at 40 msec/pt for 5000 points first at 340 nm followed by 380 nm excitation wavelengths. Traces were digitally juxtaposed. B: Effect of 1.5% (1.9 MAC) halothane on electrically stimulated Ca^{2+} transients. The shift in diastolic level during halothane exposure was not consistently observed. C: Effect of halothane on electrically stimulated Ca^{2+} transients recorded at 400 msec/pt (5000 pts) at the 340 nm excitation wavelength. Note that the traces are displaced vertically for clarity. The diastolic [Ca^{2+}]$_i$ did not shift significantly during these recordings.

to only 74 ± 4 percent of control at a concentration of 1%. These results have not indicated enflurane- or isoflurane-induced changes in the rate of rise of the Ca^{2+} transient. In addition, the inhibition of electrically stimulated Ca^{2+} transients by these agents was not easily reversed by washout of the anesthetic. Increased cell irritability was also observed. In figure 8, a cell paced by suction pipette is treated with 0.25% enflurane. After 6 minutes of superfusion with enflurane-equilibrated Tyrode's, average Ca^{2+} transient amplitude was reduced to 95% of the average control transient amplitude (P < 0.05). In addition, cell spontaneity was observed in the form of "aftertransients" (figure 8 bottom tracing). These spontaneous Ca^{2+} transients corresponded to phasic contractions of the myocyte. Recovery from enflurane exposure was minimal and required long washout periods. Similar spontaneous transients were observed during isoflurane exposure to electrically paced myocytes.

DISCUSSION

The results of these studies confirm the cardioinhibitory potential of halothane, isoflurane and enflurane. Halothane is a most potent inhibitor of the increase in cytosolic Ca^{2+} resulting from direct activation of the SR Ca^{2+} release channel (caffeine stimulation) or from indirect SR release following sarcolemmal depolarization. Isoflurane shares halothane's inhibition of the response to caffeine, high $[K^+]_o$ and electrical depolarization but, as with enflurane, induces cardiac cell irritability leading to spontaneous, rapid and phasic elevations in $[Ca^{2+}]_i$ resulting in tonic contraction of the myocyte. Enflurane was effective in blocking the depolarization-induced Ca^{2+} transients but did not affect the ability of 15 mM caffeine to release SR stores of Ca^{2+}. Both isoflurane and enflurane produced only minimal concentration-dependent inhibition of the K^+-stimulated change in $[Ca^{2+}]_i$. The absence of observed concentration effects by these anesthetics on the K^+-stimulated Ca^{2+} transients may indicate that lower concentrations of these agents must be used to reveal dose dependency. However, given that roughly equivalent MAC concentrations of halothane, isoflurane and enflurane were used; direct comparisons of the effects of these three agents on $[Ca^{2+}]_i$ is appropriate.

The evidence obtained from these experiments is consistent with the current consensus among various laboratories that halothane exerts its negative inotropic action on the heart by depleting SR stores of Ca^{2+}, thereby reducing the Ca^{2+} available for contraction.[6,7,27] The results presented here generally support those of Lynch[14] and Housmans and Murat[16] which demonstrated an order of negative inotropic potency of halothane > enflurane > isoflurane in guinea pig and ferret papillary muscles, respectively. The results of these studies indicate that at roughly equivalent MAC concentrations, halothane is more effective than isoflurane and approximately equipotent with enflurane at reducing the rise in $[Ca^{2+}]_i$ induced by 50 mM K^+, although enflurane failed to produce an obvious concentration-dependent inhibition.

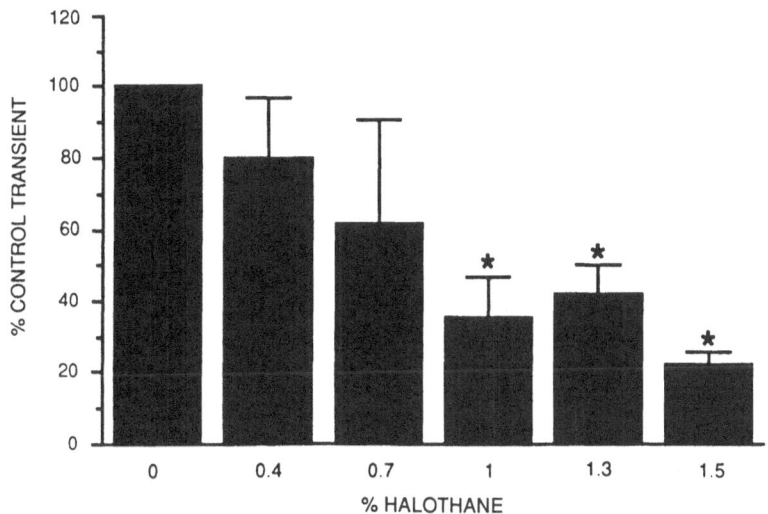

FIGURE 7. Concentration dependence of the halothane inhibition of electrically stimulated Ca^{2+} transients in single cardiac myocytes. Values represent percent of the net control transient amplitude, averaged over 15 transients (peak minus diastolic $[Ca^{2+}]_i$). Results presented as mean ± SEM. Significant differences from control are denoted by * where $P < 0.05$.

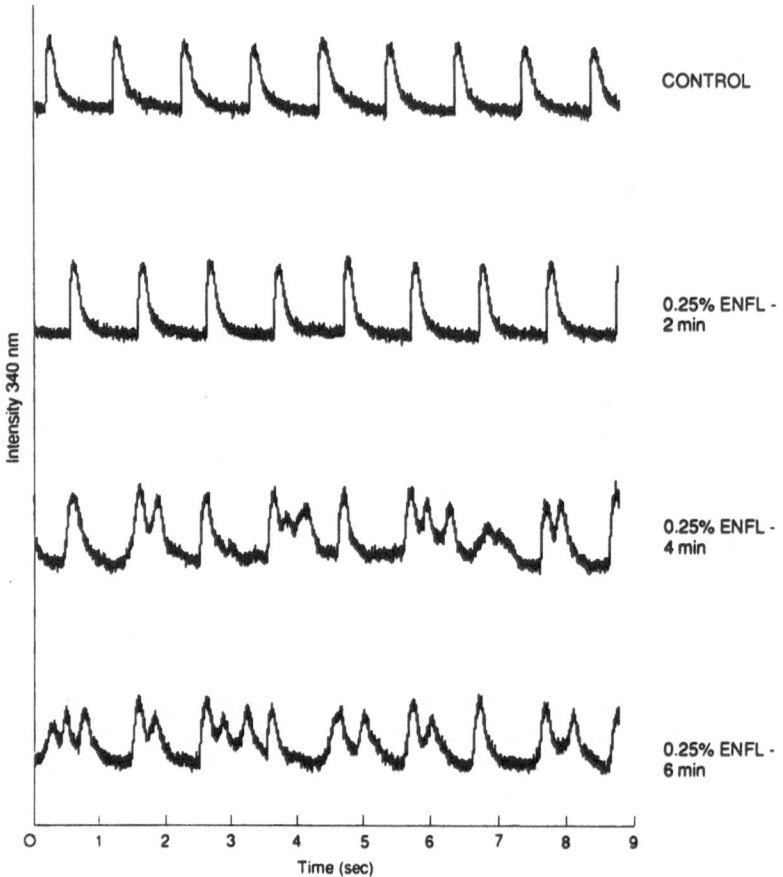

FIGURE 8. Effect of 0.25% enflurane (0.1 MAC) on electrically stimulated Ca^{2+} transients. At 6 minutes of treatment with the anesthetic (bottom trace), average transient amplitude has been significantly reduced to 95% of control ($P < 0.05$). Cell irritability has increased as evinced by the incidence of "aftertransients" which corresponded to phasic cell shortening. Transients were recorded at the 340 nm excitation wavelength at a sample rate of 4 msec/pt (5000 pt acquisition).

At equipotent MAC values, enflurane was more effective than isoflurane at inhibiting the K^+-induced Ca^{2+} transients. However, halothane and isoflurane are roughly equipotent in reducing the caffeine-stimulated Ca^{2+} transient, while enflurane appears to have no effect. While the evidence is relatively clear that halothane acts to deplete SR stores of Ca^{2+},[7,9,10,12] it remains uncertain if isoflurane's negative inotropic effect occurs by the same mechanism. Enflurane did not diminish the response of cells to caffeine, suggesting that it does not deplete SR stores of Ca^{2+}. The effect of enflurane and isoflurane on sarcolemmal Ca^{2+} channels (L-type) to reduce conductance or create partial or total blockade would reduce the amount of Ca^{2+} influx during the normal excitation-contraction cycle leading to reduced Ca^{2+}-stimulated Ca^{2+} efflux from the SR.

The evidence obtained in our laboratory indirectly suggests that enflurane and isoflurane may act on sarcolemmal L-channels to alter their activation properties. In 60% of the isoflurane experiments and 69% of the enflurane experiments using either caffeine or 50 mM K^+_o stimulation, spontaneous $[Ca^{2+}]_i$ transients and contractions were observed during washout of the anesthetic. The blockade of this spontaneity by 1 μM nitrendipine implicates the dihydropyridine-sensitive, L-type, voltage-gated Ca^{2+} channels in the initiation of the spontaneous transients. The action of nitrendipine is use dependent.[28] Therefore, enflurane and isoflurane must be inducing opening in the L-channel population. Surprisingly, the isoflurane- and enflurane-induced spontaneity was observed following either caffeine or 50 mM K^+ stimulation. However, spontaneous activity was not observed in the absence of any initial cell stimulation. Alterations in L-channel function during enflurane or isoflurane exposure may result following L-channel activation in 50 mM K^+ but the results from caffeine-stimulated cells seem to infer an anesthetic modulation of the interaction between intracellular Ca^{2+} release and L-channel activation. It is possible that enflurane and isoflurane modulation of L-channel activation is dependent on prior stimulation of those channels. By contrast, induction of spontaneity was not observed during or after exposure of single myocytes to halothane and, although halothane reduced the K^+-stimulated and electrically paced Ca^{2+} transients, it is unclear whether this effect was simply a consequence of SR Ca^{2+} depletion or a direct effect at the sarcolemma. Interactions between enflurane and the Ca^{2+} channel blocker diltiazem have been described as additive in patients receiving enflurane anesthesia.[29] Additive effects of these two agents would suggest that enflurane increases the likelihood that the Ca^{2+} channel blocker has access to the open channel. An action of halothane on membrane Ca^{2+} channels has also been demonstrated. Ikemoto et al.[11] have published voltage-clamp data from halothane-treated rat ventricular myocytes which indicate a direct effect of the agent on inward Ca^{2+} current. However, evidence obtained by Wheeler (personal communication) demonstrates that halothane's effect on sarcolemmal Ca^{2+} current may occur subsequent to its action on SR Ca^{2+} release in rat ventricular myocytes. Halothane appears to increase G-protein-mediated second messenger generation in mouse heart.[30] Elevation of cGMP due to halothane action could feed back to decrease Ca^{2+} channel contributions to the elevation of $[Ca^{2+}]_i$ during each contraction cycle (for review see Schultz et al.[31]).

The results of our studies also indicate that isoflurane reduces the caffeine-stimulated Ca^{2+} transient in single cardiac myocytes. The inhibition is less pronounced than that caused by halothane but does exhibit a similar concentration dependence. Enflurane, on the other hand, had no effect on the caffeine response. Although these evidences infer a similarity of mechanism for halothane and isoflurane, the direct action of isoflurane on SR Ca^{2+} release could not be demonstrated in this preparation.

In summary, halothane, enflurane and isoflurane exert potent negative inotropic actions on single cardiac myocytes, reducing the availability of Ca^{2+} for cellular shortening. However, the mechanisms by which these agents produce their effects differ. Halothane limits Ca^{2+} availability via depletion of SR Ca^{2+} stores and may secondarily limit entry of Ca^{2+} at the sarcolemma. Isoflurane exhibits effects similar to those of halothane regarding SR-stimulated Ca^{2+} release and, like enflurane, appears to affect the L-type Ca^{2+} channels in the sarcolemma. Both isoflurane and enflurane induce spontaneous, nitrendipine-sensitive elevations in $[Ca^{2+}]_i$ which are slow to wash out, suggesting direct and long-lasting alteration of L-channel function by these anesthetics. Although these effects of isoflurane and enflurane may not be evident in vivo or in isolated tissue preparations, they suggest a potentially damaging effect on myocardial cell viability which may further act to reduce the function of this organ in vivo. In addition, the differences between the responses of single ventricular myocytes and isolated heart or papillary muscles to halothane, enflurane and isoflurane suggest that although single cells are appropriate for examination of molecular mechanisms of anesthetic action the results of such studies must be carefully interpreted in order to compare the results with those obtained from the intact organism. The latter is a less easily controlled biological model that possesses intrinsic homeostatic mechanisms which may counteract the deleterious cardiovascular effects of anesthetic exposure.

REFERENCES

1. Z. J. Bosnjak and J. P. Kampine, Effects of halothane on transmembrane potentials, Ca^{2+} transients and papillary muscle tension in the cat, *Am J Physiol* 251:H374-381 (1986).
2. S. M. Malinconico, C. R. Hartzell and R. L. McCarl, Effect of calcium on halothane-depressed beating in heart cells in culture, *Mol Pharmacol* 23:417-423 (1983).
3. H. Komai and B. F. Rusy, Effect of halothane on rested-state and potentiated-state contractions in rabbit papillary muscle: Relationship to negative inotropic action, *Anesth Analg* 61:403-409 (1982).
4. B. F. Rusy and H. Komai, Anesthetic depression of myocardial contractility: A review of possible mechanisms, *Anesthesiology* 67:745-766 (1987).
5. M. Maze, Transmembrane signalling and the holy grail of anesthesia, *Anesthesiology* 72:959-961 (1990).
6. D. M. Wheeler, R. T. Rice, R. G. Hansfors and E. G. Lakatta, The effect of halothane on the free intracellular calcium concentration of isolated rat heart cells, *Anesthesiology* 69:578-583 (1988).
7. D. M. Wheeler, R. T. Rice and E. G. Lakatta, The action of halothane on spontaneous contractile waves and stimulated contractions in isolated rat and dog heart cells, *Anesthesiology* 72:911-922 (1990).
8. M. Katsuoka and S. T. Ohnishi, Inhalation anesthetics decrease calcium content of cardiac sarcoplasmic reticulum, *Br J Anaesth* 62:669-673 (1989).
9. J. Y. Su and W. G. L. Kerrick, Effects of halothane on caffeine-induced tension transients in functionally skinned myocardial fibers, *Pflugers Arch* 380:29-34 (1979).
10. J. S. Herland, F. J. Julian and D. G. Stephenson, Halothane increases Ca^{2+} channels of sarcoplasmic reticulum in chemically skinned rat myocardium, *J Physiol* (Lond) 426:1-18 (1990).
11. Y. Ikemoto, A. Yatani, J. Arimura, J. Yoshitake, Reduction of the slow inward current of isolated rat ventricular cells by thiamylal and halothane, *Acta Anaesthesiol Scand* 29:583-586 (1985).
12. T. E. Nelson and T. Sweo, Ca^{2+} uptake and Ca^{2+} release by skeletal muscle sarcoplasmic reticulum, *Anesthesiology* 69:571-577 (1988).
13. T. J. J. Blanck and M. Thompson, Calcium transport by cardiac sarcoplasmic reticulum: Modulation of halothane action by substrate concentration and pH, *Anesth Analg* 60:390-394 (1981).
14. C. Lynch III, Differential depression of myocardial contractility by halothane and isoflurane *in vitro*, *Anesthesiology* 64:620-631 (1986).
15. C. Lynch III and M. J. Frazer, Depressant effects of volatile anesthetics upon rat and amphibian ventriculur myocardium: Insights into anesthetic mechanisms of action. *Anesthesiology* 70:511-522 (1989).
16. P. R. Housmans and I. Murat, Comparative effects of halothane, enflurane, and isoflurane at equipotent anesthetic concentrations on isolated ventricular myocardium of the ferret: I. Contractility, *Anesthesiology* 69:451-463 (1988)
17. I. Murat, P. Lechene and R. Ventura-Clapier, Effects of volatile anesthetics on mechanical properties of rat cardiac skinned fibers, *Anesthesiology* 73:73-81 (1990).
18. J. R. Blinks, Intracellular $[Ca^{2+}]$ measurements, in: "The Heart and Cardiovascular System," H. A. Fozzard, ed., Raven Press, New York (1986) pp. 671-701.
19. A. Fabiato, Calcium-induced release of calcium from the cardiac sarcoplasmic reticulum, *Am J Physiol* 245:C1-C14 (1983).
20. R. Mitra and M. Murad, A uniform enzymatic method for dissociation of myocytes from hearts and stomachs of vertebrates, *Am J Physiol* 249:H1056-H1060 (1985).
21. A. P. Jackson, M. P. Timmerman, C. R. Bagshaw and C. C. Ashley, The kinetics of calcium binding to fura-2 and indo-1, *FEBS Lett* 216:35-39 (1987).
22. J. P. Y. Kao and R. Y. Tsien, Ca^{2+} binding kinetics of fura-2 and azo-1 from temperature-jump relaxation measurements, *Biophys J* 53:635-639 (1988).
23. G. Grynkiewicz, M. Peonie and R. Y. Tsien, A new generation of Ca^{2+} indicators with greatly improved fluorescent properties, *J Biol Chem* 260:3440-3450 (1985).

24. Q. Li, R. A. Altschuld and B. T. Stokes, Quantitation of intracellular free calcium in single adult cardiomyocytes by fura-2 fluorescence microscopy: Calibration of fura-2 ratios, *Biochem Biophys Res Comm* 147:1120-1126 (1987).

25. R. A. Haworth, Quantitation of intracellular free calcium in single myocytes by fura-2 fluorescence microscopy, *Cell Calcium* 10:263-264 (1989).

26. G. Callewaert. L. Cleeman and M. Murad, Caffeine-induce Ca^{2+} release stimulates efflux of Ca^{2+} via the Na^+-Ca^{2+} exchanger in single mammalian cardiac myocytes, *Biophys J* 55:411a (1989).

27. M. Katsuoka, K. Kobayashi and S. T. Ohnishi, Volatile anesthetics decrease calcium content of isolated myocytes, *Anesthesiology* 70:954-960 (1989)

28. K. S. Lee, E. W. Lee and R. W. Tsien, Calcium channel inhibition by nitrendipine and other agents in single dialyzed heart cells, *in:* "Nitrendipine," A. Scriabine, Ed., Urban and Schwarzenberg, Baltimore (1984) pp. 169-184.

29. C. B. Hantler, N. Winton, D. M. Learned, A. E. G. Hill and P. R. Knight, Impaired myocardial conduction in patients receiving diltiazem therapy during enflurane anesthesia, *Anesthesiology* 67:94-96 (1987).

30. Y. Vulliemoz and M. Verosky, Halothane interaction with guanine nucleotide binding proteins in mouse heart, *Anesthesiology* 69:876-880 (1988).

31. G. Schultz, W. Rosenthal, J. Herscheler and W. Trautwein, Role of G protein in calcium channel modulation, *Annu Rev Physiol* 52:275-292 (1990).

13

Effects of Volatile Anesthetics on Cardiac Sarcoplasmic Reticulum as Determined in Intact Cells

David M. Wheeler, Ana Katz, R. Todd Rice

INTRODUCTION

While it is generally agreed that the volatile anesthetics halothane, enflurane and isoflurane act at both the sarcoplasmic reticulum (SR) and sarcolemma (SL) in cardiac cells, it is less clear whether all three agents have the same qualitative effects at these sites. In many preparations, it has been demonstrated that isoflurane has a less potent negative inotropic effect than halothane or enflurane.[1-5] It has recently been demonstrated that at equi-anesthetic concentrations the agents produce a similar degree of depression of peak second inward current.[6-8] The effect of these agents at the SR is less readily measured, and evidence exists that there is a major difference between halothane's and isoflurane's action there.[9] Since it appeared that the differences in negative inotropic potency among these agents are a consequence of their actions at the SR, we sought to document differences in effect at the SR for all three commonly used agents.

In order to study the actions of the volatile anesthetics on the cardiac SR in intact cells, we have utilized isolated rat cardiac cells. The rat exhibits a cardiac excitation-contraction coupling mechanism unique among mammals. The rat cardiac contraction is heavily dependent on the function of the cardiac SR. On the basis of the ryanodine sensitivity of twitch tension in isolated cardiac muscle, Bers has concluded that nearly all of the twitch tension developed by rat cardiac muscle is dependent on SR calcium release (figure 1).[10] While all beats exhibit this property, the rested state contraction (a stimulated beat following a period of quiescence) is even more dependent on SR function than subsequent contractions. One reason behind this phenomenon is that unlike other mammals, the rat cardiac SR loads with Ca during rest.[11] Therefore, the rat cardiac contraction provides us with the opportunity to study the function of the cardiac SR without disrupting the cell.

Single isolated rat heart cells at rest also exhibit phenomena related to SR function which are helpful in the study of differences among the volatile anesthetics. These isolated cells exhibit spontaneous contractile waves when they are allowed to rest; that is, not stimulated to beat.[12,13] These spontaneous contractile waves are quite different from a normal synchronous contraction. The waves are felt to be the result

DAVID M. WHEELER, R. TODD RICE, Department of Anesthesiology and Critical Care Medicine, The Johns Hopkins Hospital, Baltimore, Maryland 21205. ANA KATZ, Department of Anesthesiology, Soroka Medical Center, Beer Sheva, Israel.

Mechanisms of Anesthetic Action in Skeletal, Cardiac, and Smooth Muscle
Edited by T.J.J. Blanck and D.M. Wheeler, Plenum Press, New York, 1991

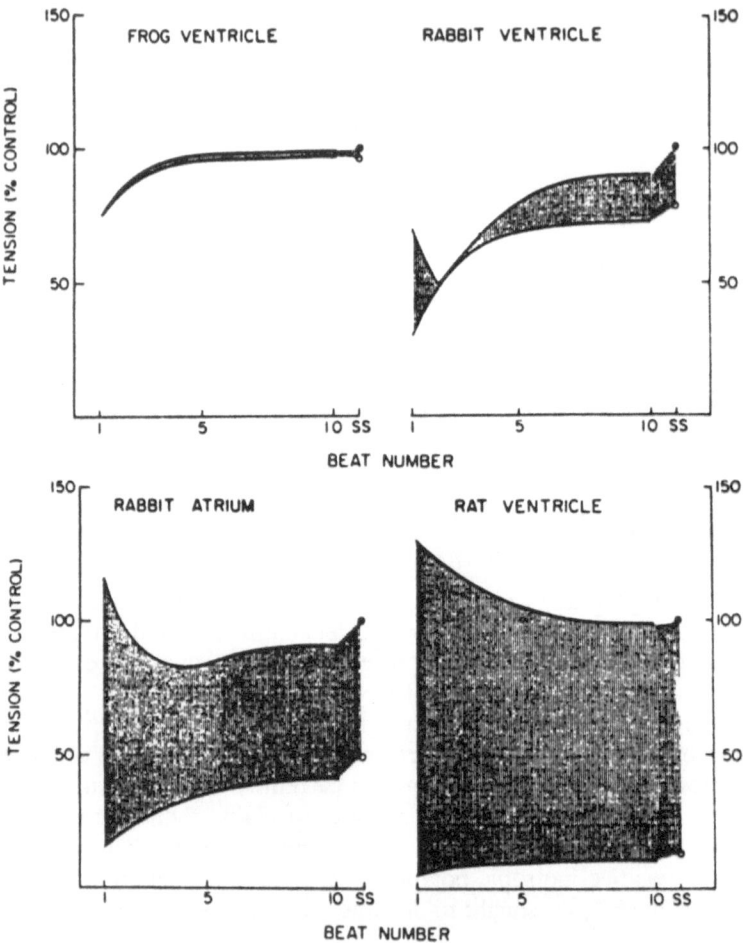

FIGURE 1. Representation of post-rest tension recovery in various cardiac tissues. The shaded areas represent the ryanodine-sensitive portion of tension development. Reprinted from D. M. Bers, *Am J Physiol* 248:H366-H381 (1985) with permission.

of a spontaneous calcium release from one region of the SR. The calcium released by that region then triggers calcium release from adjacent areas of SR, and the wave therefore propagates along the length of the cell (figure 2). While there is debate as to the functional significance of these spontaneous contractile waves, they are observed in isolated rat heart cells at physiologic extracellular calcium concentrations and can be observed in the cells of other species at elevated extracellular calcium concentrations.[14] Phenomena that likely reflect spontaneous contractile wave activity have been observed in intact rat cardiac muscle at rest as well as in the cardiac muscles of other species at elevated extracellular calcium concentrations.[15] The likelihood of occurrence of a spontaneous contractile wave appears to be related to the total cell calcium content.[13] Agents which tend to increase calcium influx, such as catecholamines, increase the frequency and amplitude of spontaneous contractile waves. On the other hand, lowering extracellular calcium tends to decrease the frequency and amplitude of spontaneous contractile waves and may eliminate them entirely. These spontaneous contractile waves are associated only with a very small depolarization of the sarcolemma[12] (figure 3); therefore, any alterations in the spontaneous contractile waves must occur independently of changes in sarcolemmal voltage-dependent channels, which would not be activated by such a small depolarization.

The series of experiments reported here focuses on concentrations of the individual volatile anesthetics which produced equivalent amounts of twitch amplitude depression, similar to an approach previously used by others.[16] If true qualitative differences in mechanisms between the volatile anesthetics exist, then these differences should be apparent at doses which produce equivalent twitch depression. If quantitative differences exist but no mechanistic differences are present, then the actions of the volatile anesthetics at equi-depressant concentrations should be similar.

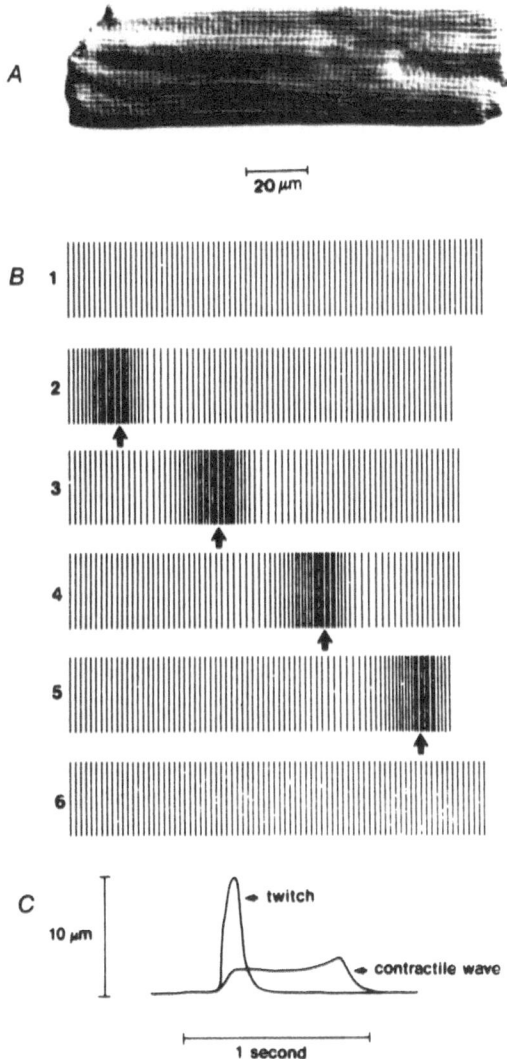

FIGURE 2. Panel A: photomicrograph of an isolated rat ventricular cell. Panel B: sequential schematic representation of localized sarcomere shortening as a spontaneous contractile wave passes along the cell. Panel C: superimposed cell length tracings (shortening upward) comparing a stimulated contraction to a spontaneous contractile wave. Adapted from M. C. Capogrossi, A. A. Kort, H. A. Spurgeon and E. G. Lakatta, *J Gen Physiol* 88:589-613 (1986) with permission.

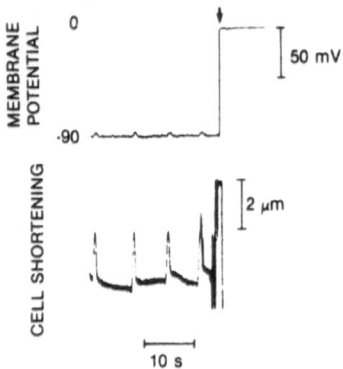

FIGURE 3. Simultaneous length and membrane potential recording of a single, unstimulated rat ventricular myocyte. The upward deflections in the lower (cell length) panel reflect spontaneous contractile waves. At the arrow, the microelectrode was withdrawn from the cell to define zero potential. Reprinted from M. C. Capogrossi, A. A. Kort, H. A. Spurgeon and E. G. Lakatta, *J Gen Physiol* 88:589-613 (1986) with permission.

One issue not addressed by the present series of experiments is a possible variable depression of myofilament sensitivity to calcium between the different anesthetics. The endpoint on which we focus, twitch amplitude, is roughly analogous to an unloaded contraction in cardiac muscle. Unloaded contractions are thought to operate at a relatively low myofilament calcium sensitivity.[4] Therefore, small changes in myofilament calcium sensitivity, such as might be produced by the volatile anesthetics, will be relatively less important in unloaded contractions or in the twitch amplitude of isolated cells.

To summarize the rationale for the experiments reported here, we hoped to demonstrate a qualitative or mechanistic difference among the three commonly used volatile anesthetics in their actions at the cardiac SR. The use of isolated rat cardiac cells provided an experimental preparation in which endpoints related to SR function could be examined in the intact cell. Should a major difference exist in the action of the agents at the cardiac SR, there should be differences in these endpoints at anesthetic concentrations that produce equal twitch depression.

METHODS

Rat heart cells were isolated from two- to three-month old Wistar rats as described in detail elsewhere.[12,17] Briefly, for each experiment a heart was perfused at constant pressure through a catheter secured in the proximal aorta. Digestion with collagenase and protease and subsequent mechanical mincing were used to disaggregate the tissue. Calcium-tolerant cells were produced by gradually introducing the disaggregated cells to physiologic calcium concentrations. Experiments were performed within six hours of cell isolation. All experiments were performed at 37°C.

The measurement of the length of a single cardiac cell was accomplished by allowing an aliquot of cells to settle in a glass-bottomed perfusion chamber. After approximately 5-10 minutes of no flow, many cells were lightly adherent to the glass. Perfusion of the chamber was then begun, and a suitable rod-shaped cell without blebs or granulations was selected for measurement. The cell length was measured by video dimension analysis or by casting the image of the cell onto a linear photodiode array. Electrical stimulation was provided by field stimulation *via* platinum wire electrodes located at either end of the perfusion chamber. During each experiment, solution continuously flowed across the perfusion chamber. The composition of the solution could be rapidly changed through the use of parallel fluid streams and a four-way diverting valve.

Intracellular calcium concentration was measured in suspensions of isolated rat heart cells using the calcium-sensitive fluorescent dye quin2 as previously described.[18] Briefly, cells were loaded with quin2 through a 30 minute exposure to a 50 μM concentration of the acetoxymethylester of quin2. The cells were then washed once, resuspended, and placed in a cuvette in a fluorometer chamber maintained at 37°C and gassed with a 95% O_2/5% CO_2 mixture to maintain pH at approximately 7.4. The fluorescence signal at wavelengths > 480 nm in response to excitation light at 333 nm was continuously recorded. Additions of solution to the cell suspension were made by syringe injection without interruption of fluorescence recording. The suspension was continuously stirred using a magnetic disk.

The volatile anesthetic concentration in the fluorometer cuvette or in the perfusion chamber was determined by sampling the aqueous solution and measurement using gas chromatography after heptane extraction. In all cases the results obtained in the presence of an anesthetic were compared to a paired control. In the case of cell length or twitch amplitude measurements, each cell served as its own control. For the intracellular calcium measurements, an aliquot of cells from the same cell suspension served as the control. Wherever possible, the control was the average of measurements before and after exposure to anesthetic or of two aliquots of cells run immediately before and after the anesthetic-treated aliquot of cells. Paired t-tests were therefore used to detect statistically significant differences from control, and analysis of variance was used for comparisons among the anesthetics.

RESULTS

In order to detect qualitative differences in mechanisms of volatile anesthetic depression of contractility, we sought to evaluate anesthetic concentrations that produced similar degrees of twitch depression. Cells stimulated to beat at 60 bpm were exposed to halothane, enflurane and isoflurane, and the twitch amplitude in the anesthetic was compared to that in control. Results are shown in table 1. The concentrations of volatile anesthetics were deliberately selected to produce similar twitch depression under these conditions. While a full dose-response curve would be necessary to make clear statements with respect to relative potency, it appears that, considering MAC fractions, more isoflurane is required to produce comparable twitch depression than either halothane or enflurane. This result is, of course, consistent with comparable results in a number of other cardiac preparations.[1-5] The depression of rested state contractions caused by the anesthetics was then evaluated. The data in

Table 1. Twitch Depression by Volatile Anesthetics

Anesthetic	Concentration			N	Twitch Amplitude at 60 beats/min (% control)
	mM	Vol %	f rat MAC		
Halothane	0.27	1.0	0.94	12	62 ± 5
Enflurane	0.67	1.9	0.87	18	62 ± 3
Isoflurane	0.52	2.2	1.4	71	56 ± 2

Twitch values are mean ± SEM, and each is significantly different from control, although there are no significant differences between anesthetics. The "f rat MAC" is the anesthetic concentration expressed as a fraction of MAC for rats.

Table 2. Anesthetic Depression of Rested State Contractions (RSC)

Anesthetic	Concentration (f rat MAC)	N	RSC Amplitude (% control)
Halothane	0.94	7	80 ± 5
Enflurane	0.87	9	79 ± 4
Isoflurane	1.4	20	82 ± 4

RSC amplitudes are expressed as mean ± SEM.

table 2 reveal that the anesthetic concentrations which produce a similar degree of twitch depression at 60 bpm also similarly depress rested state contractions. A rat heart cell at rest accumulates calcium in its SR, and the rested state contraction is extremely dependent on SR calcium. Therefore, it appears that in resting rat heart cells at 37°C each volatile anesthetic similarly depresses the amount of SR calcium releasable during a contraction.

The effect of each volatile anesthetic on spontaneous contractile waves was also measured, and the results are shown in table 3. Each agent tended to decrease the wave period (the mean time interval between waves) and decrease their amplitude. While small differences in the magnitude of these effects between the three agents may be present, it is clear that any distinction between the agents on the basis of their effect on spontaneous contractile waves is subtle at best. While these results do not reveal any clear differences between the agents they do indicate that all three agents have an effect on the sarcoplasmic reticulum in resting rat heart cells. Such an effect would not be mediated by any action on voltage-sensitive sarcolemmal channels since such channels would not be active in these resting cells.

Several groups have previously demonstrated that the abrupt addition of halothane to beating heart cells or tissue causes a short-lived increase in contractility or twitch amplitude (refs. 17, 19 and 20 and Chapter 7). This temporary increase in contractility is insensitive to the calcium channel blocker, verapamil (figure 4). Figure 5 shows an example (from a dog heart cell) suggesting that the temporary increase is sensitive to the SR-disabling drug, ryanodine. Thus, the SR, but not Ca channels, is likely to be important in the generation of the temporary increase in twitch. A possible mechanism for this effect, previously suggested by ourselves and others,[20] is that halothane enhances calcium release from the SR. With each beat, a greater

Table 3. Anesthetic Effects on Spontaneous Contractile Waves

Anesthetic	Concentration (f rat MAC)	N	Wave Period (% control)	Wave Amplitude (% control)
Halothane	0.94	24	84 ± 3	80 ± 3
Enflurane	0.87	17	80 ± 6	64 ± 3
Isoflurane	1.4	36	92 ± 5	74 ± 3

Wave properties expressed as mean ± SEM. With the exception of wave period in isoflurane, all results are significantly different from control ($P < 0.05$). The only significant difference between results in different anesthetics is for wave amplitude in halothane *vs.* wave amplitude in enflurane ($P < 0.01$).

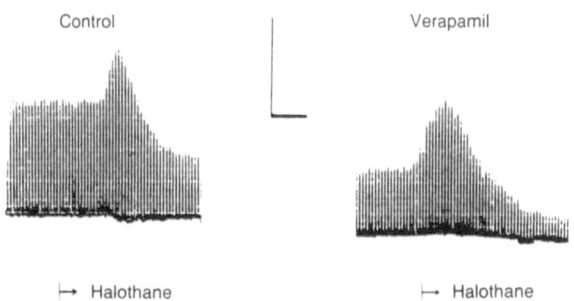

FIGURE 4. Twitch amplitude of a single rat heart cell abruptly exposed to solution containing halothane (0.47 mM or 1.7 vol%). Between panels, the halothane was washed out and the cell exposed to verapamil 1 μM. Halothane was then re-introduced in the presence of verapamil in the right panel. Calibration bars: 5 μm vertical, 10 sec horizontal.

amount of calcium is released and twitch or contractility is enhanced. With continued beating, the SR becomes depleted of calcium so that even an enhanced calcium release results in a calcium transient that is reduced from the control condition. The verapamil and ryanodine results are consistent with this interpretation. We have now studied a number of cells acutely exposed to enflurane and isoflurane to determine whether their effect is comparable to that of halothane. As shown in figure 6 and table 4, enflurane clearly produces a temporary increase in twitch amplitude when cells beating at 60 bpm are abruptly exposed to the agent. Isoflurane produces a very small temporary increase which is not significantly different from cells switched between two streams of control solution and which clearly represents a very much smaller effect than is the case with either halothane or enflurane.

We have also made comparisons among the volatile anesthetics with respect to their actions on quiescent cardiac cell suspensions loaded with the calcium indicator quin2. These experiments were also conducted at 37°C and, as with many of the single cell experiments, involved resting cells. However, quin2 loading does produce at least one important difference in comparison to cells studied without quin2. Because of the substantial intracellular quin2 concentration required to produce an adequate fluorescent signal, intracellular calcium is buffered and the free calcium is probably reduced below control. Quin2-loaded cells typically do not exhibit spontaneous contractile waves. Direct addition of volatile-anesthetic-containing solution to

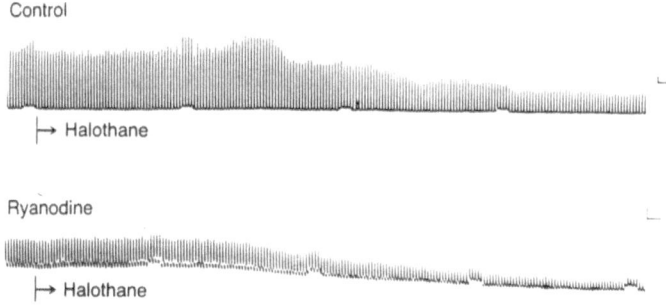

FIGURE 5. Twitch amplitude of a single dog heart cell exposed to halothane (0.47 mM or 1.7 vol%) in the absence and presence of ryanodine 1 μM. In this case, the temporary increase in twitch amplitude is delayed, probably related to the position of the cell in the chamber with a longer time required for the new solution to reach it. Calibration bars: 5 μm vertical, 5 sec horizontal.

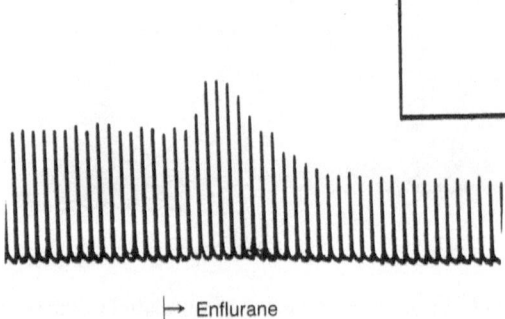

├→ Enflurane

FIGURE 6. Twitch amplitude tracing of a single rat heart cell abruptly exposed to enflurane (0.28 mM or 0.80 vol%). Calibration bars: 5 μm vetical, 10 sec horizontal.

quin2-loaded suspensions of cardiac cells produces different results among the different volatile anesthetics. As we have previously reported,[18] the abrupt addition of halothane results in a short-lived increase in the quin2 fluorescence, which measures the free intracellular calcium concentration (figure 7). This effect was dose-dependent and quite reproducible. When enflurane or isoflurane was added to these cardiac cell suspensions, the gross appearance of the quin2 fluorescence trace was typically that of a downgoing step function.[21] This result is consistent with simple dilution of the cell suspension. However, comparison of anesthetic addition to addition of the carrier solution alone revealed that small, transient increases in the quin2 fluorescence may occur, particularly with enflurane. Any such effect was small, especially when compared with that of halothane. In these same quin2-loaded cardiac cell suspensions, we have evaluated the SR calcium content of these quiescent cells by utilizing high concentrations of caffeine to release essentially all of the calcium available in the SR into the cytoplasm. The abrupt addition of caffeine produces a short-lived increase in cytoplasmic calcium concentration (figure 7). We have used the magnitude of this increase as an index of SR calcium content. Anesthetic-induced changes in the caffeine-induced fluorescence increase should reflect changes in SR calcium content that are independent of changes in voltage-sensitive sarcolemmal ion channels since these cells are quiescent. As we have previously reported, halothane reduces the SR calcium content in a dose-dependent manner. Preliminary results with enflurane and isoflurane reveal that enflurane may have a similar effect to that of halothane, but isoflurane does not appear to alter SR calcium content.

Table 4. Twitch Amplitude during Introduction of Anesthetics

Solution Change	N	Maximum Twitch Amplitude (% of max control twitch)
Control → Enflurane 1.9%	18	121 ± 5
Control → Isoflurane 2.2%	70	104 ± 1
Control → Control	33	101 ± 1

Values are mean ± SEM. The maximum twitch in enflurane is significantly different from the results from solution changes to either isoflurane or control (P < 0.001).

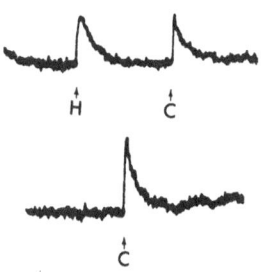

FIGURE 7. Quin2-Ca fluorescence tracings of two aliquots of cells from the same cell suspensions. At the H, halothane (final conc.: 0.55 mM or 1.9%) was added, and at the C, caffeine (final conc.: 10mM). Calibration bar: 30 sec.

DISCUSSION

The results that we have presented indicate that the relative potency of halothane, enflurane and isoflurane for depression of twitch amplitude is similar to that for depression of twitch tension in other preparations. In units of equi-anesthetic potency, more isoflurane was required to produce a similar degree of twitch depression than either enflurane or halothane. It should be noted that twitch amplitude in single cardiac cells approximates the extent of shortening of an unloaded contraction in intact cardiac muscle.

One of the bases for our experiments is the extreme dependence of rat ventricular cells on SR calcium. This dependence allows us to evaluate changes in SR function in intact cardiac cells. Initially, we expected there to be qualitative differences among the volatile anesthetics concerning their actions at the SR. We anticipated that these differences would become manifest if the relative depression of beating and resting rat heart cells were examined. However, such qualitative differences between the volatile anesthetics were not found. While isoflurane was less potent for depression of twitch amplitude at 60 bpm, it was similarly less potent than halothane or enflurane for depression of rested state contractions.

The rested state contraction represents a response dependent upon both the SR Ca content and the efficiency of SR Ca release (as well as "downstream" effects not addressed in this chapter). Spontaneous contractile waves, while not triggered specifically by sarcolemmal mechanisms, are also a complex endpoint. Wave amplitude reflects factors similar to those contributing to the rested state contraction (SR Ca content and the fraction of it released in response to each spontaneous trigger). Wave frequency may depend on both the total cell Ca load and the sensitivity, or threshhold, of the spontaneous release trigger. The SR Ca load (on which amplitude depends) is reciprocally related to wave frequency and directly to the rate of "reloading" or SR Ca uptake.

Halothane, enflurane and isoflurane change spontaneous contractile wave frequency and amplitude. While isoflurane's effect on wave frequency is not significant, major differences between the agents (at concentrations which produced equal twitch depression) are not apparent. This result agrees with the lack of difference between the agents for rested state contraction depression. However, because none of these endpoints reports a single cellular mechanism, there is a possibility that differences in action do exist between the agents, but due to offsetting effects no net difference is detected.

It is also interesting to note that the magnitude of twitch amplitude depression at 60 bpm was somewhat greater for all anesthetics than their depression of rested state contraction amplitude. This may indicate that the net depression of SR calcium content by the anesthetics is greater during steady beating than at rest. This greater depression during beating could be explained through an anesthetic-induced reduction of calcium influx with each beat, as has been observed by a number of investigators.[2,6-8,19,22]

In contrast to the observations on spontaneous contractile waves and rested state contractions, the phenomenon of a short-lived increase in twitch amplitude upon introduction of the anesthetics did reveal a major difference between the agents. An enhancement of the SR Ca release with each beat is a likely explanation for this phenomenon.[23] We conclude then that an equi-depressant dose of isoflurane is much less apt or less potent to induce calcium release from the SR in beating cells as compared to halothane or enflurane. Obviously, there is a conflict with the conclusions based on rested state contractions and spontaneous contractile waves. We will return to this problem after reviewing the results from quin2-loaded cell suspensions.

The results involving quin2-loaded rat heart cell suspensions at first glance appear to contradict some of the observations made with single cardiac cells. Specifically, the apparent lack of effect of isoflurane on the SR Ca content of quin2-loaded quiescent cells seems at odds with isoflurane's depression of rested state contractions or spontaneous contractile wave amplitude. However, it must be recalled that in quin2-loaded suspensions the resting intracellular calcium concentration, and presumably also the resting SR calcium content, is not controlled by spontaneous contractile wave activity. In cells without the Ca-buffering action of quin2, the occurrence of spontaneous contractile waves sets an upper limit for the SR Ca content at rest.[24] Thus, the contrast between the lack of a differential effect on rested state contractions (and spontaneous contractile waves) and the differences produced in SR Ca content in quin2-loaded cells may simply indicate the SR Ca content is not controlled by the same mechanism in the two experiments. Alternatively, the discrepancy could exist because the rested state contraction may not exclusively reflect changes in SR Ca content.

The short-lived increase in cytoplasmic calcium concentration evoked by halothane may reflect halothane's ability to induce spontaneous contractile wave activity in these cells (figure 8). Since halothane induces this short-lived increase in intracellular calcium whereas enflurane and isoflurane produce a much smaller increase, if any, it would appear that halothane is the only one of the three drugs that is able to produce large amounts of spontaneous contractile wave activity when abruptly added to quin2-loaded cardiac cell suspensions.

The following presents a possible synthesis of the results incorporating many of the assumptions and interpretations previously discussed. Halothane appears capable of enhancing SR Ca release in beating cells and evoking spontaneous Ca release from the SR of resting cells (both quin2-loaded and not). In both situations, SR Ca content is reduced by halothane once a steady state is achieved. The reduction of the rested state contraction by halothane reflects the summation of competing actions. The depression of SR Ca content and the reduction of the triggering signal (the slow inward current) combine to exceed the enhancement of SR Ca release, so that the net effect is a reduction of the rested state contraction. Isoflurane does not appear to enhance SR Ca release to any meaningful extent, nor does it reduce SR Ca content in quin2-loaded, quiescent cells. Its inhibition of the rested state contraction in rat cells may be due to downstream effects such as reduction of myofilament sensitivity to Ca, to slow inward current inhibition (thereby reducing the trigger for SR Ca release) or to a reduction of SR Ca content not apparent in quin2-loaded cells. Enflurane generally appears to act like halothane, with the exception that enflurane appears to be less potent for induction of spontaneous Ca release in resting cells. The above scheme of the relative action of the agents at the SR (halothane and enflurane showing significant depression, isoflurane little effect) is a agreement with that proposed by

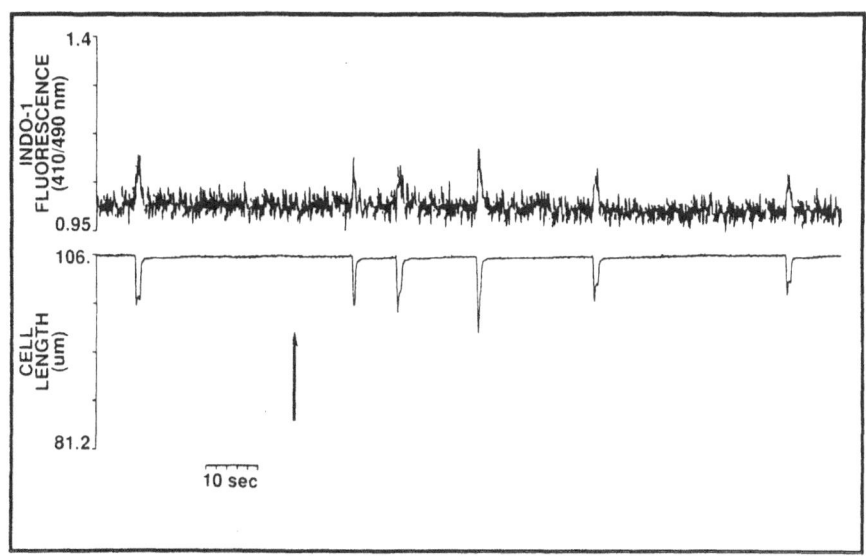

FIGURE 8. Simultaneous cell length and fluorescence ratio recording of a single rat heart cell abruptly exposed to halothane (0.65 mM or 1.2 vol %) at the arrow. This cell was loaded with the Ca-indicator indo-1 using its acetoxymethylester. Temperature was 22°C. Reprinted from D. M. Wheeler, R. T. Rice and E. G. Lakatta, *Anesthesiology* 72:911-920 (1990) with permission.

DeTraglia *et al.*[16] in an intact cardiac muscle preparation. Thus, perhaps some agreement is developing regarding the effects of volatile anesthetics on the cardiac SR. However, a number of these conclusions remain indirect, and not all of our observations fit readily into a simple scheme.

ACKNOWLEDGEMENTS

This work was supported in part by an ASA Research Starter Grant, an Andrew W. Mellon Foundation grant for faculty development, and NIH grant R29 GM39568.

REFERENCES

1. B. R. Brown Jr. and J. R. Crout, A comparative study of the effects of five general anesthetics on myocardial contractility: I. Isometric conditions, *Anesthesiology* 34:236-245 (1971).
2. C. Lynch III, Differential depression of myocardial contractility by halothane and isoflurane *in vitro*, *Anesthesiology* 64:620-631 (1986).
3. H. Komai and B. F. Rusy, Negative inotropic effect of isoflurane and halothane in rabbit papillary muscles, *Anesth Analg* 66:29-33 (1987).
4. P. R. Housmans and I. Murat, Comparative effects of halothane, enflurane, and isoflurane at equipotent anesthetic concentrations on isolated ventricular myocardium of the ferret: I. Contractility, *Anesthesiology* 69:451-463 (1988).
5. P. R. Housmans, Negative inotropy of halogenated anesthetics in ferret ventricular myocardium, *Am J Physiol* 259:H827-H834 (1990).
6. D. A. Terrar and J. G. G. Victory, Effects of halothane on membrane currents associated with contraction in single myocytes isolated from guinea pig ventricle, *Br J Pharmacol* 94:500-508 (1988).
7. D. A. Terrar and J. G. G. Victory, Isoflurane depresses membrane currents associated with contraction in myocytes isolated from guinea-pig ventricle, *Anesthesiology* 69:742-749 (1988).

8. Z. J. Bosnjak and N. J. Rusch, The effects of halothane, enflurane, and isoflurane on calcium current in isolated canine ventricular cells, *Anesthesiology* 74:340-345 (1991).

9. H. Komai and B. F. Rusy, Direct effect of halothane and isoflurane on the function of the sarcoplasmic reticulum in intact rabbit atria, *Anesthesiology* 72:694-698 (1990).

10. D. M. Bers, Ca influx and sarcoplasmic reticulum Ca release in cardiac muscle during postrest recovery, *Am J Physiol* 248:H366-H381 (1985).

11. M. J. Shattock and D. M. Bers, Rat *vs.* rabbit ventricle: Ca flux and intracellular Na assessed by ion-selective microelectrodes, *Am J Physiol* 256:C813-C822 (1989).

12. M. C. Capogrossi, A. A. Kort, H. A. Spurgeon and E. G. Lakatta, Single adult rabbit and rat cardiac myocytes retain the Ca^{2+}- and species-dependent systolic and diastolic contractile properties of intact muscle, *J Gen Physiol* 88:589-613 (1986).

13. M. C. Capogrossi, B. A. Suarez-Isla, and E. G. Lakatta, The interaction of electrically stimulated twitches and spontaneous contractile waves in single cardiac myocytes, *J Gen Physiol* 88:615-633 (1986).

14. E. G. Lakatta, M. C. Capogrossi, A. A. Kort and M. D. Stern, Spontaneous myocardial calcium oscillations: Overview with emphasis on ryanodine and caffeine, *Fed Proc* 44:2977-2983 (1985).

15. M. D. Stern, A. A. Kort, G. M. Bhatnager and E. G. Lakatta, Scattered-light intensity fluctuations in diastolic rat cardiac muscle caused by spontaneous Ca^{++}-dependent cellular mechanical oscillations. *J Gen Physiol* 82:119-153 (1983).

16. M. C. DeTraglia, H. Komai and B. F. Rusy, Differential effects of inhalational anesthetics on myocardial potentiated contractions in vitro, *Anesthesiology* 68:534-540 (1988).

17. D. M. Wheeler, R. T. Rice and E. G. Lakatta, The action of halothane on spontaneous contractile waves and stimulated contractions in isolated rat and dog heart cells, *Anesthesiology* 72:911-920 (1990).

18. D. M. Wheeler, R. T. Rice, R. G. Hansford and E. G. Lakatta, The effect of halothane on the free intracellular calcium concentration of isolated rat heart cells, *Anesthesiology* 69:578-583 (1988).

19. C. Lynch III, Halothane depression of myocardial slow action potentials, *Anesthesiology* 55:360-368 (1981).

20. H.-N. Luk, C.-I. Lin, C.-L. Chang and A.-R. Lee, Differential inotropic effects of halothane and isoflurane in dog ventricular tissues, *Eur J Pharmacol* 136:409-413 (1987).

21. D. M. Wheeler, R. T. Rice, E. G. Lakatta and R. G. Hansford, Changes in free intracellular calcium in isolated heart cells during exposure to volatile anesthetics (abstract), *Anesthesiology* 69:A69 (1988).

22. C. Lynch III, Enflurane depression of myocardial slow action potentials, *J Pharmacol Exp Therap* 222:405-409 (1982).

23. R. T. Rice, D. M. Wheeler and E. G. Lakatta, Halothane enhances Ca release from the cardiac sarcoplasmic reticulum (abstract), *FASEB J* 3:A986 (1989).

24. M. C. Capogrossi, M. D. Stern, H. A. Spurgeon and E. G. Lakatta, Spontaneous Ca^{2+} release from the sarcoplasmic reticulum limits Ca^{2+}-dependent twitch potentiation in individual cardiac myocytes, *J Gen Physiol* 91:133-155 (1988).

14

Alcohol and Anesthetic Actions
on Myocardial Contractility
Evidence for a Lipophilic/Electrophilic Sarcoplasmic Reticulum Site

Carl Lynch III

INTRODUCTION

While volatile anesthetics typically depress cardiac contractility, the actions of the presently used volatile anesthetics (halothane, enflurane, isoflurane) upon the pattern and magnitude of myocardial tension development differ significantly, suggesting different sites of action.[1-3] Previous physiological and biochemical studies have suggested multiple mechanisms of action for the volatile agents, primarily implicating altered control of Ca delivery to myofibrils (depression of Ca entry and depression of sarcoplasmic reticulum uptake and release of Ca).[4] To better understand the alterations in myocardial tension development, which may in part be related to the lipid solubility of the anesthetic agents, the aliphatic alcohols were employed as model anesthetic compounds with known physical characteristics which vary in a systematic fashion. With increasing chain length, the aliphatic alcohols possess increasing general anesthetic potency as well as increasing ability to inhibit nerve conduction.[5] The actions of alcohols can thus be compared not only with the volatile anesthetics, but also with effects of the local anesthetics (LAs), also known to selectively alter myocardial contractility.[3,6] Effects of these various anesthetic compounds were also contrasted with the effects of ryanodine, a plant alkaloid which binds with very high affinity to the Ca release channel of the sarcoplasmic reticulum (SR) and reduces the release of Ca^{2+} from the SR.[7-10]

METHODS

Isometric contractions of right ventricular papillary muscles from guinea pig heart were studied in Tyrode solution bubbled with $95\%O_2$-$5\%CO_2$ at 37°C. Contractions were studied following rest (10-15 min) and at stimulation rates of 0.1, 0.25, 0.5, 1, 2 and 3 Hz. Each stimulation frequency was maintained until a stable and unchanging response was obtained. These muscles demonstrated a "positive staircase" or "treppe." Variation in the stimulation frequency altered the rate, pattern, and duration of tension development, and these characteristics at each stimulation rate

CARL LYNCH III, Department of Anesthesiology, University of Virginia Health Sciences Center, Charlottesville, Virginia 22908.

Mechanisms of Anesthetic Action in Skeletal, Cardiac, and Smooth Muscle
Edited by T.J.J. Blanck and D.M. Wheeler, Plenum Press, New York, 1991

were very reproducible over the 1-3 hr experiments. To emphasize even further the differing patterns of tension development, the following modifications were usually employed.

Papillary muscles were partially depolarized in 26 mM K Tyrode solution, and their contractility was enhanced by increasing intracellular cyclic-AMP via stimulation of adenylate cyclase, either by β-adrenergic stimulation with addition of isoproterenol (10^{-7} M) or directly by application of 1 μM forskolin (also with 5 μM metoprolol, to prevent any effect due to release of endogenous norepinephrine). Muscles were stimulated by 0.5-1 msec pulses (after rest and at rates of 0.1, 0.25, 0.5, 1, 2, and 3 Hz) and demonstrated propagated slow action potentials. Accompanying contractions displayed a remarkably prominent late peaking tension after rest and at stimulation rates up to 0.5 Hz, with rapid tension development becoming prominent at 2-3 Hz. Typical control responses at the differing stimulation rates are shown superimposed in figure 1. When the tension signal was differentiated, a maximum early rate of tension development (dT_E/dt) could be defined approximately 20-40 msec following the stimulus; dT_E/dt increased dramatically at ≥ 1 Hz stimulation rate. A maximum late tension development (dT_L/dt) could be defined 100-120 msec after the stimulus, being prominent after rest and at 0.1-0.25 Hz stimulation rates; dT_L/dt disappeared at ≥ 2 Hz, the stimulation rate at which the early rate of tension development became dominant. In a few experiments, intracellular impalements were maintained and action potential (AP) characteristics (amplitude, duration, maximum rate of depolarization) were monitored.

Muscles in 26 mM K Tyrode with 0.1 μM isoproterenol were also studied in double stimulus experiments. A rested state (RS) contraction was elicited by a stimulus followed by a second stimulus after an interval of 250, 300, 350, 400, 500 or 600 msec. The second contraction (C2) showed rapid early tension development similar to that at 2-3 Hz.

Some muscles were superfused with low Na Tyrode (40 mM Na, 200 mM sucrose substitution), which resulted in rapid tension development after rest and at 0.1 Hz, similar to that seen at 2-3 Hz in the previous setting. In this setting the decreased external [Na^+] decreased the depletion of intracellular Ca stores which normally occurs during rest and at low stimulation rates via Na/Ca exchange.

The straight chain aliphatic alcohols 1-octanol, 1-heptanol, 1-hexanol, 1-pentanol, 1-butanol, as well as ethanol and benzyl alcohol, were employed in this study. Each alcohol was applied by direct addition of the appropriate volume into a perfusate reservoir. Following control studies, the alcohol solutions were applied for 20-30 minutes, which was found to be sufficient time for an unchanging effect; during the course of the 20-30 min exposure, no attempt was made to replace any alcohol which may have been volatilized into the atmosphere. In studies of local anesthetics and ryanodine, the final drug concentration was achieved by addition of appropriate aliquots from concentrated stock solutions of each drug.

RESULTS

Figure 1A shows the pattern of contractions accompanying slow action potentials at various frequencies as indicated beside each tracing. Clearly, 10 mM 1-butanol caused relatively uniform depression of tension development all stimulation rates, which was largely reversible with washout of the alcohol. In contrast, subsequent application of 200 μM 1-octanol caused selective depression of the late-peaking component of tension. Washout of octanol typically resulted in partial recovery toward control behavior (50-80% of control). Figure 1B shows the average depression in the

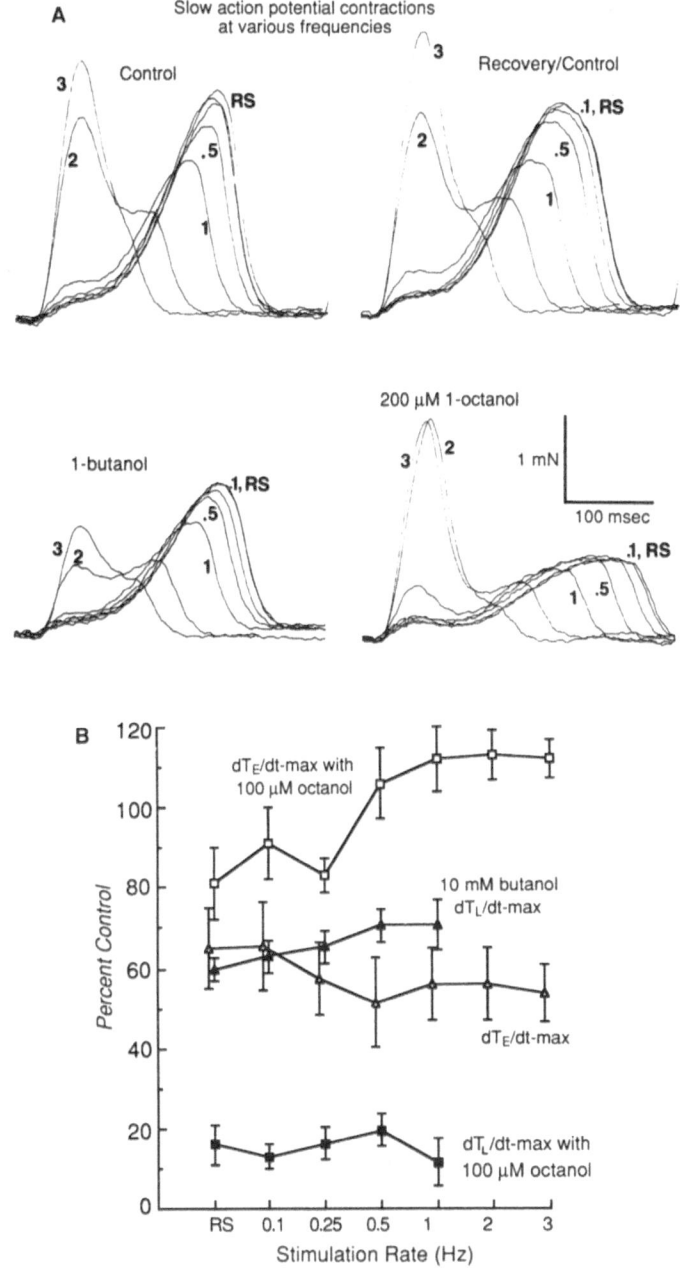

FIGURE 1. Effects of 10 mM 1-butanol and 100 μM 1-octanol on contractions of guinea pig papillary muscle in 26 mM Tyrode solution with 0.1 μM isoproterenol. Numbers beside tracing denote stimulation rate (in Hz). RS indicates rested state contraction. A: Tension tracings for steady state contractions observed at various stimulation rates are superimposed. Following application of 1-butanol, the alcohol was washed out and octanol was applied. B: Average effects (n = 4) of 100 μM 1-octanol (square symbols) and 10 μM 1-butanol (triangular symbols) on maximum rates of early ($[dT_E/dt]_{max}$, open symbols) and late tension development ($[dT_L/dt]_{max}$, filled symbols) of slow action potential contractions.

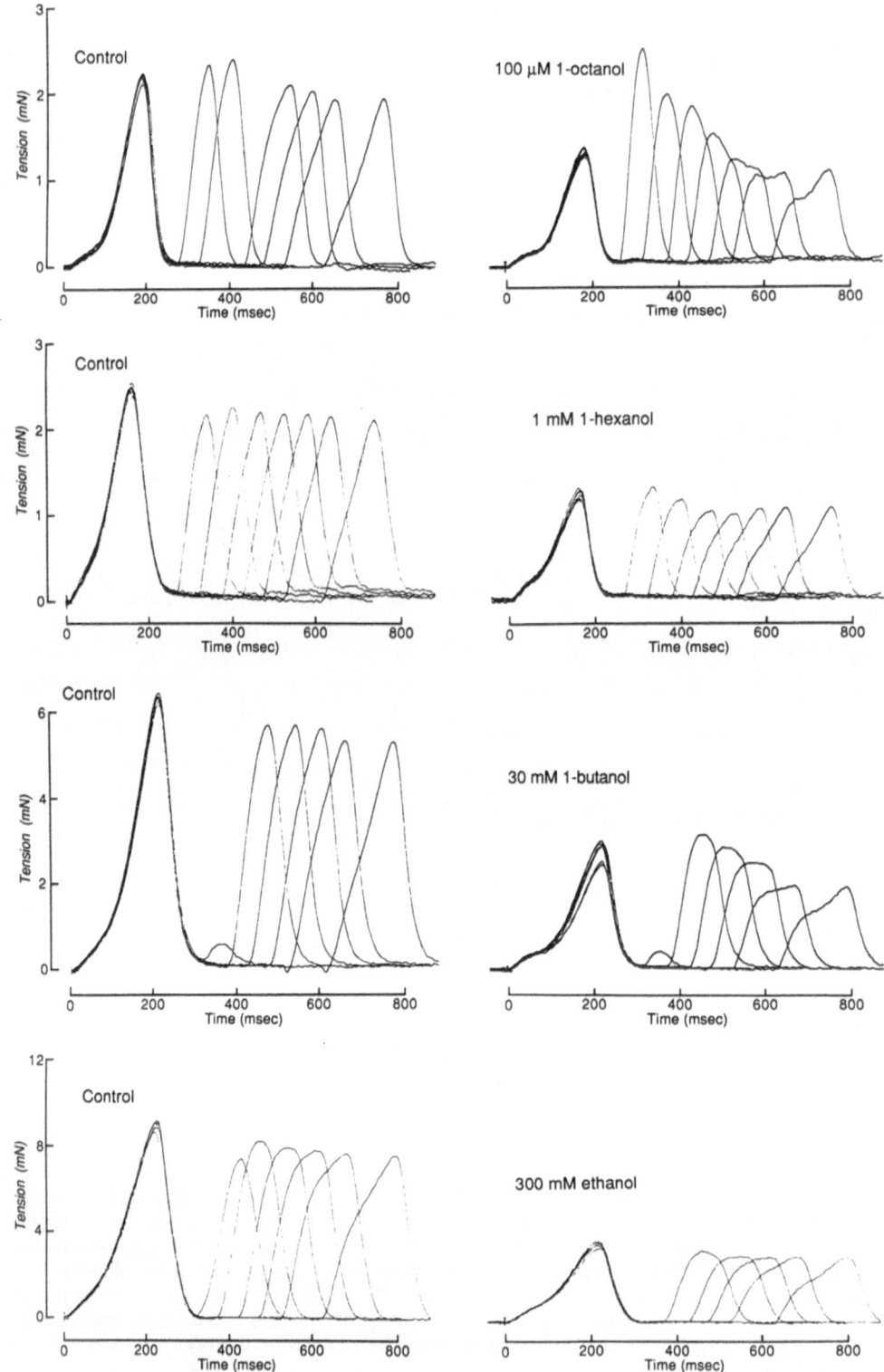

maximum rates of early (dT_E/dt) and late (dT_L/dt) tension development caused by octanol (n = 6) and butanol (n = 5).

Figure 2 shows the results of experiments employing double stimulation after rest for various alcohols studied in separate muscles. The tracings obtained for the various stimulus intervals are superimposed; the rested state contractions (RSCs), which typically showed very little variation, are superimposed. 100 μM 1-octanol, 1 mM 1-hexanol, 30 mM 1-butanol, and 300 mM ethanol showed roughly similar depression of the late-peaking RSC; depression equivalent to that caused by 100 μM octanol required a ten- to thirty-fold increase in alcohol concentration for each decrease in alkyl chain length by two carbon atoms. The rapidly developing tension of the second contraction (C2) was selectively spared by 1-octanol, while the other alcohols showed moderate depression of the early tension development associated with C2.

The early maximum rate of tension development ($[dT_E/dt]_{max}$) of C2 contractions 400-500 msec following a RSC (double stimulation experiments as in figure 2) was tabulated in the presence of the various alcohols. The ($dT_E/dt)_{max}$ in the presence of the various alcohol concentrations employed are plotted in figure 3A as a fraction of the control value. The effects of 1-octanol, 1-heptanol and benzyl alcohol are clearly distinct from those of ethanol, 1-butanol, 1-pentanol, and 1-hexanol. Octanol, heptanol and benzyl alcohol had either no effect on or enhanced the ($dT_E/dt)_{max}$ of the C2 contractions, while the shorter chain length alcohols from hexanol to ethanol caused significant dose-dependent depression of ($dT_E/dt)_{max}$.

The effect of the various alcohol concentrations on the late maximum rate of tension development (dT_L/dt) of the RSC was likewise tabulated as a fraction of the observed control value. The results are plotted in figure 3B. Each alcohol showed a similar dose-dependence (slope), with a 3-5 fold increase in concentration causing a 25-35% increase in the amount of depression (measured as percent of control).

In two experiments, intracellular impalements were maintained and slow APs monitored during application of 1-octanol. There was no alteration caused by 200 μM octanol in the slow AP maximum rate of depolarization ($[dV/dt]_{max}$), a qualitative measure of ionic flux through the calcium channel. The higher concentration of octanol (500 μM), which obliterated the late peaking RSC tension while having minimal effect on early tension (see figure 3), caused a 5-10% decline in ($dV/dt)_{max}$. In both instances, the slow AP duration was increased by 5-10%.

The selective alteration of late tension development caused by 1-octanol was also verified in contractions in normal Tyrode solution. Figure 4A shows the tension tracings in normal Tyrode under control conditions and in the presence of 100 μM 1-octanol. Although there is little depression of the maximum initial rate of tension development, the loss of a late component of tension results in marked depression of peak tension after rest up to 0.5 Hz. The average depression of contractility for five muscles is shown in figure 4B, with peak tension showing greater depression than ($dT/dt)_{max}$. In contrast, hexanol, butanol and ethanol produced relatively similar depression of peak tension and ($dT/dt)_{max}$ at all frequencies.

FIGURE 2. Effects of various alcohols upon a pair of contractions elicited after rest from papillary muscles in 26 mM Tyrode solution with 0.1 μM isoproterenol. The stimulation interval was varied from 250, 300, 350, 400, 450, 500 and 600 msec. The rested state contraction (RSC) showed a typical 100 msec delay after the stimulus (time 0) before maximum tension development, while the second contraction (C2) showed immediate early tension development following stimulation. For each panel, tension tracings for the pairs of contractions are superimposed at the time of the first stimulus, so that the RSCs are superimposed, while the C2 obtained with differing stimulation intervals are not.

159

Effects upon the rapid early tension development were also studied in the low Na medium. Results obtained for depressant effects upon the $(dT/dt)_{max}$ of the contractions in low Na Tyrode solution are listed in table 1. These results were very similar to those for $(dT_E/dt)_{max}$ observed for 2-3 Hz stimulation rates or for C2 of the double stimulation experiments.

The selective depression of late tension by octanol, heptanol and benzyl alcohol is reminiscent of effects previously described for the local anesthetics.[3,6] Figure 5 shows the effects of tetracaine upon action potentials and accompanying contractions.

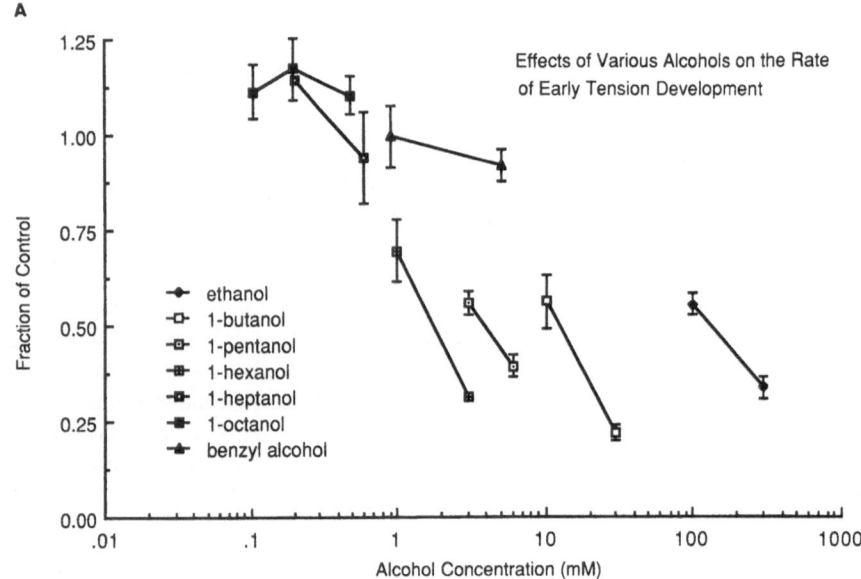

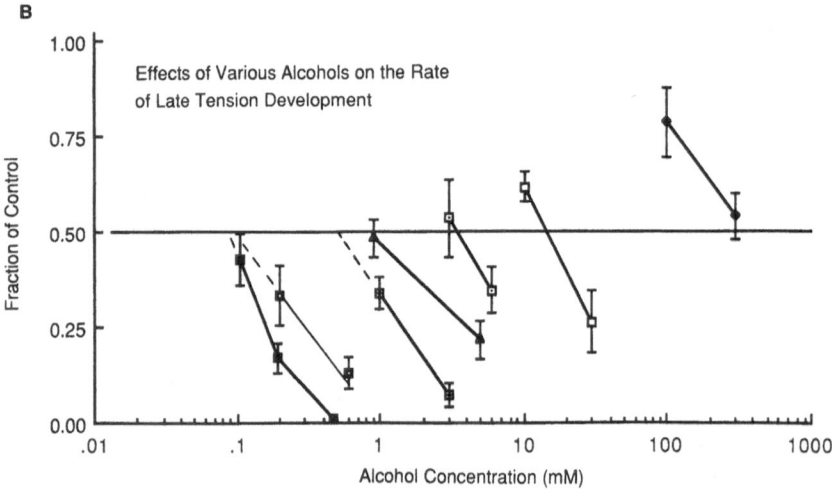

FIGURE 3. Effects of various alcohols upon early or late tension development of slow action potential contractions. The results, expressed as the fraction of the control response prior to alcohol application, are the average of 3 experiments with each alcohol except for heptanol and benzyl alcohol (n = 2). A: Effects upon the maximum rate of early tension development ($[dT_E/dt]_{max}$) of the second contractions (C2) which followed by 400-500 msec a rested state contraction (RSC) as shown in figure 2. B: Effects upon the maximum rate of late tension development ($[dT_1/dt]_{max}$) of a late-peaking RSC.

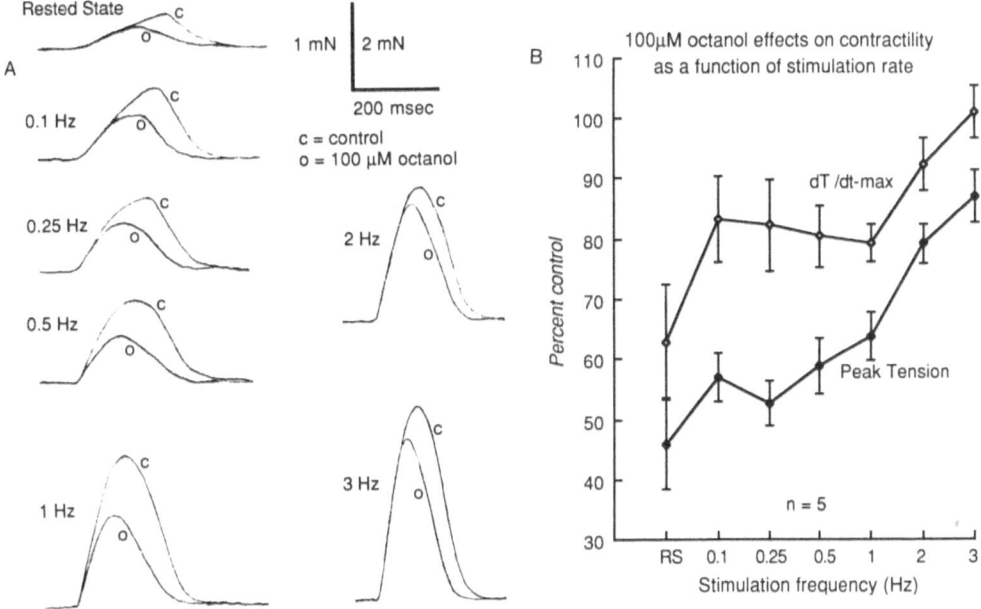

FIGURE 4. Effects of 1-octanol on contractions in normal Tyrode solution. A: Contractions after rest (rested state) and at the indicated stimulation rates. Tracings for control and in the presence of 100 μM 1-octanol at each frequency are superimposed. Note the loss of the late component of tension. B: Plot of contractile depression at each frequency (by percent of control response at that frequency). At all points, peak tension was depressed significantly more than $(dT/dt)_{max}$ by paired t-test (n = 5).

Although there is no alteration in the slow AP $(dV/dt)_{max}$, the late peaking tension is selectively depressed by 10 μM tetracaine, with minimal effect upon the rapid early tension development. Similar effects on contractions are observed for procaine, as shown in figure 6. This muscle was studied both at varied stimulation rates as well as using double stimulation after rest. Procaine caused profound depression of late RSC tension with enhancement of the early tension development. Accompanying slow

Table 1. Effects of alcohols on early tension development of papillary muscles in 40 mM Na Tyrode solution

| Alcohol | Conc (mM) | N | Rested state contraction | | 0.1 Hz stimulation rate | |
			Peak tension (% cont)	dT/dt_{max} (% cont)	Peak tension (% cont)	dT/dt_{max} (% cont)
Ethanol	300	3	30 ± 3	34 ± 2	34 ± 3	37 ± 4
Butanol	10	3	59 ± 7	63 ± 5	54 ± 4	67 ± 9
Pentanol	5	2	62 ± 4	69 ± 1	65 ± 6	73 ± 7
Hexanol	1	4	71 ± 4	85 ± 3	73 ± 5	86 ± 3
Octanol	0.2	6	117 ± 12	131 ± 17	118 ± 9	119 ± 4
Benzyl alcohol	5	1	115	136	110	115

161

action potentials (not shown) showed no depression of $(dV/dt)_{max}$. The effect of procaine was removed with washout (not shown) and 0.5 μM ryanodine was then applied. This drug resulted in its typical reduction of the early rate of tension development, especially prominent at 2-3 Hz, yet the late peaking tension development of the RSC remained. Interestingly, the subsequent re-application of 200 μM procaine resulted in more modest depression of the RSC late tension.

DISCUSSION

Tension development by guinea pig ventricular myocardium may be separated into two components. One component is the slowly developing or delayed component present at low stimulation frequencies, which is greatly enhanced (10-50X) by β-adrenergic stimulation. This tension development is selectively depressed by local anesthetics (LAs), and also by octanol, heptanol, and benzyl alcohol. The other is the rapidly developing or early component of tension present at short stimulation intervals, high stimulation rates (2-3Hz), or in low Na medium. This component is sensitive to ryanodine. Ryanodine appears to bind to the Ca release channel of both skeletal and cardiac SR with high selectivity (K_D < 30 nM).[7] This channel is also thought to represent the "foot" protein which couples the junctional SR to the t-tubular membrane.[8,9] Single channel recordings of skeletal muscle SR channels suggest that this high conductance channel is opened by ryanodine to a low conducting state.[10] It is likely that ryanodine depresses tension development by permitting Ca to leak from the SR and also by preventing the channel from achieving a high conductance state.

The late peak tension of the rested state contractions appears to involve function of the SR,[11] and multiple lines of evidence implicate an action of LAs on the SR.[6,12,13] However, local anesthetics at higher concentrations have been reported to depress sarcolemmal Ca currents,[14] and blockade of Ca entry is also demonstrated to depress late peak tension.[11] Can local anesthetic blockade of Ca entry explain depressed late tension? While 10 μM tetracaine dramatically depressed late tension,

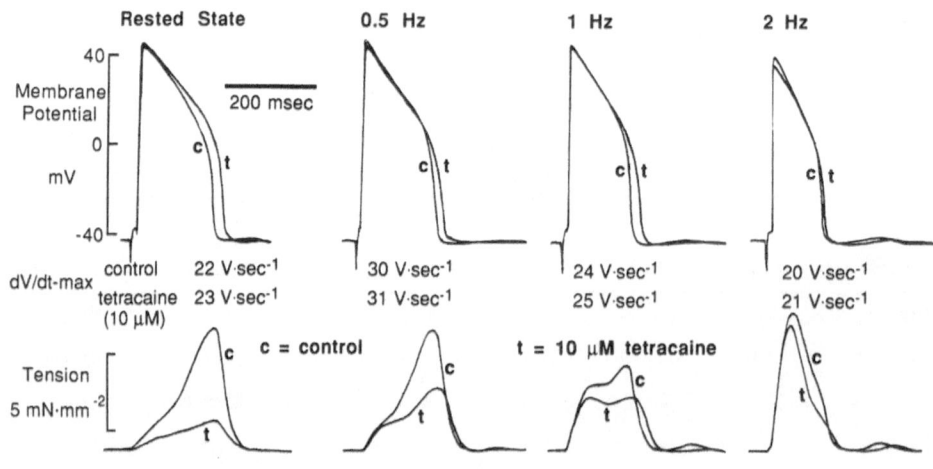

FIGURE 5. Effects of tetracaine on slow action potentials and accompanying contractions. Tension and action potential tracings are superimposed for control conditions and after application of 10 μM tetracaine. Incomplete recovery was noted upon washout of drugs.

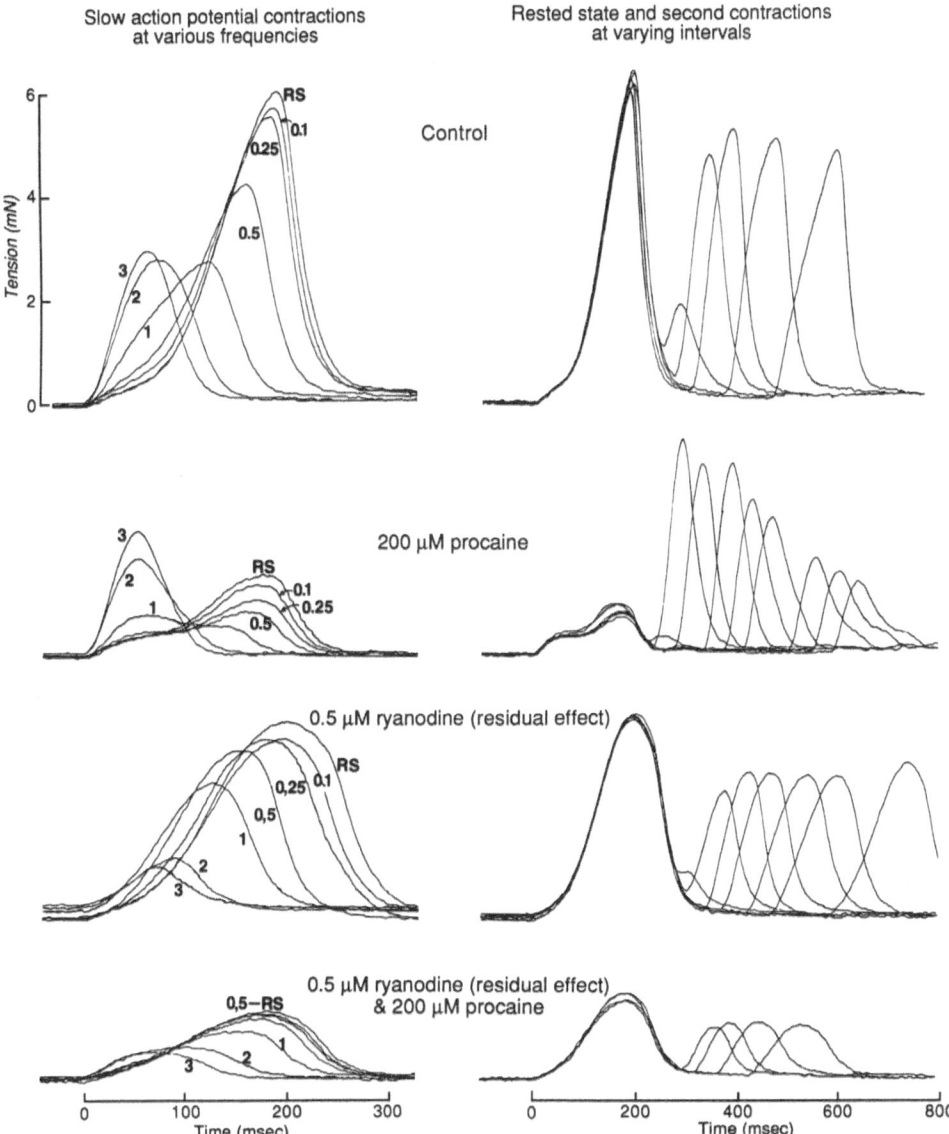

FIGURE 6. Effects of procaine and ryanodine upon the slow AP contractions. The results of stimulation at different rates (rested state up to 3 Hz) are shown on the left. The results employing double stimulation after rest, with intervals of 250, 300, 350, 400, 450, 500 and 600 msec, are shown on the right panels (some intervals were omitted for clarity; note change in time scale from left panels). The rested state contraction (RSC) showed a typical delay after the stimulus (time 0) before maximum tension development, while the second contraction (C2) showed immediate early tension development following stimulation. For each panel, tension tracings for the pairs of contractions are superimposed at the time of the first stimulus, so that the RSCs are superimposed, while the C2s obtained with differing stimulation intervals are not.

it caused no depression of slow AP $(dV/dt)_{max}$. This lack of action is not inconsistent with the previous report of a K_D for depression of high threshold calcium channel currents by tetracaine of 80 μM,[14] eight-fold higher than the greatest concentration employed here. Late tension was also markedly reduced by concentrations of procaine and lidocaine which showed no depression of slow AP $(dV/dt)_{max}$, and by 10 μM bupivacaine, which showed slight depression slow AP $(dV/dt)_{max}$.[6] It also seems unlikely that local anesthetics would significantly depress calcium current, yet have minimal effect on 2-3 Hz contractions. Concentrations of nifedipine, diltiazem and verapamil which depress 2-3 Hz slow AP contractions to 50 percent of control uniformly decrease the slow AP $(dV/dt)_{max}$ to 30-50 percent of control (unpublished observations).

The failure of local anesthetics to depress the ryanodine-sensitive second contraction (C2) in the double stimulation rested state experiments is difficult to rationalize if the late-peaking RSC is depressed by local anesthetics *via* blockade of

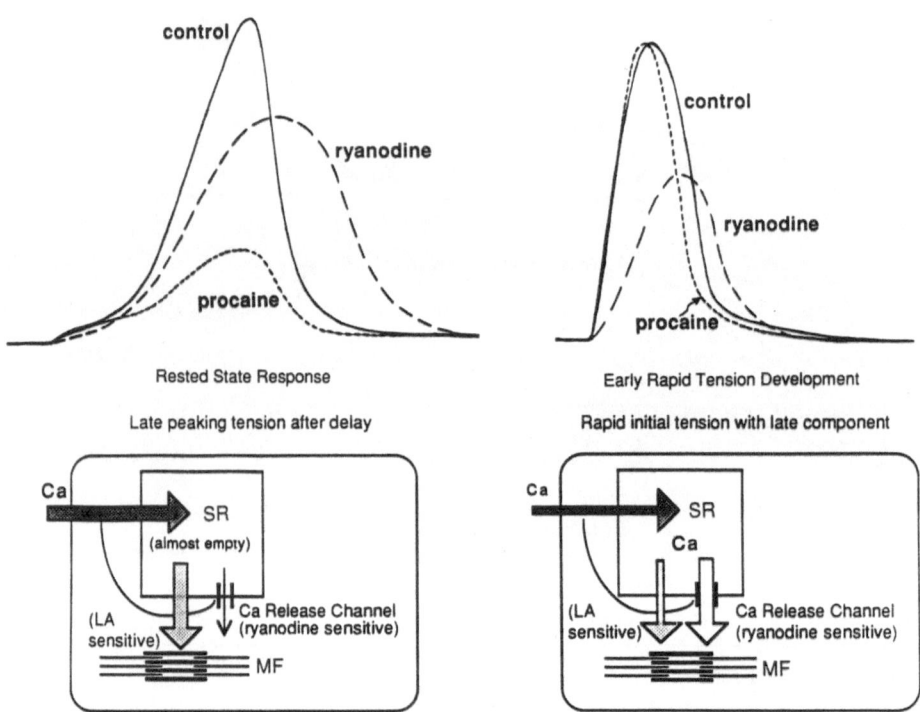

FIGURE 7. Schematic of proposed Ca^{2+} fluxes responsible for late tension development following rest (rested state response) and rapid tension development following a short stimulation interval. Considerations apply primarily to myocardium in which Ca^{2+} influx and sarcoplasmic reticulum (SR) Ca^{2+} uptake has been enhanced by increased cyclic-AMP. After rest, the SR pools are almost empty. Upon depolarization, any residual Ca^{2+} may be released to activate a small amount of tension, Ca^{2+} enters through the sarcolemma, is accumulated into an SR pool, and with continued depolarization and Ca^{2+} influx, the accumulated Ca^{2+} is released by a local anesthetic sensitive pathway to the myofibrils (MF) to activate tension. Relaxation is mediated by uptake into the SR. Following a short period of repolarization, the large store of accumulated Ca^{2+} is rapidly released to activate rapid tension development. Since this SR Ca^{2+} release is ryanodine sensitive, it presumably represents flux through the Ca release channel (foot protein). The separate efflux paths from the SR could represent distinct channels or different states of the same channel.

Ca^{2+} entry. The ryanodine-sensitive C2 which rapidly follows a RSC represents rapid SR release of Ca^{2+} accumulated from the immediately preceding RSC. This Ca^{2+} must have entered during the previous depolarization (the RSC), been accumulated in the SR to produce relaxation, and released to activate C2. The local anesthetics depressed the RSC, yet occasionally enhanced dT$_E$/dt and peak tension of the C2 above control levels. Therefore, it seems unlikely that local anesthetic blockade of initial Ca^{2+} entry accounts for the depressed late peak of the RSC. The LAs thus appear to alter the control of some component of SR Ca^{2+} stores, but it must be a mechanism which is pharmacologically distinct from that perturbed by ryanodine. Figure 7 suggests a scheme of Ca fluxes which may explain the observed changes in the patterns of tension development caused by the local anesthetics and ryanodine. It is possible that both pathways could represent the same release channel in different states (*e.g.*, open *versus* closed).

In addition to the differential pharmacologic sensitivity to LAs and ryanodine, these two components of tension development (both apparently related to SR Ca^{2+} release) also show differential sensitivity to the alcohols. Octanol, heptanol, and benzyl alcohol depress the late, LA-sensitive component, with little or no depression of the rapid, ryanodine-sensitive component. Like the LAs, octanol's effect occurs in the absence of any apparent depression of Ca entry, as measured by slow AP dV/dt. The arguments for a specific LA-sensitive pathway through the SR, which leads to late tension development, also applies to the larger alcohols. Similarly, isoflurane also demonstrates marked depression of late tension development, although it is usually accompanied some modest depression of (dT$_E$/dt)$_{max}$. Its additional effects are perhaps related to its depression of Ca currents.[6,15]

The shorter chain alcohols typically depress early and late tension to a similar extent. It is unclear whether the depression of late tension by these smaller alcohols represents an action analogous to that of the LAs and larger alcohols, or whether it is due to a separate process. For example, down regulation of Ca channels and Ca^{2+} entry by 100 nM nifedipine depresses slow AP (dV/dt)$_{max}$, (dT$_E$/dt)$_{max}$ and (dT$_L$/dt)$_{max}$ to similar extents (~50%, unpublished observations), and it is unknown whether the shorter chain alcohols have a similar action.

However, the effect of all the alcohols on late tension may be due to a specific effect on the postulated LA-sensitive SR pathway. To further characterize these actions, the concentration of each alcohol which causes approximately 50% depression the the late rate of tension development of RSCs was estimated from figure 3B. This 50% depressant concentration is plotted as a function of the octanol-water partition coefficient for each alcohol in figure 8. The ability of any aliphatic alcohol to depress the late tension correlates closely with its lipophilic character. For the aliphatic alcohols the correlation was excellent (r^2 = 0.979) and is similar to that for nerve blocking potency.[5] Depression of tension is probably unrelated to Na channel blockade *per se*, since in all cases of partially depolarized muscle, Na channels are inactivated and probably contribute little to function.

Also plotted on the same graph is the 50% depressant concentration calculated in the same manner for a number of local anesthetics, where the octanol-water partition coefficient was estimated for pH 7.4.[16] For a given partition coefficient, the local anesthetics appear approximately 10 times more potent than the alcohols. Benzyl alcohol is also more potent than the aliphatic alcohols than might be expected based upon its octanol-H$_2$O coefficient. The obvious common feature of the local anesthetics and benzyl alcohol is the presence of an aromatic ring. The presence of the more polarizable aryl moiety further enhances the ability of lipophilic substances to alter late tension, suggesting that the site of action is a lipophilic site which possesses some affinity for electron-rich molecules.

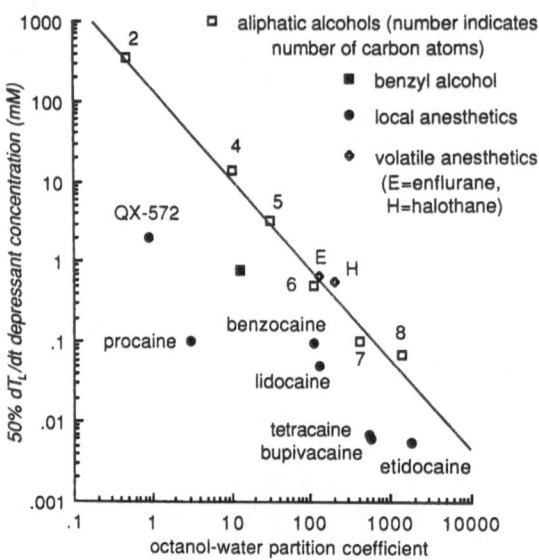

FIGURE 8. The concentration of each alcohol on anesthetic which caused a 50% depression the late rate of tension development of RSCs (estimated from Figure 3) is plotted versus its octanol-water partition coefficient.[5] For the aliphatic alcohols the correlation was excellent (dotted line, $r^2 = 0.985$). A similar correlation of nerve blocking concentration to octanol-water partition is observed. However, the slope of the relation is less steep and an almost ten-fold greater concentration of 1-octanol and 1-hexanol is required for nerve blockade. Enflurane and halothane behave in a manner similar to that for the aliphatic alcohols (data from ref. 3). Also plotted in the same manner is the 50% depressant concentration of local anesthetics (data from ref. 6 and unpublished data), where the octanol-water partition coefficient represents the measured partitioning between octanol and buffer at pH 7.4.[16] The local anesthetics possess approximately ten-fold greater potency than the alcohols, as does benzyl alcohol.

Other than an artificially developed model, the depression of which correlates with anesthetic potency and octanol-water partitioning, does the depression of the late component have any physiologic, pharmacologic or clinical significance? Clearly, in the case of octanol as well as for isoflurane[1] and local anesthetics[6], contractions in normal Tyrode solution demonstrate greater depression of peak tension than of $(dT/dt)_{max}$. This effect suggests that even in contractions unenhanced by increased cyclic-AMP, the time-tension interval is reduced due to this loss of late tension. This may account for less obvious differences in myocardial depression among anesthetics *in vivo* and in the clinical setting.

ACKNOWLEDGEMENTS

This work was supported in part by NIH grant R01 GM31144. The author thanks Martha J. Frazer and Jacqueline Washington for their excellent technical assistance.

REFERENCES

1. C. Lynch III, Differential depression of myocardial contractility by halothane and isoflurane in vitro, *Anesthesiology* 64: 620 (1986).
2. M. C. DeTraglia, H. Komai and B. F. Rusy, Differential effects of inhalational anesthetics on myocardial potentiated-state contractions in vitro, *Anesthesiology* 68: 534 (1988).

3. C. Lynch III, Differential depression of myocardial contractility by volatile anesthetics in vitro: comparison with uncouplers of excitation-contraction coupling, *J Cardiovasc Pharmacol* 15:155 (1990).

4. B. F. Rusy and H. Komai, Anesthetic depression of myocardial contractility: a review of possible mechanisms, *Anesthesiology* 67:745 (1987).

5. L. L. Firestone, J. C. Miller and K. W. Miller, Appendix: Tables of physical and pharmacological properties of anesthetics, *in*: "Molecular and Cellular Mechanisms of Anesthetics," S. H. Roth and K. W. Miller, ed., Plenum Medical Book Co., New York (1986).

6. C. Lynch III, Depression of myocardial contractility in vitro by bupivacaine, etidocaine, and lidocaine, *Anesth Analg* 65:551 (1986).

7. M. Michalek, P. Dupraz and V. Shoshan-Rarmatz, Ryanodine binding to sarcoplasmic reticulum membrane; comparison between cardiac and skeletal muscle, *Biochim Biophys Acta* 939:587 (1988).

8. M. Inui, A. Saito and S. Fleischer, Purification of the ryanodine receptor and identity with the feet structures of junctional terminal cisternae of sarcoplasmic reticulum from fast skeletal muscle, *J Biol Chem* 262:1740 (1987).

9. F. A. Lai, L. P. Erickson, E. Rousseau, Q.-Y. Liu and G. Meissner, Purification and reconstitution of the calcium release channel from skeletal muscle, *Nature* 331:315 (1988).

10. K. Nagasaki and S. Fleischer, Ryanodine sensitivity of the calcium release channel of sarcoplasmic reticulum, *Cell Calcium* 9:1 (1988).

11. M. Reiter, W. Vierling and K. Seibel, Where is the origin of activator Ca^{++} in cardiac ventricular contraction?, *Basic Res Cardiol* 79:1 (1984).

12. B. K. Chamberlain, P. Volpe and S. Fleischer, Inhibition of calcium-induced Ca^{++} release from purified cardiac sarcoplasmic reticulum vesicles, *J Biol Chem* 259:7547 (1984).

13. S. D. Prabhu and G. Salama, The heavy metal ions Ag^+ and Hg^{2+} trigger calcium release from cardiac sarcoplasmic reticulum, *Arch Biochem Biophys* 277:47 (1990).

14. E. Carmeliet, M. Morad, G. V. d. Heyden and J. Vereecke, Electrophysiological effects of tetracaine in single guinea-pig ventricular myocytes, *J Physiol (Lond)* 376:143 (1986).

15. D. A. Terrar and J. G. G. Victory, Isoflurane depresses membrane currents associated with contractions in myocytes isolated from guinea-pig ventricle, *Anesthesiology* 69:742 (1988).

16. G. R. Strichartz, V. Sanchez, G. R. Arthur, R. Chafetz and D. Martin, Fundamental properties of local anesthetics. II. Measured octanol:buffer partition coefficients and pK_a values of clinically used drugs, *Anesth Analg* 71:158 (1990).

15

Volatile Anesthetics and Second Messengers in Cardiac Tissue

Yvonne Vulliemoz

INTRODUCTION

The depression of myocardial contractility induced by volatile anesthetics is well documented by *in vivo* and *in vitro* studies. These agents' direct myocardial depressant effect has been attributed to a decreased availability of free intracellular calcium to the contractile proteins and to a decreased sensitivity of the contractile proteins to activation by calcium. The volatile anesthetics have been shown to depress slow inward calcium currents in the sarcolemma, as well as inhibit calcium uptake and release by the sarcoplasmic reticulum.[1] All these processes are modulated by hormones, neurotransmitters and other endogenous factors which regulate calcium movements either directly or indirectly *via* a second messenger system. Therefore, the effect of volatile anesthetics on calcium homeostasis and myocardial contractility may be due either to a direct interaction of the anesthetic with proteins regulating calcium movements or may be secondary to an action of the anesthetic on metabolic pathways modulating myocardial contractility.

The results of our earlier studies *in vitro* and *in vivo* have shown that volatile anesthetics alter the cyclic nucleotide systems in the heart. Halothane and isoflurane, at clinically used concentrations, produced a dose-dependent, reversible inhibition of the stimulatory action of β-adrenergic catecholamines on adenylate cyclase.[2-4] Since cAMP mediates the positive inotropic effect of catecholamines by modulating the concentration of free intracellular calcium,[5] these results suggested that the well documented antagonism between volatile anesthetics and catecholamines on myocardial contractility is related to the decrease in cAMP induced by the anesthetics. However, halothane has no effect on basal, non-stimulated myocardial adenylate cyclase activity, and in mice pretreated with propranolol it does not decrease myocardial cAMP content.[2,3] These observations indicated that the decrease in contractile force induced by volatile anesthetics in the absence and in the presence of β-adrenergic stimulation involves cAMP independent mechanisms.

Phospholipids, which are important structural constituents of the cell membrane, are precursors of active intracellular constituents—autacoids or second messengers—many of which affect cardiac function. Perturbation of the cell membrane, whether by chemical, mechanical or hormonal stimuli, initiates the activation of phospholipases

YVONNE VULLIEMOZ, Departments of Anesthesiology and Pharmacology, College of Physicians and Surgeons, Columbia University, New York, New York 10032.

Mechanisms of Anesthetic Action in Skeletal, Cardiac, and Smooth Muscle
Edited by T.J.J. Blanck and D.M. Wheeler, Plenum Press, New York, 1991

and the release of bioactive compounds.[6,7] Volatile anesthetics interact with cell membranes and affect their physical and chemical properties.[8] Because of their lipophilicity, high concentrations of volatile anesthetics may accumulate in these membranes and affect phospholipid metabolism by producing conformational changes in the structure of specific proteins, *e.g.*, enzymes, or in their environment, altering the availability or the access of the substrate to the enzyme.

Phospholipids can be metabolized by two main pathways.[6,9] Phospholipase A_2 catalyzes the release of arachidonic acid, which, though bioactive itself, is rapidly converted to active compounds of the prostaglandin and leukotriene family[6] (figure 1). Unsaturated fatty acids, their peroxide and hydroxyacid derivatives are potent activators of guanylate cyclase[10] and halothane has been shown *in vivo* to stimulate cGMP content in the heart.[11] Phosphoinositides, a small pool of phospholipids, are metabolized by phospholipase C to inositol phosphates and diacylglycerol. These metabolites play an important role in the regulation of free intracellular calcium.[7] These enzymes have been shown, in certain cells, to be coupled to a guanine nucleotide binding protein (G protein)[12] and volatile anesthetics influence many hormone/transmitter functions dependent on a G protein.

The studies presented in this paper examined the role of the phospholipid-arachidonic acid pathway and that of pertussis toxin-sensitive G proteins in the negative inotropic effect of halothane and isoflurane, as well as the effect of halothane on phosphoinositide turnover in rat ventricular and atrial tissue.

METHODS AND PROCEDURES

The protocol for these studies was approved by the Institutional Animal Care and Use Committee.

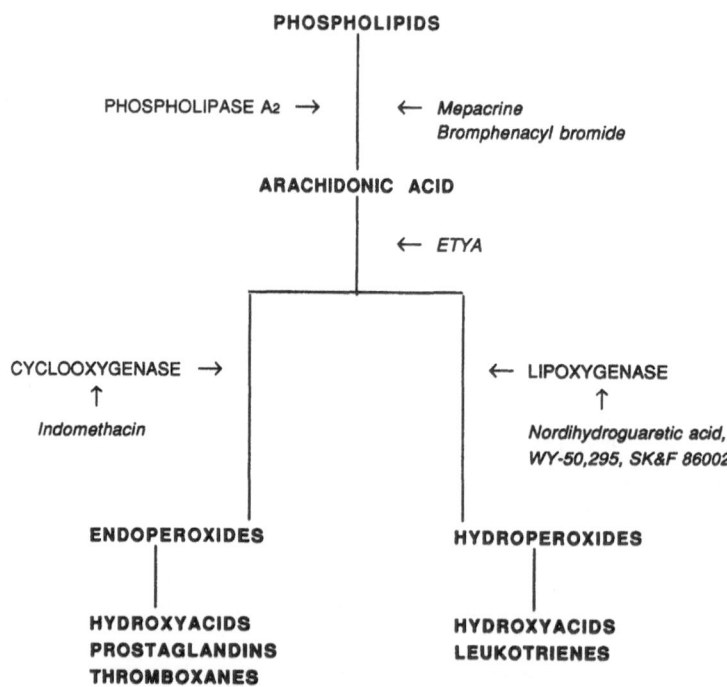

FIGURE 1. Schema of the arachidonic acid cascade. Selective inhibitors are indicated in italics.

Contractility Studies

Sprague Dawley adult male rats were killed by decapitation. Left ventricular papillary muscles or left atria were excised and mounted in 10 ml tissue baths and superfused, at 33°C, with Krebs-Ringer bicarbonate buffer equilibrated with 95% O_2 and 5% CO_2 to maintain the pH at 7.45 ± 0.05. The composition of the buffer was the following (in mM): NaCl, 119; KCl, 4.8; $MgCl_2$, 1.2; $CaCl_2$, 2.5; KH_2PO_4, 1.2; $NaHCO_3$, 24.9; glucose, 10. One end of the muscle was attached to a Statham force displacement transducer and the other to a moveable mount controlled by a micrometer used to stretch the muscle to the peak of its length-tension curve. The muscles were stimulated by silver point electrodes at 1 Hz with square-wave pulses 5 msec in duration. Stimulus voltage was maintained at approximately 10% over threshold. Isometric contractions were recorded. Preparations were allowed to equilibrate until a stable active developed tension was obtained. Drug solutions, freshly prepared, were diluted with buffer to the desired final concentrations. The concentrations used in this study are those shown, in other studies, to act on the specific pathways. Halothane or isoflurane was vaporized from an anesthetic machine equipped with a Vernitrol or Drager vaporizer carried in a humidified gas mixture of 95% O_2 and 5% CO_2. The buffer was equilibrated with the anesthetic gas mixture for at least 20 min prior to superfusing the muscle preparation. The anesthetic concentration in the gas was monitored with an EMMA gas analyzer, regularly calibrated with standard gas mixtures. All the tubing of the system was teflon tubing.

The muscles were superfused with buffer, either alone or containing the drug under study, until the contraction reached a steady level. The intensity of the contraction was taken as control value. The muscles were then exposed to the same buffer solution equilibrated with a known concentration of anesthetic. Cumulative concentration responses to halothane or isoflurane were performed in the absence or in the presence of propranolol, atropine, mepacrine, 4-bromphenacyl bromide, 5,8,11,14-eicosatetraynoic acid (ETYA), indomethacin, and nordihydroguaiaretic acid. The effects of WY-50295 and SK&F 86002, were determined on the muscle response to a single concentration of anesthetic, close to the ED_{50}.

Pertussis Toxin Treatment

Seventy-two hours before the experiment, rats were treated with pertussis toxin, 50 μg/kg iv, or with an equal volume of saline. This dose of pertussis toxin has been shown, in our own and other laboratories, to produce a maximal inactivation of the G proteins in rat heart.[13] Pertussis toxin treatment has been shown to prevent the inhibitory action of carbachol on the tension of the left atria.[14] Therefore, in this group of experiments, the response to carbachol was also tested in the left atria to insure the efficacy of the treatment. Cumulative concentration responses to halothane and isoflurane were obtained in heart papillary muscles from pertussis toxin- and saline-treated rats. In order to eliminate possible sensitivity differences to pertussis toxin between atria and ventricles, a concentration response to halothane was also obtained in the left atria.

Measurement of Inositol Phosphates Formation

Right atria and ventricles of rat hearts were preincubated in Krebs-Ringer-Hepes buffer (1.25 mM Ca^{2+}), pH 7.5, for 30 to 60 min at 37°C. The tissues were then resuspended in 1 ml fresh buffer containing ^{3}H-myoinositol, 40-50 μCi/ml. After 60 min of incubation, halothane or isoflurane, carried in humidified air, was blown over the surface of the assay buffer in each test tube, 10 min later lithium chloride and EDTA were added to a final concentration of 10 mM and 40 μM, respectively, followed 10 min later by norepinephrine or incubation buffer. In some experiments the anesthetics were added directly to the buffer dissolved in a 22% glycerol-buffer solution (ratio 1:10). The reaction was stopped 20 min after the last addition. The tissue was quickly rinsed in incubation buffer and transferred to a mixture of chloroform, methanol, 6N HCl (1:2:0.006 by volume). After extraction, the water soluble inositol phosphates were separated by Dowex anion exchange chromatography, as described by Berridge.[15] The radioactivity in the inositol phosphate eluates was counted by liquid scintillation spectometry.

Statistical Analysis

Analysis of variance and Student's t tests were used to compare the effect of halothane and isoflurane in the different experimental conditions. ED_{50} values and slopes were calculated separately for each muscle by regression analysis. Significance was set at $P < 0.05$.

RESULTS AND DISCUSSION

Effect of Inhibitors of the Arachidonic Acid Cascade

The involvement of arachidonic acid or a metabolite in the negative inotropic response to halothane and isoflurane was examined by exposing the muscle to selective inhibitors prior to superfusion with the anesthetics. Figure 1 is a simplified diagram of the arachidonic acid cascade. Arachidonic acid release from phospholipids is catalyzed by phospholipase A_2,[6] an enzyme which is inhibited by mepacrine and bromphenacyl bromide.[16,17] Arachidonic acid is rapidly oxidized to highly reactive unstable prostaglandin endoperoxides and to hydroperoxides by the action of the cyclooxygenase complex and lipoxygenases, respectively;[6] ETYA, a false substrate, prevents the metabolism of arachidonic acid.[18] Indomethacin is an inhibitor of the cyclooxygenase and nordihydroguaiaretic acid, WY-50295 and SK&F 86002 are inhibitors of lipoxygenases.[19-21] The cyclooxygenase products are further metabolized to prostaglandins, thromboxanes and hydroxyacids, the lipoxygenase products to leukotrienes.[6] All these metabolites, including arachidonic acid, are biologically active and many affect cardiac function directly. Leukotriene C4 and D4, for example, have a profound depressant effect on myocardial contractility; prostaglandins elicit negative and positive inotropic effects.[22-24] The heart is also very sensitive to lipid peroxidation; lipid-derived free radicals have been shown to decrease myocardial contractile force.[25]

Both anesthetics produced a concentration-dependent decrease in the tension of the papillary muscle. Halothane was more potent than isoflurane, an observation in agreement with published data.[26-28] The ED_{50} values for halothane and isoflurane were 1.21 ± 0.24 and 3.22 ± 0.46 vol%, respectively. This effect was not altered by atropine or propranolol, as illustrated by the overlapping concentration-response curves to halothane in figure 2. Thus the effects of the anesthetics observed in this preparation are independent of any influence due to endogenous acetylcholine or norepinephrine.

The inhibitors did not alter the potency or the efficacy of the anesthetics to decrease the tension of the papillary muscle. The concentration-response curves for the effect of halothane or isoflurane in the absence and in the presence of 4-bromphenacyl bromide, ETYA or indomethacin were identical (figure 3). The

172

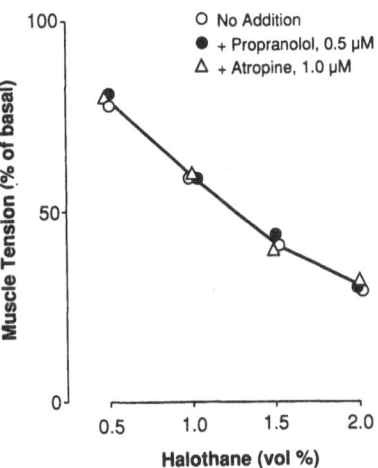

FIGURE 2. Effect of halothane on the contractile force of rat heart papillary muscle in the absence and presence of propranolol or atropine. Values are mean ± SEM; n = 17, 4 and 3, respectively.

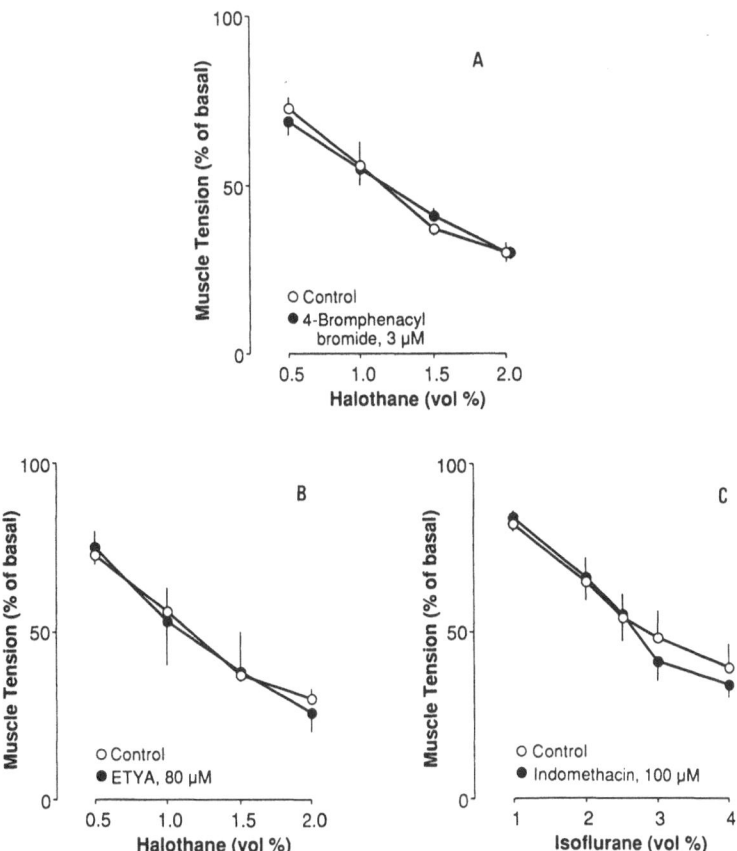

FIGURE 3. A: Effect of halothane on the contractile force of rat heart papillary muscle in the absence and presence of 4-bromphenacyl bromide. N = 5 and 3. B: Effect of halothane on the contractile force of rat heart papillary muscle in the absence and presence of ETYA. N = 5 and 4. C: Effect of isoflurane on the contractile force of rat heart papillary muscle in the absence and presence of indomethacin. N = 6 and 6. Points are means ± SEM throughout.

results of these experiments are summarized in table 1, and show that the decrease in contractile force induced by halothane 1 vol% or isoflurane 2.5 vol% was not diminished nor prevented by any of the inhibitors. Thus, the depressant effect of the anesthetics on myocardial contractility is not mediated by arachidonic acid or an active metabolite of the cyclooxygenase or lipoxygenase pathway.

Effect of Halothane on Phosphoinositide Turnover

The action of volatile anesthetics on the uptake and release of calcium from the sarcoplasmic reticulum has been proposed as one of the mechanisms by which the anesthetics modulate free intracellular calcium levels.[29-32] Many hormones and transmitters that increase cytosolic calcium by mobilizing calcium from the sarcoplasmic reticulum[5] hydrolyze phosphoinositides.[6] The primary step in the action of the agonists is the hydrolysis of polyphosphoinositides by a specific membrane bound phospholipase C. This reaction results in the formation of inositol triphosphate (IP_3), a second messenger which mediates the release of calcium from the sarcoplasmic reticulum.[7,33-35] IP_3 is then rapidly metabolized to inositol biphosphate and monophosphate directly or through a complex network of enzymatic reactions.[36] Lithium, which inhibits myoinositol-1-phosphatase, promotes the accumulation of inositol monophosphate, thereby amplifying the receptor-mediated changes in phosphoinositide turnover.[37] In different heart preparations, stimulation of α_1-adrenoceptors by norepinephrine or phenylephrine induces a rapid, transient increase in IP_3 and the accumulation of inositol monophosphate in the presence of lithium.[35,38-40] This increase in polyphosphoinositide breakdown correlates with the positive inotropic and chronotropic effect of α_1-adrenergic agonists. In addition, the stimulatory effect of norepinephrine on polyphosphoinositide breakdown in the heart is mediated by a pertussis toxin-insensitive G protein.[40] As we will see below, the negative inotropic effect of halothane is not impaired by pertussis toxin. Therefore, the effect of halothane on norepinephrine-induced phosphoinositide turnover was examined in heart ventricular and atrial tissue.

Table 1. Decrease in Contractile Force Induced by Halothane and Isoflurane in the Absence and in the Presence of Inhibitors of the Arachidonic Acid Cascade

Inhibitor	Decrease in Tension	
	Halothane 1 vol%	Isoflurane 2.5 vol%
None	44±2 (8)	46±7 (8)
4-Bromphenacyl bromide,[a] 3 μM	45±5 (3)	49±4 (4)
Mepacrine,[a] 10 μM	50 (2)	50±3 (3)
Indomethacin,[b] 100 μM	47±2 (3)	45±6 (4)
Eicosatetraynoic acid,[b,c] 80 μM	47±13 (3)	-----
Nordihydroguaiaretic acid,[c] 50 μM	45±3 (6)	59 (2)
WY - 50295,[c] 10 μM	45±2 (4)	-----
SK&F - 86002,[c] 10 μM	39 (2)	-----

Notes: [a] phospholipase A_2 inhibitors; [b] cyclooxygenase inhibitors; [c] lipoxygenase inhibitors. Values are mean ± SEM; (number) = number of papillary muscles.

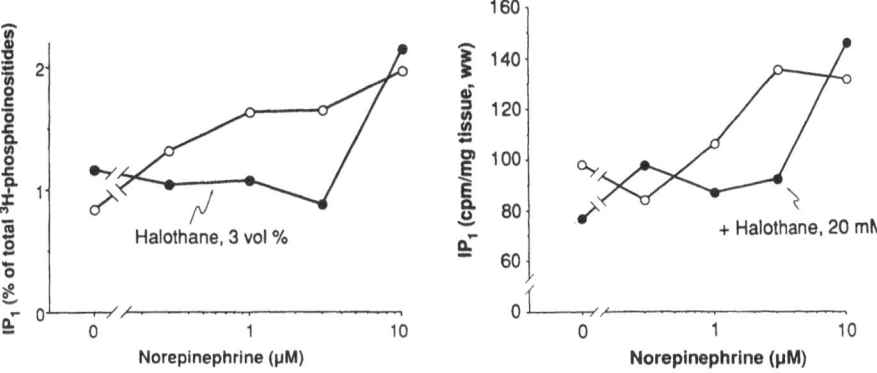

FIGURE 4. Effect of halothane on norepinephrine-stimulated phosphoinositides turnover in rat heart ventricle (left panel) and atria (right panel). Values are means of duplicate samples from one representative experiment.

Norepinephrine produced a dose-dependent increase in IP_1 formation in ventricles; in the atria, norepinephrine had a biphasic effect, producing a decrease at the lower concentration and a concentration-dependent increase at the higher concentrations. In both tissues, halothane shifted the norepinephrine concentration-response curve to the right (figure 4). At the lower concentration of norepinephrine, halothane inhibited the norepinephrine response, while at higher concentrations inositol phosphate release was the same or even greater in the presence of halothane than in its absence, suggesting a competitive type of interaction between norepinephrine and halothane. However, analysis of the data by Lineweaver-Burk plots indicates a non-competitive interaction.

General anesthetics, including volatile anesthetics, interact with different classes of receptor systems linked to many effectors. In these systems they have been shown to modulate cellular functions by a non-specific interaction with membrane lipid bilayers, or by interacting with specific hydrophobic sites on proteins, altering the activity of these molecules by an allosteric mechanism or/and by competing for endogenous ligands.[8] Even though the limited number of points on the concentration-response curves do not allow for a detailed analysis of the data presented in this study, such a dual action of halothane could account for the present results. In an earlier study halothane, as well as diethylether, has been shown to have a biphasic effect on acetylcholine-stimulated phosphatidylinositol turnover in brain membranes.[41] At low concentrations the anesthetics enhanced the acetylcholine-induced incorporation of ^{32}P into phosphatidylinositol and phosphatidic acid (a product of phosphatidylinositol hydrolysis) and inhibited it at higher concentrations. These results indicate a complex interaction between the anesthetics and the phosphoinositide system.

These results, though preliminary, are the first to demonstrate an effect of halothane on the phosphoinositide turnover in the heart. Future studies will establish the site and type of interaction of halothane and other volatile anesthetics with this biochemical pathway and the relevance of this system to their negative inotropic effect.

The Role of G Proteins in Anesthetic-Induced Myocardial Depression

Recent studies, *in vivo* and *in vitro*, have indicated that G proteins may be a target for the action of volatile anesthetics[11,42] and other general anesthetics.[43,44] In the heart at least four G proteins have been identified.[45] Gs mediates receptor stimulation of adenylate cyclase and directly activates dihydropyridine-sensitive calcium channels, leading to increased rate and contractility.[12,46] Gi mediates receptor inhibition of adenylate cyclase.[12,47] Gk is directly coupled to potassium channels activated by acetylcholine and adenosine and mediates their negative inotropic effects.[12,14,48] The function of Go is still undetermined; in neurons, Go activation inhibits calcium currents.[12] Gi, Gk and Go represent the majority of the G proteins in the heart.[45] Their activation is prevented by the bacterial toxin, pertussis toxin.[45,47] Therefore, in this study, we have investigated the involvement of pertussis toxin-sensitive G proteins in anesthetic-induced myocardial depression by comparing the response to halothane and isoflurane in papillary muscles and atria from animals pretreated with saline or with pertussis toxin. Since pretreatment with pertussis toxin abolishes the increase in potassium conductance and the negative inotropic response to acetylcholine in atrial preparations from rat and guinea pig hearts,[14,48] the efficacy of the treatment was assessed by testing the effect of carbachol on the contractile force of the atria in the same preparations.

In agreement with data reported in the literature, carbachol produced a concentration-dependent decrease in the contractile force of the atria from saline treated rats (ED_{50}: 0.33 ± 0.10 μM). The carbachol effect was abolished in tissue from pertussis toxin-treated rats (figure 5), indicating that the treatment was effective. In contrast, the pertussis toxin treatment did not prevent or diminish the decrease in contractile force induced by halothane or isoflurane in atria and papillary muscles (figures 5 and 6). The concentration-response curves for the effect of isoflurane in tissue preparations from pertussis toxin- or saline-treated rats are identical (figure 6). In both atria and papillary muscles, the depressant effect of halothane is significantly more pronounced in tissues from pertussis toxin-treated rats than in saline-treated rats. However, both concentration-response curves are parallel, as demonstrated by the similarity of the slopes (table 2). This indicates that the treatment with pertussis toxin did not affect the response of the muscle to halothane *per se*. The greater decrease in contractile force may reflect an action of pertussis toxin on mechanisms regulating calcium homeostasis which have a different sensitivity for halothane and isoflurane, as observed in a number of studies.[29-32]

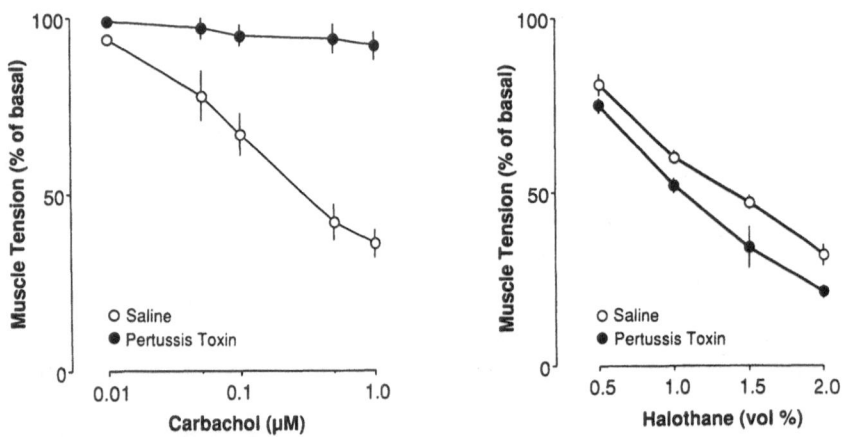

FIGURE 5. Effect of halothane and carbachol on the contractile force of heart left atria from rats pretreated with pertussis toxin. Values are mean ± SEM; n = 8 and 5 in preparations exposed to halothane; n = 6 and 5 in preparations exposed to carbachol, from rats treated with saline or pertussis toxin, respectively.

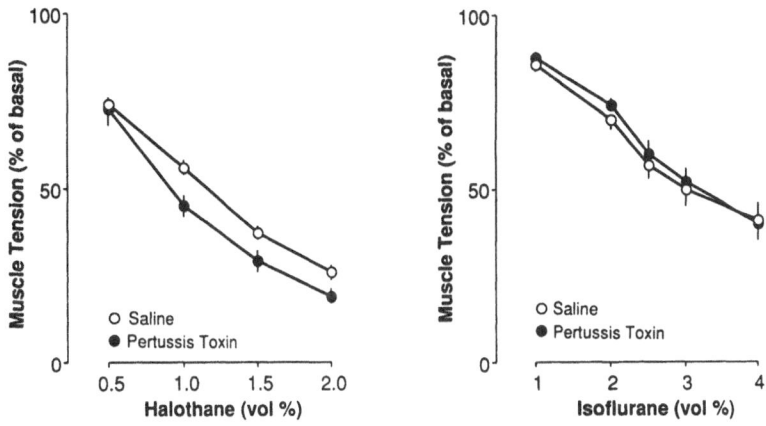

FIGURE 6. Effect of halothane and isoflurane on the contractile force of heart papillary muscles from rats pretreated with pertussis toxin. Values are mean ± SEM; n = 6 and 8 for halothane and isoflurane, respectively, without and with pretreatment with pertussis toxin.

The above results indicate that the mechanism(s) associated with the depressant effect of the anesthetics are not dependent on a functional pertussis toxin-sensitive G protein, and that they differ from that mediating the response to the muscarinic agonist, which is associated with an increase in potassium conductance.[48] The involvement of potassium channels in the action of the anesthetics has been suggested in studies with experimental models.[49-51] The present results, and those of work in progress showing that the decrease in contractile force induced by halothane and isoflurane is not blocked by the potassium channel antagonists 4-aminopyridine, tetraethylammonium chloride or glibenclamide,[52] suggest that the effect of halothane or isoflurane is not related to an action of the anesthetic on the regulation of potassium conductance.

The results of recent studies, showing that halothane, isoflurane and enflurane decrease the specific binding of the calcium antagonist nifedipine,[53] indicate that the anesthetics interfere with the functioning of the calcium channel. Our earlier studies have indicated that, in the heart, halothane decreases the stimulatory action of isoproterenol on cAMP formation, probably at the level of Gs, the protein which couples the β-adrenoceptor to adenylate cyclase. This G protein, which is not pertussis

Table 2. Characteristics of the Myocardial Depressant Effect of Halothane and Isoflurane

	ED_{50} (vol%)	Slope
Saline treated		
Halothane - Papillary muscle	1.21 ± 0.24	−83 ± 8
Halothane - Atria	1.31 ± 0.07	−71 ± 7
Isoflurane - Papillary muscle	3.22 ± 0.46	−83 ± 23
Pertussis toxin treated		
Halothane - Papillary muscle	0.96 ± 0.02*	−108 ± 17
Halothane - Atria	0.99 ± 0.05*	−73 ± 8
Isoflurane - Papillary muscle	3.19 ± 0.39	−111 ± 22

Asterisks (*) indicate significant differences from corresponding value in preparations from saline-treated rats.

toxin sensitive, has been shown recently to directly regulate the opening of the calcium channel.[46] Thus, it may be that the anesthetic also inhibits the opening of calcium channels by interacting with Gs.

The myocardial depressant effect of both volatile anesthetics and muscarinic agonists is due, in part, to their antagonism of the action of β-adrenergic catecholamines.[14,26] This effect is related to their ability to decrease the stimulatory effect of the catecholamines on cAMP formation.[2-4,14] However, the decrease in cAMP induced by the volatile anesthetics is not blocked by pertussis toxin and is thought to reflect an action at the level of Gs. Muscarinic agonists, in contrast, reduce cAMP levels by activating Gi, the pertussis toxin-sensitive G protein which mediates receptor inhibition of adenylate cyclase.[14] Thus, the mechanisms underlying the anti-adrenergic and the intrinsic effects of volatile anesthetics and muscarinic agonists on myocardial contractile force are different.

In summary, the potent depressant effect of halothane and isoflurane on myocardial contractility involves neither arachidonate metabolites nor a functional pertussis toxin-sensitive G protein. Halothane inhibits the stimulatory effect of norepinephrine on phosphoinositide hydrolysis.

ACKNOWLEDGEMENTS

This work was supported by NIH grant GM-34483. WY-50295 was kindly supplied by Dr. B. Weichmann, Wyeth-Ayerst Research Laboratories, Princeton, NJ. SK&F 86002 was provided by Dr. D. E. Griswold of Smith, Kline and French Laboratories, King of Prussia, PA.

REFERENCES

1. B. F. Rusy, H. Komai, Anesthetic depression of myocardial contractility: A review of possible mechanisms, *Anesthesiology* 67:745-766 (1987).
2. Y. Gangat, Y. Vulliemoz, M. Verosky, P. Danilo, K. Bernstein, L. Triner, Action of halothane on myocardial adenylate cyclase of rat and cat, *Proc Soc Exp Biol Med* 160:154-159 (1979).
3. Y. Vulliemoz, M. Verosky, L. Triner, Effect of halothane on myocardial cyclic AMP and cyclic GMP content of mice, *J Pharmacol Exp Ther* 236:181-186 (1986).
4. Y. Vulliemoz, M. Verosky, L. Triner, Myocardial cyclic nucleotides in response to volatile anesthetics, *Fed Proc* 41:1303 (1982).
5. A. M. Katz, Role of the contractile proteins and sarcoplasmic reticulum in the response of the heart to catecholamines, *Adv Cyclic Nucleotide Res* 11:303-343 (1979).
6. S. Moncada, R. J. Flower, J. R. Vane, Prostaglandins, prostacyclin, thromboxane A2 and leukotrienes, *in* "The Pharmacological Basis of Therapeutics," 7th edition, A. G. Gilman, L. S. Gilman, T. W. Rall, F. Murad, eds., Macmillan Publishing Company, New York (1985) pp. 660-673.
7. M. J. Berridge, Inositol triphosphate and diacylglycerol: Two interacting second messengers, *Ann Rev Biochem* 56:159-193 (1987).
8. S. H. Roth and K. W. Miller, eds., "Molecular and Cellular Mechanisms of Anesthetics," Plenum Press, New York (1986).
9. E. G. Lapetina, Regulation of arachidonic acid production: Role of phospholipases C and A, *Trends in Pharmacol Sci* 3:115-118 (1982).
10. F. Murad, W. P. Arnold, C. K. Mittal, T. M. Braughler, Properties and regulation of guanylate cyclase and some proposed functions of cyclic GMP, *Adv Cyclic Nucleotide Res* 11:175-204 (1979).
11. Y. Vulliemoz, M. Verosky, Halothane interaction with guanine nucleotide binding proteins in mouse heart, *Anesthesiology* 69:876-880 (1988).
12. L. Birnbaumer, J. Codina, R. Mattera, A. Yatani, N. Scherer, M. J. Toro, A. M. Brown, Signal transduction by G proteins, *Kidney Int* 32 (Suppl. 23):S14-S37 (1987).
13. H. M. Han, R. B. Robinson, J. P. Bilezikian, S. F. Steinberg, Developmental changes in guanine nucleotide regulatory proteins in the rat myocardial α_1-adrenergic receptor complex, *Circ Res* 65:1763-1773 (1989).

14. M. Endoh, M. Maruyama, T. Iijima, Attenuation of muscarinic cholinergic inhibition by islet-activating protein in the heart, *Am J Physiol* 249:H309-H320 (1985).

15. M. J. Berridge, R. M. C. Dawson, C. P. Downes, J. P. Heslop, R. F. Irvine, Changes in the levels of inositol phosphates after agonist-dependent hydrolysis of membrane phosphoinositides, *Biochem J* 212:473-482 (1983).

16. R. J. Flower, G. J. Blackwell, The importance of phospholipase A2 in prostaglandin biosynthesis, *Biochem Pharmacol* 25:285-291 (1976).

17. M. F. Roberts, R. A. Deems, T. C. Mincey, E. A. Dennis, Chemical modification of the histidine residue in phospholipase A2, *J Biol Chem* 252:2405-2411 (1977).

18. R. J. Flower, Drugs which inhibit prostaglandin biosynthesis, *Pharmacol Rev* 26:33-65 (1974)

19. J. R. Vane, Inhibition of prostaglandin synthesis as a mechanism of action for aspirin-like drugs, *Nature (New Biology)* 231:232-235 (1971).

20. M. Hamberg, On the formation of thromboxane B2 and 12L-hydroxy-5,8,10,14-eicosatetraynoic acid (12L0-20:4) in tissues from guinea pig brain, *Biochim Biophys Acta* 432:651-654 (1976).

21. D. E. Griswold, P. J. Marshall, E. F. Webb, R. Godfrey, J. Newton Jr, M. J. Dimartino, H. M. Sarau, J. G. Gleason, G. Poste, N. Hanna, SK&F 86002: A structurally novel anti-inflammatory agent that inhibits lipoxygenase and cyclooxygenase-mediated metabolism of arachidonic acid, *Biochem Pharmacol* 36:3463-3470 (1987).

22. Y. Hattori, R. Levi, Negative inotropic effect of leukotrienes: Leukotrienes C4 and D4 inhibit calcium-dependent contractile responses in potassium-depolarized guinea-pig myocardium, *J Pharmacol Exp Ther* 230:646-651 (1984).

23. G. Allan, R. Levi, The cardiac effects of prostaglandins and their modification by the prostaglandin antagonist N-0164, *J Pharmacol Exp Ther* 214:45-59 (1980).

24. L. Sterin-Borda, L. Canga, A. Pissani, A. L. Gimeno, Inotropic effecy of PGE_1 and PGE_2 on isolated rat atria: Influence of adrenergic mechanisms, *Prostaglandins* 20:825-837 (1980).

25. D. K. Basu, M. Karmazyn, Injury to rat hearts produced by an exogenous free radical generating system. Study into the role of arachidonic acid and eicosanoids, *J Pharmacol Exp Ther* 242:673-685 (1987).

26. C. Lynch, III, Differential depression of myocardial contractility by halothane and isoflurane in vitro, *Anesthesiology* 64:620-631 (1986).

27. W. J. Wolf, M. B. Neal, B. P. Mathew, D. E. Bee, Comparison of the in vitro myocardial depressant effects of isoflurane and halothane anesthesia, *Anesthesiology* 69:660-666 (1988).

28. P. R. Housmans, I. Murat, Comparative effects of halothane, enflurane, and isoflurane at equipotent anesthetic concentrations on isolated ventricular myocardium of the ferret. I. Contractility, *Anesthesiology* 69:451-463 (1988).

29. E. S. Casella, N. D. A. Suite, Y. I. Fisher, T. J. J. Blanck, The effect of volatile anesthetics on the pH dependence of calcium uptake by cardiac sarcoplasmic reticulum, *Anesthesiology* 67:386-390 (1987).

30. J. Y. Su, W. G. L. Kerrick, Effects of halothane on caffeine-induced tension transients in functionally skinned myocardial fibers, *Pflugers Arch* 380:29-34 (1979).

31. H. Komai, B. F. Rusy, Negative inotropic effects of isoflurane and halothane in rabbit papillary muscle, *Anesth Analg* 66:29-33 (1987).

32. D. M. Wheeler, R. T. Rice, R. G. Hansford, E. G. Lakatta, The effect of halothane on the free intracellular calcium concentration of isolated rat heart cells, *Anesthesiology* 69:578-583 (1988).

33. M. J. Berridge, Rapid accumulation of inositol triphosphate reveals that agonists hydrolyse polyphosphoinositides instead of phosphatidylinositol, *Biochem J* 212:849-858 (1983).

34. T. M. Nosek, M. F. Williams, S. T. Zeigler, R. E. Godt, Inositol triphosphate enhances calcium release in skinned cardiac and skeletal muscle, *Am J Physiol* 250:C807-811 (1986).

35. H. Otani, H. Otani, D. K. Das, Alpha-1 adrenoceptor-mediated phosphoinositide breakdown and inotropic response in rat left ventricular papillary muscles, *Circ Res* 62:8-17 (1988).

36. P. W. Majerus, T. M. Connolly, H. Deckmyn, T. S. Ross, T. E. Bross, H. Ishii, V. S. Bansal, D. B. Wilson, The metabolism of phosphoinositide-derived messenger molecules, *Science* 234:1519-1526 (1986).

37. M. J. Berridge, C. P. Downes, M. R. Hanley, Lithium amplifies agonist dependent phosphatidylinositol responses in brain and salivary glands, *Biochem J* 206:587-595 (1982).

38. J. Poggioli, J. C. Suplice, G. Vassort, Inositol phosphate production following α_1-adrenergic, muscarinic or electrical stimulation in isolated rat heart, *FEBS Letters* 206:292-298 (1986).

39. J. Scholz, B. Schaefer, W. Schmitz, H. Scholz, M. Steinfath, M. Lohse, U. Schwabe, J. Puurunen, Alpha-1 adrenoceptor-mediated positive inotropic effect and inositol triphosphate increase in mammalian heart, *J Pharmacol Exp Ther* 245:327-335 (1988).

40. S. F. Steinberg, L. M. Kaplan, T. Inouye, J. I. Fang Zhang, R. B. Robinson, Alpha-1 adrenergic stimulation of 1,4,5-inositol triphosphate formation in ventricular myocytes, *J Pharmacol Exp Ther* 250:1141-1148 (1989).

41. J. C. Miller, Anesthetics and phospholipid metabolism, *in:* "Molecular Mechanisms of Anesthesia," B. R. Fink, ed., Raven Press, New York (1975) pp. 439-447.

42. R. S. Aronstam, B. L. Anthony, R. L. Dennison, Halothane effects on muscarinic acetylcholine receptor complexes in rat brain, *Biochem Pharmacol* 35:667-672 (1986).

43. A. J. Robinson-White, S. M. Muldoon, L. Elson, D. M. Collado-Escobar, Evidence that barbiturates inhibit antigen-induced responses through interactions with a GTP-binding protein in rat basophilic leukemia (RBL-2H3) cells, *Anesthesiology* 72:996-1004 (1990).

44. C. Okuda, M. Miyazaki, K. Kuriyama, Alterations in cerebral β-adrenergic receptor-adenylate cyclase system induced by halothane, ketamine and ethanol, *Neurochem Int* 6:237-244 (1984).

45. N. M. Scherer, M. J. Toro, M. L. Entman, L. Birnbaumer, G-protein distribution in canine cardiac sarcoplasmic reticulum and sarcolemma: Comparison to rabbit skeletal muscle membranes and to brain and erythrocyte G-proteins, *Arch Biochem Biophys* 259:431-440 (1987).

46. A. Yatani, Y. Imoto, J. Codina, S. L. Hamilton, A. M. Brown, L. Birnbaumer, The stimulatory G protein of adenylyl cyclase, Gs, also stimulates dihydropyridine-sensitive Ca^{2+} channels, *J Biol Chem* 263:9887-9895 (1988).

47. T. Katada, M. Ui, Direct modification of the membrane adenylate cyclase system by islet-activating protein due to ADP-ribosylation of a membrane protein, *Proc Natl Acad Sci* (USA) 79:3129-3133 (1982).

48. P. J. Pfaffinger, J. M. Martin, D. D. Hunter, N. M. Nathanson, B. Hille, GTP-binding proteins couple cardiac muscarinic receptors to a K channel, *Nature* 317:536-538 (1985).

49. R. A. Nicoll, D. V. Madison, General anesthetics hyperpolarize neurons in the vertebrate central nervous system, *Science* 217:1055-1057 (1982).

50. I. S. Segal, J. Tinklenberg, R. W. Aldrich, M. Maze, Decreased potassium channel conductance, encoded by the SHAKER locus in drosophila, increases volatile anesthetic requirements, *Anesthesiology* 71:A641 (1989).

51. P. W. L. Tas, H. G. Kress, K. Koschel, Volatile anesthetics inhibit the ion flux through Ca^{2+}-activated K^+ channels of rat glioma C6 cells, *Biochim Biophys Acta* 983:264-268 (1989).

52. N. S. Cook, The pharmacology of potassium channels and their therapeutic potential, *TIPS* 9:21-28 (1988).

53. B. Drenger, T. J. J. Blanck, Volatile anesthetics depress the binding of calcium channel blocker to purified cardiac sarcolemma, *Anesthesiology* 69:A16 (1989).

16

The Effects of Volatile Anesthetics on the Calcium Sensitivity of Cardiac Myofilaments

Isabelle Murat, Renée Ventura-Clapier

INTRODUCTION

Volatile anesthetics mainly depress myocardial contractility by their actions on sarcoplasmic reticulum function and on sarcolemmal ionic currents.[1] A direct effect of volatile anesthetics on myocardial contractile proteins was first described by Merin in 1974,[2] but this was only observed at rather high anesthetic concentrations. Skinned fiber preparations represent an unique model for studying the contractile apparatus itself. Studies on detergent-treated skinned fibers of various animal species have provided evidence for a volatile-anesthetic-induced decrease in both the calcium sensitivity and the maximal developed tension of cardiac myofilaments. These effects are dose-dependent, reversible and quantitatively equivalent for the three currently used anesthetics, halothane, enflurane, and isoflurane.[3,4,5] This chapter will review the physiological properties of the contractile proteins and the experimental studies on volatile anesthetic effects.

REGULATORY PROTEINS

The generation of contractile force in cardiac muscle ultimately depends upon a cyclic interaction between the regulatory proteins actin and myosin, which are organized in a lattice of thick (myosin) and thin (actin) filaments in the myocardial cell. This interaction takes place between the myosin heads and the individual actin molecules of the myofilaments, and involves the myosin ATPase enzyme located in the head regions of the myosin molecule. During contraction, conformational changes in actin-myosin crossbridges take place. The instantaneous developed force depends on the number of actively cycling crossbridges, on the individual force generated by each crossbridge, as well as on their attachment and detachment rates.[6] The number of actively cycling crossbridges depends on the amount of calcium ions bound to Ca^{2+}-binding sites of troponin C (TnC).[7] The increase in intracellular free calcium concentration from 10^{-7} to 10^{-5} M produces an increase in force production and ATPase activity, both reaching a plateau at calcium concentrations of about 10^{-5} M.

ISABELLE MURAT, Department of Anesthesia, Hopital Saint-Vincent de Paul, 82 av. Denfert Rochereau, F-75674 Paris, France. RENEE VENTURA-CLAPIER, Laboratoire de Physiologie Cellulaire Cardiaque, INSERM U-241, Universite Paris-Sud, F-91405 Orsay, France.

Mechanisms of Anesthetic Action in Skeletal, Cardiac, and Smooth Muscle
Edited by T.J.J. Blanck and D.M. Wheeler, Plenum Press, New York, 1991

The relationship between parameters is sigmoïdal[8,9] (see figure 1). Many factors may change the tension/pCa (pCa = $-\log_{10}[Ca^{2+}]$) relationship. Indeed, calcium sensitivity of cardiac myofilaments depends on sarcomere length,[10,11,12,13] temperature,[14] pH,[15,16,17] ionic strength and viscosity,[18] intracellular inorganic phosphate[19,20] and $MgATP^{2-}$ concentration,[21] intracellular cAMP and cGMP concentrations (which are responsible for intracellular phosphorylation),[22,23] myosin light chain phosphorylation,[24] and also actin-myosin attachment.[7] Cardiac TnC differs from skeletal TnC, in that it has only three calcium binding sites instead of four. This difference in molecular structure is responsible for the differential effects of length changes on muscle Ca^{2+} sensitivity.[25] Indeed, the calcium sensitivity of cardiac muscle has been demonstrated to be modified to a greater extent by changes in sarcomere length than that of skeletal muscle, indicating that Starling's law of the heart definitely operates at a molecular level.[10] The role of the other sub-units of the troponin system was recently outlined by several workers.[26,27,28] Changes in troponin T and tropomyosin isoforms seem to be associated with changes in Ca^{2+} sensitivity,[27] while modifications in troponin I isoforms may influence the sensitivity of cardiac myofilaments to pH changes, especially acidosis.[26] On the other hand, changes in myosin isoforms are not associated with changes in myocardial Ca^{2+} sensitivity.[29] The number of factors potentially affecting the Ca^{2+} sensitivity of myofilaments explains the importance of carefully controlling experimental conditions to assess physiological or pharmacological changes.

STUDIES ON MYOFIBRILLAR PREPARATIONS

Several authors have attempted to demonstrate an effect of volatile anesthetics on myocardial contractile proteins. Brodkin et al.[30] in 1967 observed that halothane decreased rat myofibrillar ATPase activity, but only at very high anesthetic concentrations (20-100 mM). Myofibrillar ATPase activity was decreased by 50% for halothane concentrations of about 40 mM, which is roughly 100 times higher than

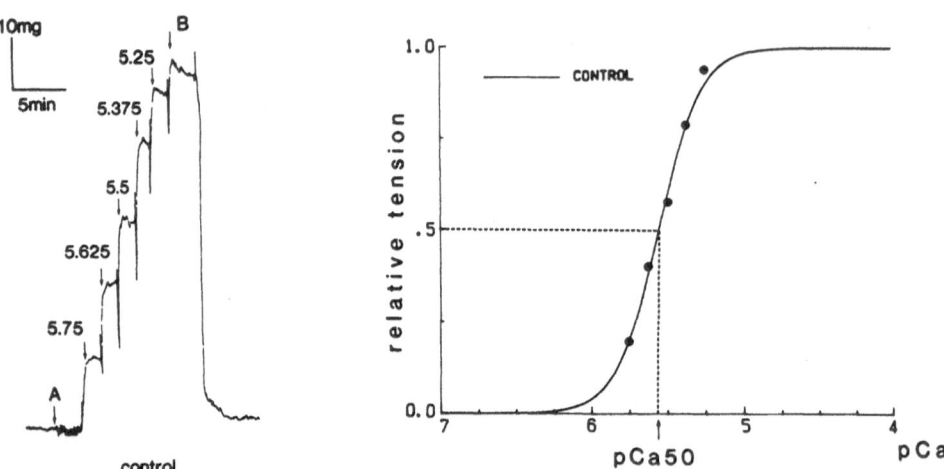

FIGURE 1. Left: Force changes of one rat left ventricular skinned fiber during stepwise exposure to solutions of increasing Ca^{2+} concentrations ranging from the relaxing solution A (pCa = 9) to the activating solution B (pCa = 4.5). Numbers indicate pCa values (pCa = $-\log_{10}[Ca^{2+}]$). Right: Sigmoïdal tension/pCa curve of the same fiber. Tension obtained in the relaxing solution A represents 0%, maximal activated tension obtained in solution B represents 100%. The formulas used to assess calcium sensitivity are as follows: relative force = $[Ca]^{n_H}/(K + [Ca]^{n_H})$ and $pCa_{50} = (-\log_{10}K)/n_H$, where n_H represents the Hill coefficient.

those used in clinical practice. However, in these experiments, Ca^{2+} concentration was not controlled. In 1973, Leuwenkroon-Strösberg et al.[31] reported that halothane (3-10%) was capable of producing conformational changes in myosin heads without modifying Ca^{2+}-activated myosin ATPase. Merin et al.[2] were the first to produce evidence for a direct effect of halothane on the physiologically active enzyme, actomyosin ATPase (actin-activated myosin ATPase, hydrolyzed by $MgATP^{2-}$). It was observed that halothane concentrations greater than 0.9 mM (about 2%, v/v) produced a reduction in maximal actomyosin ATPase activity together with a reduction in ATPase Ca^{2+} sensitivity. The effect of halothane on maximal actomyosin ATPase activity was completely antagonized by 10^{-3} to 10^{-2} M Ca^{2+}. Ohnishi et al.[32] also observed a 36 percent depression of actomyosin ATPase activity in dog heart myofibrils by 4.4 mM halothane at a pCa of 6.5. The most recent report of the action of volatile anesthetics on myofibrillar ATPase is that of Pask et al.[33] using a bovine myofibrillar preparation. Halothane and isoflurane produced a similar depression of enzyme activity (close to that reported by both Merin[2] and Ohnishi[32] with halothane), but surprisingly, enflurane had no effect on this preparation. A critical review of these studies was recently published by Rusy & Komai,[1] pointing out several drawbacks of these investigations. First, as anesthetics were directly added into the bathing solutions, the corresponding gas phase concentration was unknown. Second, it was suggested that the isolation process used to obtain myofibrils may be responsible for such relative insensitivity of the latter to Ca^{2+} and drugs.

STUDIES ON SKINNED FIBERS

The term "skinned fibers" refers to a number of different preparations in which the surface membrane of a muscle cell is removed or made permeable to small molecules.[34] Two different preparations have been used to assess the effects of volatile anesthetics on myocardial Ca^{2+} sensitivity. Su and coworkers[35,36,37,38] used mechanically disrupted fibers, while we[3,4,5] used fibers skinned with a nonionic detergent, triton X-100 (1%). Both mechanically disrupted and detergent treated fibers are multicellular preparations. In the first technique, large amounts of sarcolemma are retained in the preparations. The other cellular membranes including sarcoplasmic reticulum, T-tubules, and mitochondria are presumably intact, and these preparations can be used to study the drug effects on uptake and release capacities of sarcoplasmic reticulum.[36,37] The function of the sarcoplasmic reticulum can, however, be blunted by high EGTA concentrations. With nonionic detergents, all cellular membranes are destroyed while the contractile proteins are basically unaffected by this procedure.[39] An open question when using skinned fibers is represented by the ability of the preparation to respond to intracellular phosphorylation, as described in hyperpermeable fibers.[22,23] However, detergent treated rat or rabbit skinned fibers are insensitive to phosphatase treatment (unpublished data), indicating that the potential effects of volatile anesthetics will not result from changes in the level of intracellular phosphorylation in this particular preparation.

The effects of halothane, enflurane and isoflurane were studied in detergent treated rat cardiac fibers at a temperature of 22°C.[3] Left ventricular fibers were mounted between two stainless steel hooks, one connected to a transducer, and they were studied at a constant sarcomere length of 2.1 to 2.2 μm, as assessed by laser diffraction before each series of experiments. Control measurements involved the measurement of maximal developed force in an activating solution in which pCa was set at 4.5, pH 7.1, and ionic strength 0.16 M, and the generation of a tension/pCa curve by exposing the fibers to solutions of increasing calcium concentrations (from pCa 5.875 to pCa 4.5) and measuring developed force at each pCa (as in figure 1). This set of measurements was then obtained in a randomly designed study using three to four different anesthetic concentrations of each of the three anesthetics on the same cardiac muscle. For that purpose, experimental solutions were equilibrated in separate chambers by bubbling for 15 min with the chosen anesthetic, using calibrated anesthetic vaporizers. The anesthetic concentration in the gas phase was monitored with an infrared analyzer. During experiments, solutions were vigorously stirred at high speed to facilitate diffusion of Ca, substrates and anesthetic molecules into the muscle. Control measurements were again performed at the end of the experimental studies on each fiber.

The results can be summarized as follows: 1) volatile anesthetics produced a dose-dependent and reversible decrease in maximal activated tension; this decrease ranged between 5% and 15% for clinically relevant anesthetic concentrations; 2) volatile anesthetics also decreased myocardial calcium sensitivity in a dose-dependent (between 0.5 and 2 MAC) and reversible fashion (figure 2); 3) these effects were observed at clinically relevant anesthetic concentrations and ranged from 0.05 to about 1.5 pCa unit; and 4) the effects of the three anesthetics used, halothane, enflurane, and isoflurane, were identical for equianesthetic concentrations expressed in MAC multiples.

A similar effect of volatile anesthetics was observed in different animal species: rat,[3,4] hamster,[4] rabbit,[5] and ferret (unpublished data). In addition, the effects of equipotent anesthetic concentrations of the three anesthetics were quantitatively identical in adult rats,[3,4] as described above, and in adult ferrets (unpublished data). The effects of equipotent anesthetic concentrations of halothane (1%, v/v) and isoflurane (1.5%, v/v) were also of same magnitude in the same age-group of rabbits,[5] whatever the age of the animal (from near-term fetuses to adults).

Halothane effects were also assessed in two types of cardiomyopathies:[4] the genetic dilated cardiomyopathy of the Syrian hamster UM-X7.1 and the streptozotocin-induced diabetic cardiomyopathy in rats. Although differences in

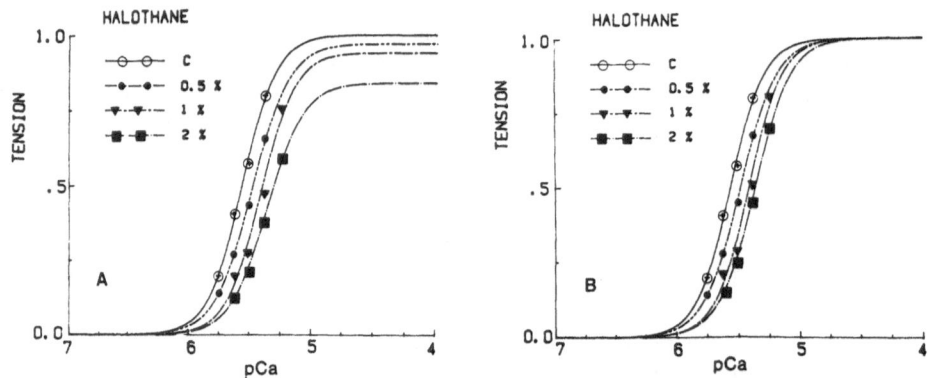

FIGURE 2. Family of tension/pCa curves obtained in the same fiber in control conditions (C) and with increasing halothane concentrations (0.5, 1, and 2%). Panel A. Experimental points and computed curves. Maximal developed tension decreased in a dose-dependent fashion (for these curves 100% represent maximal activation in anesthetic-free solution obtained at the same time). The curves are shifted in a dose-dependent fashion to the right. Panel B. Same curves as in panel A but normalized to maximal tension in control B solution. The shift to the right remains obvious with increasing anesthetic concentrations, pCa_{50} decreases in a dose-dependent fashion, but the slope of each curve (Hill coefficient) remains unchanged. This indicates a decrease in the apparent Ca^{2+} sensitivity of the contractile proteins with increasing halothane concentration. From Murat, Ventura-Clapier and Vassort, *Anesthesiology* 69:892-899 (1988) with permission.

mechanical properties of these two different pathologic hearts were observed in the absence of anesthetics, the effects of halothane on both maximal activated tension and myocardial calcium sensitivity were identical to those observed in their respective control hearts.

Developmental changes in effects of halothane and isoflurane on contractile proteins were studied in rabbits.[5] Different age-groups of animals were used: near-term fetuses (24 h before birth), 1-day-old newborn (within the 24 h after birth), immature (3-, 8-, and 17-day old) and young adult (8-11 months) rabbits. In control conditions, age-related changes in maximal tension (which increased with the age of the animal) and calcium sensitivity (which was lower in newborn and immature hearts than in adult hearts) were observed. During anesthetic exposure, both calcium sensitivity and maximal developed tension decreased significantly in all age-groups of animals, both anesthetics having a similar effect in animals of identical age (figure 3). However, calcium sensitivity decreased significantly more in newborn animals (0.192 and 0.196 pCa unit for halothane and isoflurane, respectively) compared to adults (0.122 and 0.137 pCa units, respectively). By contrast, fetuses were less sensitive to the myocardial depressant effects of anesthetics than were newborn animals. In addition, maximal myofibrillar ATPase activities were measured in the absence of anesthetics and in the presence of 1% halothane in each age-group of rabbits.[5] No significant changes were observed during anesthetic exposure (0.63 mM) in any age-group of rabbits. These results differ from those reported by Krane and Su,[40] who did not find differences in the effects of halothane on mechanical properties of mechanically disrupted cardiac skinned fibers of 1- to 4-day-old and adult rabbits. These discrepancies between the two studies may arise from the method used to obtain skinned fibers, as described above, as well as from differences in experimental conditions. Indeed, their fibers were studied just above the slack length, whereas our fibers were stretched by 20% above the slack length, which corresponds to a sarcomere length of 2.1 to 2.2 μm. As myocardial Ca^{2+} sensitivity depends on sarcomere length,[10,11,12,13] it was suggested that difference in length may alter the functional state of the myofibrils, possibly explaining the discrepancies between the two studies.

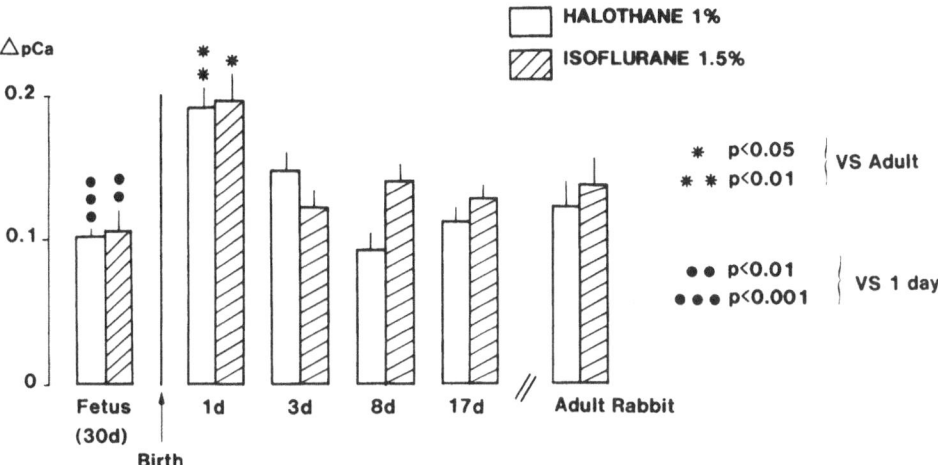

FIGURE 3. Changes in calcium sensitivity according to the age and anesthetic exposure (halothane 1% and isoflurane 1.5%) in rabbits of various ages. Stars indicate a greater decrease in pCa_{50} in newborn rabbits than in adults in the presence of anesthetic. Circles indicate significant changes between 30-day-old fetuses and newborn rabbits. From Murat, Hoerter and Ventura-Clapier, *Anesthesiology* 73:137-145 (1990) with permission.

In addition to their effects on calcium sensitivity of cardiac myofilaments, volatile anesthetics decreased maximal developed force in a dose-dependent and reversible fashion in all the species studied.[3,4,5] To further investigate the possible mechanisms involved in the decrease of force production, stiffness measurements[41,42] were performed in detergent-treated rat cardiac skinned fibers at defined levels of activation with the use of quick length changes of 0.3 to 4% of initial muscle length in the absence and in the presence of 2 MAC of halothane, enflurane or isoflurane.[43] The following results suggest that these anesthetics have multiple sites of action on cardiac myofibrillar proteins: 1) they decreased active stiffness indicating a decreased number of attached force-generating crossbridges; 2) they increased the stiffness/force ratio suggesting that the individual force developed by each crossbridge was decreased during anesthetic exposure; and 3) they increased the time-constant of force recovery, which is consistent with the decreased rate of ATP hydrolysis described by Merin.[2]

MECHANISMS OF ACTION

The mechanisms by which volatile anesthetics affect the regulatory proteins are not fully understood. However, some hypotheses may be suggested from cardiac skinned fibers experiments. Tension/pCa relationships reflect attachment to and detachment of calcium from troponin C binding sites.[7] Only the low-affinity Ca^{2+} binding site plays a role in force regulation. The two high-affinity binding sites are saturated at a very low calcium concentration ($\approx 10^{-7}$ M), and their binding capacities are not modified by halothane.[44] However, the lack of effect of halothane on isolated TnC does not necessarily imply an absence of effect on the troponin system, as calcium binding depends on both the relative interactions of the three subunits and on the actin-myosin interactions.[27] The hypothesis of an involvement of the native troponin system in the effects of volatile anesthetics on myocardial Ca^{2+} sensitivity is further supported by the results of the experiments performed on immature animals.[5] A greater decrease in Ca^{2+} sensitivity was observed in neonatal muscles than in adult muscles. The greater effect of volatile anesthetics on newborn animals may be related to changes in the troponin system that are known to occur around birth.[16,17,45,46,47,48,49] The molecular structure of troponin C does not seem to change in the developing heart,[50] but calcium binding to myofibrils from 4-day-old rabbit hearts was found to be lower than that observed in 22-day-old and adult preparations.[51] It was recently suggested that developmental changes in troponin I isoforms may be responsible for the decreased sensitivity of neonatal cardiac muscle to acidosis,[16,17] whereas developmental changes in troponin T isoforms would better correlate to changes in Ca^{2+} sensitivity.[51] In addition, the effect of volatile anesthetics on myocardial Ca^{2+} sensitivity seems to be independent of changes in myosin composition, as halothane has a similar effect in normal and diabetic rats,[4] while in the latter rats a shift from the normal V1 to the slow V3 myosin isoform is usually observed. Furthermore, a similar decrease in myocardial calcium sensitivity was observed in left ventricular fibers of rat heart, in which the isozymic profile is exclusively V1; and in those of rabbit heart, where only the V3 isoform is expressed in adult animals.[52] Therefore, the greater sensitivity to anesthetics of cardiac contractile proteins from newborn animals may be related to subtle changes in the composition of the troponin-tropomyosin complex.

Mechanisms by which volatile anesthetics decrease maximal tension seem to involve processes of attachment and detachment of actomyosin crossbridges, leading to a decrease in both the number of crossbridges involved in force generation and in the amount of force developed by individual crossbridges. Maximal ATPase activity did not change in the presence of 1% halothane (≈ 0.6 mM). This agrees closely with the data of Merin,[2] who observed a decrease in maximal ATPase activity only for halothane concentrations greater than 0.9 mM. Thus, for clinical anesthetic concentrations, the decrease in maximal force is not correlated with changes in ATPase activity, but it is probably related to changes in actin-myosin interactions, leading to a modification of energy transduction at the crossbridge level.[53]

RELATIVE CONTRIBUTION TO THE NEGATIVE INOTROPIC EFFECT

The hypothesis of an effect of clinical concentrations of volatile anesthetics on myocardial contractile proteins is strongly supported by experimental studies done in intact papillary muscles.[54,55] Indeed, in intact ferret papillary muscles, it was observed that clinical anesthetic concentrations of halothane, enflurane or isoflurane (0.5 to 2 MAC) prolonged the duration of lengthening of an isotonic twitch, yet they abbreviated the duration of isometric relaxation.[54] The latter was taken as possibly resulting from reduced myofibrillar Ca^{2+} sensitivity, while the former would more likely be the consequence of reduced sarcoplasmic reticulum function. Indeed, calcium sensitivity is higher during force production than during shortening and isometric relaxation mainly reflects the calcium responsiveness of myocardial contractile proteins while isotonic relaxation mainly depends on calcium uptake capacities of the sarcoplasmic reticulum.[56] This effect on isometric relaxation was dose-dependent and identical for the three anesthetics studied, whereas the effects of isoflurane on both myocardial contractility and relaxation were significantly less pronounced than those of both halothane and enflurane.[54,55] This is consistent with our unpublished data in normal ferret right ventricular skinned fibers where a similar reduction of calcium sensitivity was observed for 1 MAC of each of the three agents. Other evidence for an effect of volatile anesthetics on myocardial contractile proteins in intact preparations is provided by aequorin experiments.[57] A decrease in the aequorin signal during isometric contraction was observed in the presence of halothane (0.5%). When the bathing calcium concentration was increased to permit twitch tension to return to preanesthetic control value, the peak aequorin signal was higher than that measured in control conditions, suggesting that myocardial calcium sensitivity was decreased in the presence of halothane. Similar results were obtained with enflurane and isoflurane.[57]

Therefore, the action of volatile anesthetics on myocardial contractile proteins may participate in the negative inotropic effects of these agents, but it cannot explain the differential effect of volatile anesthetics on myocardial contractility. In living tissues of most species, even during a positive inotropic intervention, the force development is far from maximal, so that the force can easily be varied by increasing or decreasing the free Ca^{2+} concentration.[58] Therefore, for calcium concentrations likely to be reached during the time-course of normal activation, the effects of volatile anesthetics on intracellular calcium content, myofibrillar calcium sensitivity and maximal activated force may play a role. In rabbit papillary muscles, studied at a temperature of 37°C, Berman et al.[59] found that 2% halothane decreased maximal activated force generated by myofilaments to 79 percent of control, while twitch tension was reduced to 58 percent of control, both experiments being performed in the presence of ryanodine to suppress the function of sarcoplasmic reticulum. However, in intact ferret papillary muscles (at a temperature of 30°C), 2 MAC isoflurane reduced twitch tension to about 25% of control, while 1.5 MAC halothane decreased it to less than 20%. Therefore, the effect of volatile anesthetics on maximal activated force can only account for a small fraction of the profound depression of contractility. On the other hand, any decrease in the rise in activator Ca caused by volatile anesthetics would be magnified by the decreased calcium sensitivity of the contractile proteins.

CONCLUSION

Clinical concentrations of volatile anesthetics decrease myocardial calcium sensitivity. This effect is dose-dependent (between 0.5 and 3 MAC), reversible, identical and constant in every species studied to date, but more pronounced in newborn animals than in adult animals. In addition, volatile anesthetics decrease maximal developed tension, an effect that only participates to a moderate extent in the overall negative inotropic effects of these agents. The studies performed during development and in various adaptive and pathologic models favor the hypothesis that the site of action involved in these effects mainly lies at the level of the thin filament. These agents may modify the troponin system, and also actin-myosin interactions.

REFERENCES

1. B. F. Rusy, H. Komai, Anesthetic depression of myocardial contractility: A review of possible mechanisms, *Anesthesiology* 67:745-766 (1987).
2. R. G. Merin, T. Kumazawa, C. R. Honig, Reversible interaction between halothane and Ca^{2+} on cardiac adenosine triphosphatase: Mechanism and significance, *J Pharmacol Exp Ther* 190:1-14 (1974).
3. I. Murat, R. Ventura-Clapier, G. Vassort, Halothane, enflurane, and isoflurane decrease calcium sensitivity and maximal force in detergent treated rat cardiac fibers, *Anesthesiology* 69:892-899 (1988).
4. I. Murat, V. I. Veksler, R. Ventura-Clapier, Effects of halothane on cardiac skinned fibers from cardiomyopathic animals, *J Mol Cell Cardiol* 21:1293-1304 (1989).
5. I. Murat, J. Hoerter, R. Ventura-Clapier, Developmental changes in effects of halothane and isoflurane on contractile properties of rabbit cardiac skinned fibers, *Anesthesiology* 73:137-145 (1990).
6. A. F. Huxley, R. M. Simmons, Proposed mechanism of force generation in striated muscle, *Nature* (Lond) 233:533-538 (1971).
7. P. A. Hofmann, F. Fuchs, Evidence for a force-dependent component of calcium binding to cardiac troponin C, *Am J Physiol* 253:C541-C546 (1987).
8. M. J. Holroyde, S. P. Robertson, J. D. Johnson, R. J. Solaro, J. D. Potter, The calcium and magnesium binding sites on cardiac troponin and their role in the regulation of myofibrillar adenosine triphosphatase, *J Biol Chem* 255:11688-11693 (1980).
9. R. J. Solaro, R. M. Wise, J. S. Shiner, F. N. Briggs, Calcium requirements for cardiac myofibrillar activation, *Circ Res* 34:525-530 (1974).
10. A. Babu, E. Sonnenblick, J. Gulati, Molecular basis for the influence of muscle length on myocardial performance, *Science* 240:74-76 (1988).
11. S. M. Harrison, C. Lamont, D. J. Miller, Hysteresis and the length dependence of calcium sensitivity in chemically skinned rat cardiac muscle, *J Physiol* (Lond) 401:115-144 (1988).
12. M. G. Hibberd, B. R. Jewell, Calcium- and length-dependent force production in rat ventricular muscle, *J Physiol* (Lond) 329:527-540 (1982).
13. J. C. Kentish, H. E. D. J. Ter Keurs, L. Ricciardi, J. J. J. Buck, M. I. M. Noble, Comparison between the sarcomere length-force relations of intact and skinned trabeculae from rat right ventricle. Influence of calcium concentrations on these relations, *Circ Res* 58:755-768 (1986).
14. S. M. Harrison, D. M. Bers, Influence of temperature on the calcium sensitivity of the myofilaments of skinned ventricular muscle from the rabbit, *J Gen Physiol* 93:411-428 (1989).
15. A. Fabiato, F. Fabiato, Effects of pH on the myofilaments and the sarcoplasmic reticulum of skinned cells from cardiac and skeletal muscles, *J Physiol* (Lond) 276:233-255 (1978).
16. R. J. Solaro, P. Kumar, E. M. Blanchard, A. F. Martin, Differential effects of pH on calcium activation of myofilaments of adult and perinatal dog hearts: Evidence for developmental differences in thin filament regulation, *Circ Res* 58:721-729 (1986).
17. R. J. Solaro, L. A. Lee, J. C. Kentish, D. G. Allen, Effects of acidosis on ventricular muscle from adult and neonatal rats, *Circ Res* 63:779-787 (1988).
18. J. C. Kentish, The inhibitory effects of monovalent ions on force development in detergent skinned ventricular muscle from guinea-pig, *J Physiol* (Lond) 352:353-374 (1984).
19. J. C. Kentish, The effects of inorganic phosphate and creatine phosphate on force production in skinned muscles from rat ventricle, *J Physiol* (Lond) 370:585-604 (1986).

20. H. Mekhfi, R. Ventura-Clapier, Dependence upon high-energy phosphates on the effects of inorganic phosphate on contractile properties in chemically skinned rat cardiac fibres, *Pflügers Arch* 411:378-385 (1988).

21. P. M. Best, S. K. Bolitho Donaldson, W. G. L. Kerrick, Tension in mechanically disrupted mammalian cardiac cells: Effects of magnesium adenosine triphosphate, *J Physiol* (Lond) 265:1-17 (1977).

22. G. B. McClellan, S. Winegrad, The regulation of the calcium sensitivity of the contractile system in mammalian cardiac muscle. *J Gen Physiol* 72:737-764 (1978).

23. S. Winegrad, G. McClellan, R. Horowits, M. Tucker, E. R. Lin, A. Weisberg, Regulation of cardiac contractile proteins by phosphorylation, *Fed Proc* 42:39-44 (1983)

24. I. Morano, F. Hofmann, M. Zimmer, J. C. Ruegg, The influence of P-light chain phosphorylation by myosin light chain kinase on the calcium sensitivity of chemically skinned heart fibres, *FEBS Lett* 189:221-224 (1985).

25. A. Babu, W. Lehman, J. Gulati, Characterization of the Ca^{2+}-switch in skeletal and cardiac muscles, *FEBS Lett* 251:177-182 (1989).

26. S. P. Robertson, J. D. Johnson, M. J. Holroyde, E. G. Kranias, J. D. Potter, R. J. Solaro, The effect of troponin I phosphorylation on the Ca^{2+}-binding properties of the Ca^{2+}-regulatory site of bovine cardiac troponin, *J Biol Chem* 257:260-263 (1982).

27. F. H. Schachat, M. S. Diamond, P. W. Brandt, Effect of different troponin T-tropomyosin combinations on thin filament activation, *J Mol Biol* 198:551-554 (1987).

28. A. S. Zot, J. D. Potter, Reciprocal coupling between troponin C and myosin crossbridge attachment, *Biochemistry* 28:6751-6756 (1989).

29. R. Ventura-Clapier, H. Mekhfi, P. Oliviero, B. Swynghedauw, Pressure overload changes cardiac skinned fibers mechanics in rats not in guinea pigs, *Am J Physiol* 254:H517-H524 (1988).

30. W. E. Brodkin, A. H. Goldberg, H. L. Kayne, Depression of myofibrillar ATPase activity by halothane, *Acta Anaesthesiol Scand* 11:97-101 (1967).

31. E. Leuwenkroon-Strösberg, L. H. Laasberg, J. Hedley-White, Myosin conformation and enzymatic activity: Effect of chloroform, diethyl ether and halothane on optical rotatory dispersion and ATPase, *Biochim Biophys Acta* 295:178-186 (1973).

32. T. Ohnishi, G. S. Pressman, H. L. Price, A possible mechanism of anesthetic-induced myocardial depression, *Biochem Biophys Res Comm* 57:316-322 (1974).

33. H. T. Pask, P. J. England, C. Prys-Roberts, Effects of volatile inhalational anaesthetic agents on isolated bovine cardiac myofibrillar ATPase, *J Mol Cell Cardiol* 13:293-301 (1981).

34. P. M. Best, Cardiac muscle function: Results from skinned fiber preparations, *Am J Physiol* 244:H167-H177 (1983).

35. J. Y. Su, W. G. L. Kerrick, Effects of halothane on Ca^{2+}-activated tension development in mechanically disrupted rabbit myocardial fibers, *Pflügers Arch* 375:111-117 (1978).

36. J. Y. Su, W. G. L. Kerrick, Effects of halothane on caffeine-induced tension transients in functionally skinned myocardial fibers, *Pflügers Arch* 380:597-604 (1979).

37. J. Y. Su, W. G. L. Kerrick, Effects of enflurane on functionally skinned myocardial fibers from rabbits, *Anesthesiology* 52:385-389 (1980).

38. J. Y. Su, J. G. Bell, Intracellular mechanism of action of isoflurane and halothane on striated muscle of the rabbit, *Anesth Analg* 65:457-462 (1986).

39. D. J. Miller, H. Y. Elder, G. L. Smith, Ultrastructural and X-ray microanalysis studies of EGTA- and detergent-treated heart muscle, *J Mol Cell Motility* 6:525-540 (1985).

40. E. J. Krane, J. Y. Su, Comparison of the effects of halothane on skinned myocardial fibers from newborn and adult rabbits: I. Contractile proteins, *Anesthesiology* 70:76-81 (1989).

41. L. E. Ford, A. F. Huxley, R. M. Simmons, Tension responses to sudden length change in stimulated frog muscle fibers near slack length, *J Physiol* (Lond) 269:441-515 (1977).

42. L. E. Ford, A. F. Huxley, R. M. Simmons, The relation between stiffness and filament overlap in stimulated frog muscle fibers, *J Physiol* (Lond) 311:219-249 (1981).

43. I. Murat, P. Lechene, R. Ventura-Clapier, Effects of volatile anesthetics on mechanical properties of rat cardiac skinned fiber, *Anesthesiology* 73:73-81 (1990).

44. E. S. Casella, T. J. J. Blanck, The effect of halothane on the binding of calcium by cardiac troponin C, *Biophys J* 53:583a (1988).

45. P. A. W. Anderson, G. E. Moore, R. N. Nassar, Developmental changes in the expression of rabbit left ventricular troponin T, *Circ Res* 63:742-747 (1988).

46. J. P. Jin, J. J. C. Lin, Rapid purification of mammalian cardiac troponin T and its isoform switching in rat hearts during development, *J Biol Chem* 263:7309-7315 (1988).

47. L. Saggin, S. Ausoni, L. Gorza, S. Sartore, S. Schiaffino, Troponin T switching in the developing rat heart, *J Biol Chem* 263:18488-18492 (1988).
48. L. Saggin, L. Gorza, S. Ausoni, S. Schiaffino, Troponin I switching in the developing heart. *J Biol Chem* 264:16299-16302 (1989).
49. L. S. Tobacman, R. Lee, Isolation and functional comparison of bovine cardiac troponin T isoforms, *J Biol Chem* 262:4059-4064 (1987).
50. J. M. Wilkinson, Troponin C from rabbit slow skeletal and cardiac muscle is a product of a single gene, *Eur J Biochem* 103:179-188 (1980).
51. J. J. McAuliffe, L. Gao, R. J. Solaro, Changes in myofibrillar activation and troponin C Ca^{2+} binding associated with troponin T isoform switching in developing rabbit heart, *Circ Res* 66:1204-1216 (1990).
52. A. M. Lompré, J. J. Mercadier, C. Wisnewsky, P. Bouveret, C. Pantaloni, A. D'Albis, K. Schwartz, Species- and age-dependent changes in the relative amounts of cardiac myosin isoenzymes in mammals. *Dev Biol* 84:286-290 (1981).
53. E. Eisenberg, T. L. Hill, Muscle contraction and free energy transduction in biological systems, *Science* 227:999-1006 (1985).
54. P. R. Housmans, I. Murat, Comparative effects of halothane enflurane and isoflurane at equipotent anesthetic concentrations on isolated ventricular myocardium of the ferret: I. Contractility, *Anesthesiology* 69:451-463 (1988).
55. P. R. Housmans, I. Murat, Comparative effects of halothane enflurane and isoflurane on isolated ventricular myocardium of the ferret: II. Relaxation, *Anesthesiology* 69:464-471 (1988).
56. D. L. Brutsaert, S. U. Sys, Relaxation and diastole of the heart, *Physiol Rev* 69:1228-1315 (1989).
57. P. R. Housmans, L. A. Wanek, E. G. Carton, Halothane (H), enflurane (E) and isoflurane (I) decrease myofibrillar Ca^{2+} responsiveness in intact mammalian ventricular muscle, *Biophys J* 57:554A (1990).
58. S. Ebashi, M. Endo, Calcium ion and muscle contraction, *Prog Biophys Mol Biol* 18:125-183 (1968).
59. M. R. Berman, E. S. Casella, T. J. J. Blanck, Evidence for an effect of halothane on the maximum calcium-activated force generated by cardiac myofilaments, *Anesthesiology* 71:A250 (1989).

17

Evidence for a Halothane-Induced Reduction in Maximal Calcium-Activated Force in Mammalian Myocardium

Michael R. Berman, Eugenie S. Casella, Thomas J.J. Blanck

INTRODUCTION

In addition to their general anesthetic effects, the halogenated volatile anesthetics (*e.g.*, halothane, enflurane, isoflurane) are known to significantly depress cardiac contractility. In the late 1960's, Goldberg and Ullrick[1] reported a dose-dependent and reversible depression of twitch force by halothane (as delivered by a calibrated vaporizer) in the range of 0.1% to 2.35%. They argued that ". . . halothane seems to exert its cardiac effects primarily by decreasing the intensity of the active state . . . ." In a comparative study of the effects of several general anesthetic agents on cardiac contractility, Brown and Crout[2] reported similar results for halothane as well as for methoxyflurane. They, too, suggested that these agents exerted their cardio-depressive effects *via* a diminution of the active state of cardiac muscle. Although the concept of "active state" is no longer useful, it is known to be associated with the calcium transient; that is, the rapid rise and fall of intracellular free calcium concentration seen subsequent to electrical stimulation.

Price[3] measured twitch force in the presence of 0.5% halothane as a function of extracellular calcium concentration and found that the attenuation of twitch force could be at least partially reversed by increasing extracellular calcium concentration. He suggested that ". . . the apparent interference with the effectiveness of externally supplied Ca^{2+} could represent either an inhibition of Ca^{2+} transport from the exterior into the cytoplasm or an interference with the actions of Ca^{2+} on the contractile proteins . . . ." Our understanding of the mechanism of force generation in the heart, and of the critical role played by calcium, has increased tremendously since the appearance of these earlier reports. Even so, we remain convinced, as is demonstrated by the recent review by Rusy and Komai,[4] that volatile anesthetic agents depress cardiac contractility primarily by a derangement of cellular calcium handling.

Figure 1 is a schematic drawing representing the three primary sites of cellular calcium handling; interference by volatile agents at any one (or more) of these sites would result in a depression of myocardial contractility. It is now well accepted (see,

MICHAEL R. BERMAN, EUGENIE S. CASELLA, THOMAS J. J. BLANCK, Department of Anesthesiology and Critical Care Medicine, The Johns Hopkins Hospital, Baltimore, Maryland 21205.

Mechanisms of Anesthetic Action in Skeletal, Cardiac, and Smooth Muscle
Edited by T.J.J. Blanck and D.M. Wheeler, Plenum Press, New York, 1991

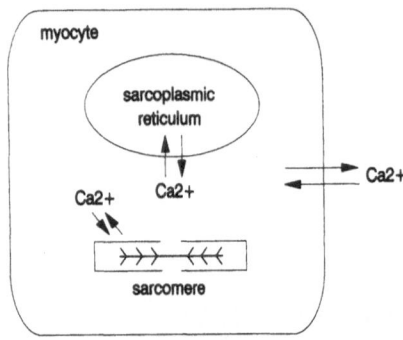

FIGURE 1. Schematic of a cardiac myocyte, showing the three major sites at which volatile anesthetics might derange cellular calcium handling. They are the sarcolemma (reduction of entry and/or enhancement of removal of Ca^{2+}), sarcoplasmic reticulum (reduced release and/or enhanced uptake of Ca^{2+}), and myofilaments (reduced binding and/or sensitivity to Ca^{2+} and/or reduced crossbridge force generation).

e.g., recent review by Rusy and Komai[4]) that the volatile anesthetics do reduce calcium entry across the sarcolemma. This unquestionably reduces contractility by diminishing the amount of Ca^{2+} available to the crossbridges; the fundamental force generators in the heart. There is recent evidence[5-7] for a direct effect of volatile agents on sarcoplasmic reticulum function, although the extent to which this contributes to the overall depression of contractility is not yet clear.

To investigate whether volatile agents had a direct effect on the myofilaments, Su and Kerrick[8] conducted studies using mechanically "skinned" myocardial bundles. In this way they were able to functionally remove the sarcolemma, thereby leaving a preparation with only sarcoplasmic reticulum and myofilaments remaining functional. They found a dose-dependent depression of maximal calcium-activated force (MCAF) and a reduction of myofilament calcium sensitivity. More recently, Murat *et al.*[9] performed similar experiments, in this instance using chemically "skinned" bundles, in which both the sarcolemma and the sarcoplasmic reticulum were not able to modulate myoplasmic calcium concentration. Their results were similar to those reported earlier;[8] the volatile anesthetic agents reduced maximal calcium-activated force generation and myofilament calcium sensitivity in a dose-dependent, reversible fashion.

There are, however, two problems which arise when interpreting the results of any experiments performed using "skinned" preparations. Most obvious is the degradation of the preparation with repeated activations.[9] This is unavoidable; it is most likely due to calcium-activated proteases which "eat" the muscle. A more subtle concern is the effect of skinning on the micro-environment within the cell and/or surrounding the structures (*e.g.*, the myofilaments and regulatory proteins) being studied. The very purpose of "skinning" is to render membrane structures hyper-permeable, so that the concentrations of low molecular weight moieties (*e.g.*, Ca^{2+}, Mg^{2+}, H^+) within the cell can be set by controlling their concentrations in the bathing medium.[10] However, this also allows the diffusion out of the cell of low molecular weight proteins which play an important role in the modulation of muscle contractile function.[11]

A third problem in interpreting the results of the particular skinned muscle experiments cited above is that they were performed at 20°C, with anesthetic concentrations in the bathing solution given in volume percent,[8] or at 22°C, with anesthetic concentrations in the bathing solution given in mM.[9] In the latter paper,[9] the measured anesthetic concentrations were related to the minimal alveolar concentration,[12] MAC, for anesthesia at 37°C. It is known[12] that volatile anesthetic potency varies inversely with temperature; thus, the anesthetic concentrations used in these studies might well be effectively higher than the range of clinical interest.

Therefore, we have re-investigated the effects of halothane on maximal calcium-activated force (MCAF) generation with the restrictions that the experiments be performed at 37°C, using a preparation with intact membranes.

METHODS

Since halothane can affect force generation *via* action at the sarcolemma and at the sarcoplasmic reticulum as well as at the myofilaments (see figure 1), we eliminated these first two sites of action to concentrate on the myofilaments. Sarcoplasmic reticulum was removed as a confounding factor by adding ryanodine to the superfusate. This drug has been shown to render the sarcoplasmic reticulum non-functional.[13,14] Ryanodine has the added advantage of having no effect on myofilament calcium-activated force generation.[15] Limitation of calcium entry to the myoplasm by effects of halothane at the sarcolemma was countered by performing experiments at a high level of extracellular calcium concentration, so that during tetany the myofilaments would be saturated with calcium;[16] therefore, tetanic force would be the maximal calcium-activated force. By so doing we also eliminated the confounding possibility that halothane would affect measured force by altering myofilament calcium sensitivity,[8,9] since the myofilaments would be saturated with calcium. In this way, we believe we have established a preparation in which tetanic force is the maximal calcium-activated force.

With the approval of the Animal Care and Use Committee of the School of Medicine, all experiments were performed on isolated right ventricular papillary muscles (n = 8) from the hearts of 5 to 6 pound male New Zealand white rabbits. Anesthesia was induced by injecting 50 mg/kg of pentobarbital *via* an ear vein. Heparin (1000 units) was given to aid the wash-out of blood from the heart following dissection. Hearts were rapidly removed and perfused through the aortic stump with a solution consisting of (in mM) NaCl 108, KCl 5, $MgCl_2$ 1, HEPES 5, $NaCH_3CO_2$ 20, dextrose 10 and $CaCl_2$ 2.5, maintained at 37°C and vigorously bubbled (flow rate $\geq$ 1 L/min) with 100% oxygen. The right ventricle was opened and, if a suitable papillary muscle was found, a small (12 mm long, 0.005 inch diameter) tungsten hook was tied at the tendinous end. The muscle was removed from the heart by cutting the tendon distal to the tie and by removing a small piece of ventricular wall at the base of the muscle.

Muscles were mounted horizontally in a chamber superfused with solution as described above. The hook at the tendinous end was attached to a force transducer; the wall end was placed in a Teflon clamp attached to the shaft of a linear motor, which for these experiments was maintained in its isometric mode. The force transducer was mounted on an X-Y micromanipulator for precise axial alignment of the muscle; the linear motor was mounted on a Z-axis micrometer for adjusting muscle length. Two platinum wire electrodes, one on either side of the muscle, ran the length of the chamber.

Muscles were stimulated (30/min, 5 msec duration, supramaximal) and incrementally stretched to L_{max}, the length at which twitch force exhibited a maximum. Muscle length at L_{max} ranged from 3.0 to 5.2 mm (3.88 $\pm$ 0.97 mm, mean $\pm$ S.D.); cross-sectional area ranged from 0.20 to 0.50 mm^2 (0.35 $\pm$ 0.10 mm^2).

Following equilibration at L_{max}, calcium sufficient to bring total extracellular concentration to 20 mM and 1 μM ryanodine were added to the superfusate reservoir.

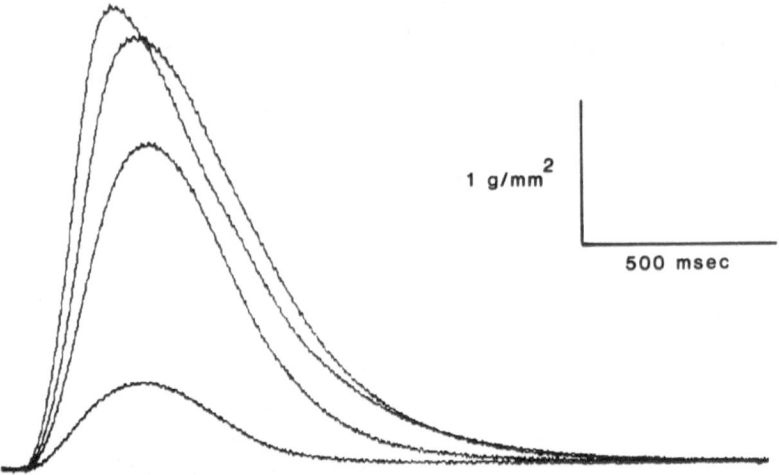

FIGURE 2. Reduction of twitch force by increasing concentrations of halothane in a muscle pre-treated with ryanodine (1 μM) and in saturating (20 mM) extracellular calcium. Halothane concentrations (in vol%), as measured in the chamber fluid by gas chromatography, were 0, 0.24 ± 0.05, 0.60 ± 0.04, and 1.34 ± 0.05 and pertain to the twitches in order of decreasing amplitude.

After equilibration, chamber fluid was sampled and twitches and tetani were recorded. Halothane was introduced into the superfusate reservoir *via* a vaporizer at dial settings of 1%, 2% and 3%. Following equilibration at each concentration of halothane chamber fluid was sampled and twitches and tetani were recorded.

Twitches were recorded in sets of four, at a sampling rate of 500 Hz. The four traces were subsequently point averaged to improve signal-to-noise ratio. Tetani were induced by a 2 sec train of 10 Hz, 50 msec duration, supramaximal stimuli. Data were recorded at 500 Hz.

At each concentration of halothane 1 ml of fluid was withdrawn from the muscle chamber and immediately vigorously mixed with 2 ml heptane in a Teflon sealed vial. A 0.5 ml aliquot of the resulting mixture was analyzed for halothane by

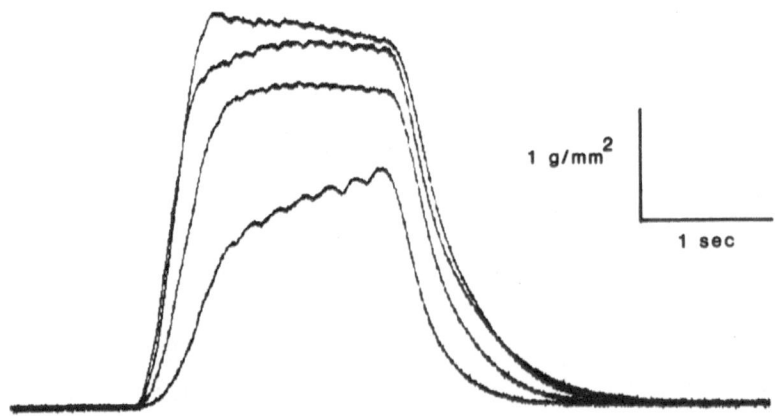

FIGURE 3. Reduction of tetanus force (MCAF) by increasing concentrations of halothane in a muscle pre-treated with ryanodine (1 μM) and in saturating (20 mM) extracellular calcium. Halothane concentrations (in vol%), as measured in chamber fluid by gas chromatography, were 0, 0.24 ± 0.05, 0.60 ± 0.04, and 1.34 ± 0.05 and pertain to the traces in order of decreasing amplitude.

gas chromatography. For our calculations we assumed halothane had a density of 1.86 g/ml, a molecular weight of 197.4 g/mole and a gas/saline partition coefficient of 1.43 (at 37°C).

RESULTS

The combination of high (20 mM) extracellular calcium and 1 μM ryanodine resulted in a reduction of twitch amplitude and a prolongation in twitch time course. This is as would be expected, since, with the sarcoplasmic reticulum rendered non-functional by ryanodine the only significant pathway for cellular calcium exchange is the sarcolemma.

Twitch force decreased steeply with the addition of halothane to the superfusate (figure 2, figure 4). Tetanic force (figure 3, figure 4), a direct measure of maximum calcium-activated force (MCAF), decreased as well, although not as steeply as did twitch force.

DISCUSSION

The results of these experiments (figure 4) demonstrate that halothane causes a small, but statistically significant ($P > 0.05$) diminution in tetanic force; which we interpret as being the maximal calcium-activated force. These results are consistent with earlier reports in the literature.[8,9] Direct comparison is made difficult by differences in anesthetic potency at low (20°C,[8] 22°C[9]) vs. high (37°C) temperatures. In a sense, then, our work is merely confirmatory; we have, however, added the new information that 1) earlier results[8,9] were not artifacts of the skinned muscle preparation and 2) the diminution of MCAF by halothane is present at a physiologically relevant temperature and anesthetic concentration.

There are several possible confounding factors which might call our results into question. The most obvious is the effect of the damaged ends of the muscles on

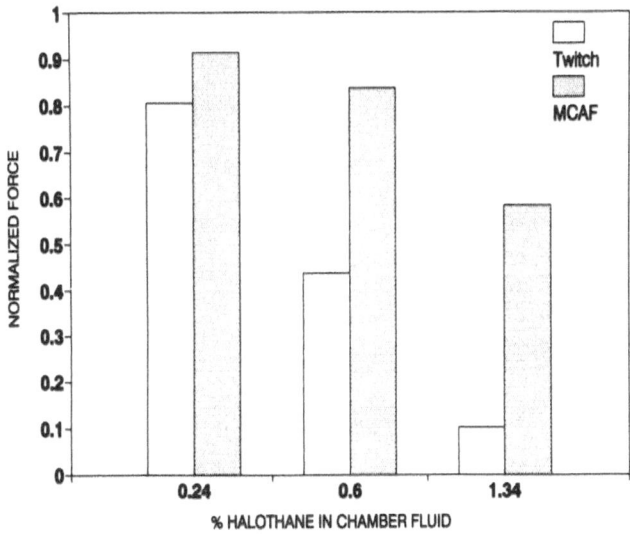

FIGURE 4. Summary (n = 8) of effects of halothane on twitch force and on maximum calcium activated force (MCAF) in ryanodized (1 μM) muscles in saturating (20 mM) extracellular calcium. Note that for rabbit, 1 MAC halothane ≈ 0.82%.[12]

measured tetanic force. Since these experiments were done in a muscle length isometric mode, it is certain that, under tetanic stimulation, the undamaged central portion of the muscles shortened to an unknown degree during contraction. Even in maximally activated cardiac muscle, force generation is somewhat sensitive to sarcomere length,[17] decreasing as sarcomeres shorten and travel down the ascending limb of the force-length curve. Therefore, "high" forces (control, low concentrations of halothane) could be underestimated to a greater extent than "low" forces (high halothane concentrations). If so, it may be that 1) there is actually little or no diminution in MCAF at low halothane concentrations and 2) the percent reduction in MCAF at mid and high halothane concentrations may be greater that shown in figure 4.

In order to interpret our results, we have assumed that the measured tetanic force is, in fact, the maximal calcium-activated force. To ensure this we performed our experiments at an extracellular calcium concentration of 20 mM, well above the saturating level of 5 mM reported by DeTraglia et al.[16] Since those measurements were made in the absence of halothane, we briefly tested the sufficiency of tetanic force by giving isoproterenol (10 μM) in the presence of 3% halothane. Twitch force increased, indicating additional calcium was coming across the sarcolemma; tetanic force did not change, which we interpreted as indicating that tetani were saturating, even at our highest halothane concentrations. In hindsight we should have added the calcium channel agonist BAY-K 8644 ($\approx$ 1 μM) to our superfusate to further ensure saturation of the tetanus force.[18]

Even if the interpretation of our results is not complicated by the factors discussed above, we must still question the significance to clinicians of the observation that halothane reduces maximal calcium-activated force, if for no other reason than that myofilaments in the beating heart are never maximally calcium activated. Further, as seen in figures 2 and 4, the depressive effect of halothane on twitch force via limitation of calcium availability to the myofilaments is far more severe than its effect on MCAF. Although our results support the hypothesis that halothane interferes with the fundamental force generating interaction of myosin with actin, it appears that at clinically relevant concentrations the contribution of this mechanism to the overall cardio-depressant effect of halothane is quite small.

ACKNOWLEDGEMENTS

Supported by grants HL-38488 (MRB) and GM-30799 (TJJB) from the National Institutes of Health.

REFERENCES

1. A. H. Goldberg and W. C. Ullrick, Effects of halothane on isometric contractions of isolated heart muscle, Anesthesiology 28:838 (1967).
2. B. R. Brown and J. R. Crout, A comparative study of the effects of five general anesthetics on myocardial contractility: Isometric conditions, Anesthesiology 34:236 (1971).
3. H. L. Price, Calcium reverses myocardial depression caused by halothane: Site of action, Anesthesiology 41:576 (1974).
4. B. F. Rusy and H. Komai, Anesthetic depression of myocardial contractility: a review of possible mechanisms, Anesthesiology 67:745 (1987).
5. E. S. Casella, N. D. A. Suite, Y. I. Fisher and T. J. J. Blanck, The effect of volatile anesthetics on the pH dependence of calcium uptake by cardiac sarcoplasmic reticulum, Anesthesiology 67:386 (1987).
6. H. Komai and B. F. Rusy, Direct effect of halothane and isoflurane on the function of the sarcoplasmic reticulum in intact rabbit atria, Anesthesiology 72:694 (1990).

7. P. R. Housmans, Negative inotropy of halogenated anesthetics in ferret ventricular myocardium, *Am J Physiol* 259:H827 (1990).

8. J. Y. Su and W. G. L. Kerrick, Effects of halothane on Ca^{2+} activated tension development in mechanically disrupted rabbit myocardial fibers, *Pflugers Arch* 375:111 (1978).

9. I. Murat, R. Ventura-Clapier and G. Vassort, Halothane, enflurane, and isoflurane decrease calcium sensitivity and maximal force in detergent treated rat cardiac fibers, *Anesthesiology* 69:892 (1988).

10. A. Fabiato and F. Fabiato, Calculator programs for computing the composition of the solutions containing multiple metals and ligands used for experiments in skinned muscle cells, *J Physiol* (Paris) 75:463 (1979).

11. S. Winegrad, Regulation of cardiac contractile proteins: Correlations between physiology and biochemistry, *Circ Res* 55:565 (1984).

12. E. I. Eger, "Anesthetic Uptake and Action," Williams and Wilkins, Baltimore (1974).

13. D. M. Bers, Ca influx and sarcoplasmic reticulum Ca release in cardiac muscle activation during postrest recovery, *Am J Physiol* 248:H366 (1985).

14. E. Marban and W. G. Wier, Ryanodine as a tool to determine the contributions of calcium entry and calcium release to the calcium transient and contraction of cardiac Purkinje fibers, *Circ Res* 56:133 (1985).

15. J. Y. Su, Effects of ryanodine on skinned myocardial fibers of the rabbit, *Pflugers Arch* 411:132 (1988).

16. M. C. DeTraglia, H. Komai, D. Redon and B. F. Rusy, Isoflurane and halothane inhibit tetanic contractions in rabbit myocardium in vitro, *Anesthesiology* 70:837 (1989).

17. D. G. Allen and J. C. Kentish, The cellular basis of the length-tension relation in cardiac muscle, *J Mol Cell Cardiol* 17:821 (1985).

18. D. T. Yue, Intracellular $[Ca^{2+}]$ related to rate of force development in twitch contraction of the heart, *Am J Physiol* 252:H760 (1987).

18

Mechanisms of Negative Inotropy of Halothane, Enflurane and Isoflurane in Isolated Mammalian Ventricular Muscle

Philippe R. Housmans

The volatile anesthetics halothane, enflurane, and isoflurane are potent myocardial depressants. They cause a concentration-dependent negative inotropic effect in humans, experimental animals, and in cardiac tissues isolated from a variety of mammalian species.[1] Their negative inotropic effect has been attributed to interference with 1) transsarcolemmal Ca^{2+} influx; 2) Ca^{2+} uptake and release from the sarcoplasmic reticulum; and 3) Ca^{2+} sensitivity of the contractile proteins. The purpose of this communication is to review the relative importance of the decrease in intracellular Ca^{2+} availability and of the decrease in myofibrillar Ca^{2+} sensitivity brought about by these anesthetics. Our conclusions will be illustrated by recent studies in intact ventricular muscle, *i.e.*, the right ventricular papillary muscle of the ferret.

The concentration-dependent myocardial depressant effect of halothane is illustrated in figure 1 for isometric twitch contractions at the optimal muscle length L_{max}. Enflurane and, to a lesser extent, isoflurane are potent negative inotropic agents as well in this preparation.[2,3] The time course of the isometric twitch is also slightly modified by these anesthetics. However, to fully appreciate the subtle changes in twitch time course, we have electronically adjusted all force traces of the isometric twitches before and during exposure to 0.5, 1.0, and 1.5 MAC halothane to the same height as shown in figure 2. Halothane (and enflurane and isoflurane) caused an abbreviation in the time to peak force and caused a faster relative decline of force; that is, force declined sooner from its peak value in twitches in the presence of halothane than in its absence (figure 2B). However, it was still uncertain whether the changes in time course of the isometric twitch were not at least in part a consequence of the decrease in twitch amplitude. Therefore, to dissociate effects of anesthetics proper from those of twitch amplitude *per se* on the time course of the twitch, we have compared time to peak force (TPF) and time to half isometric relaxation (RTH) in twitches of equal amplitude: without anesthetic in low $[Ca^{2+}]_o$ *versus* anesthetic in control $[Ca^{2+}]_o$ (2.25 mM) in the same muscles.[4] Results of this type of comparison are shown for halothane in figure 3, where time to peak force (TPF) and time from TPF to 50% isometric force decline (RTH) are plotted as a function of the

PHILIPPE R. HOUSMANS, Department of Anesthesiology, Mayo Medical School and Mayo Foundation, Rochester, Minnesota 55905.

Mechanisms of Anesthetic Action in Skeletal, Cardiac, and Smooth Muscle
Edited by T.J.J. Blanck and D.M. Wheeler, Plenum Press, New York, 1991

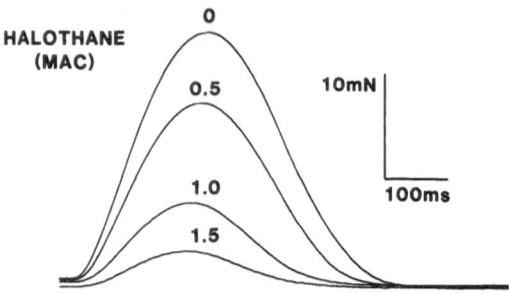

FIGURE 1. Force traces of four superimposed isometric twitch contractions at L_{max} of a ferret right ventricular papillary muscle in control (0) conditions and during exposure to 0.5, 1.0, and 1.5 MAC halothane. Stimulus interval 4 seconds, $[Ca^{2+}]_o$ 2.25 mM, 30°C.

corresponding peak twitch amplitude (developed force) for a protocol involving 1) changing $[Ca^{2+}]_o$ from 0.45 mM to 2.25 mM in 0.45 mM increments (open circles) and 2) exposure to halothane 0-1.5 MAC in 0.25 MAC increments in $[Ca^{2+}]_o$ 2.25 mM (filled circles). At lower twitch amplitudes a clear divergence in time course variables is apparent between twitches of equal amplitude in low $[Ca^{2+}]_o$ versus those in anesthetic. Halothane, enflurane and isoflurane have, therefore, specific effects on

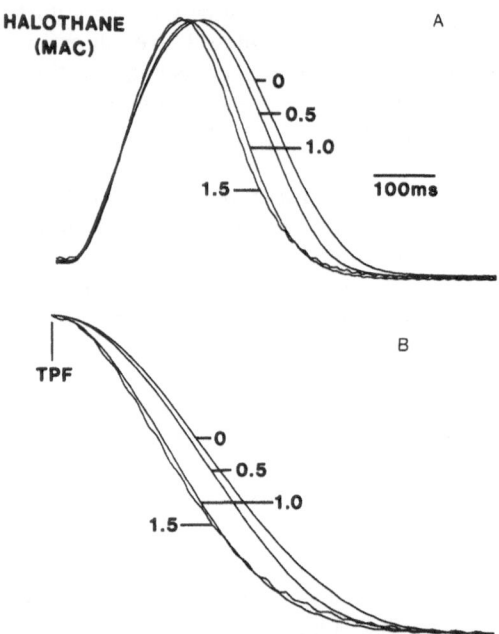

FIGURE 2. Panel A: Force traces of four superimposed isometric twitches at L_{max} of a right ventricular ferret papillary muscle in control conditions (0) and during exposure to 0.5, 1.0, and 1.5 MAC halothane. The force traces were scaled vertically to the same peak amplitude to facilitate comparison of time course among twitches. Panel B: Force traces of isometric relaxation of the twitches shown in panel A. Normalized force traces were shifted horizontally and made to superimpose at their respective time to peak force to allow for easy comparison of the time course of isometric relaxation. Ferret right ventricular papillary muscle, stimulus interval 4 seconds, $[Ca^{2+}]_o$ 2.25 mM, 30°C.

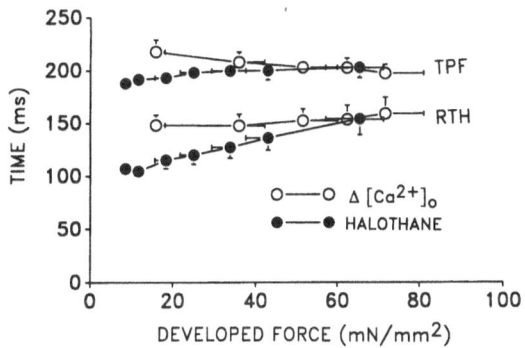

FIGURE 3. Dependence of time to peak force (TPF) and of time to 50% isometric relaxation (RTH) on peak developed force in isometric twitches in 2 different conditions: changes in $[Ca^{2+}]_o$ from 0.45 to 2.25 mM in 0.45 mM increments (o) and of halothane concentration from 0 to 1.5 MAC in 0.25 MAC increments (●) in $[Ca^{2+}]_o$ 2.25 mM. Values (mean ± S.E.) were from the same muscles (n = 8). Differences between TPF and RTH curves were significant (P < 0.01) by Hotelling's T^2 test for differences between (slope,intercept) values of least squares linear regressions among $[Ca^{2+}]_o$ and [anesthetic] experiments. (Modified after P. R. Housmans, *Am J Physiol* 259:H827-H834 (1990), with permission of the American Physiological Society.)

TPF and RTH, *i.e.*, an abbreviation of isometric contraction and relaxation. These effects are separate from those that exist as a consequence of the concomitant reduction in twitch amplitude. A reduction in RTH and therefore accelerated decline in force during isometric relaxation could result from either (or a combination of) 1) an abbreviation of the intracellular Ca^{2+} transient; 2) a decreased apparent affinity of troponin C for Ca^{2+} in its low-affinity "Ca^{2+}-specific" site II, specifically an increase in k_{off}; and 3) an increase in cross-bridge turnover, which would result in a shorter average cross-bridge cycle duration.

It is unlikely that volatile anesthetics cause the Ca^{2+} transient to fall more rapidly than in control conditions since the time course of the Ca^{2+}-induced luminescence in aequorin-injected cat[5] and ferret[6] papillary muscle was not abbreviated. Furthermore, halothane, enflurane and isoflurane have been reported to increase the permeability of the sarcoplasmic reticulum (SR), to reduce the amount of releasable Ca^{2+} from the SR, and to decrease the total Ca^{2+} content of cardiomyocytes by 10-70%.[7-10] The effects of volatile anesthetics on Ca^{2+} release from the SR is quantitatively much more important than the stimulation of SR Ca^{2+} uptake,[11] so that it is not surprising that the time course of the intracellular Ca^{2+} transient is hardly changed.

On the basis of currently available data, it is difficult to distinguish between effects of volatile anesthetics on the Ca^{2+} affinity of cardiac troponin C and on kinetics of actomyosin cross-bridge interactions. Halothane did not change the Ca^{2+} affinity of bovine cardiac troponin C over a pCa range of 9 to 3,[12] yet it remains to be established whether this is the case in myofilaments in their native environment as well. In rabbit papillary muscle in Ba^{2+} contracture, halothane, enflurane, and isoflurane did not change actin-myosin kinetics, yet the total number of cross-bridges was decreased.[13] A change in apparent Ca^{2+} sensitivity of the contractile apparatus depends, among other factors, on the temporal relationship between the time a cross-bridge consumes to complete a cycle and the time Ca^{2+} stays bound to troponin C. If the cycle rate is decreased, a smaller fraction of troponin C Ca^{2+}-binding sites must be occupied to keep a given fraction of cross-bridges active.[14] The acceleration of isometric relaxation

201

in intact[2,3] and skinned[15] cardiac fibres can therefore be accounted for by a decreased affinity of troponin C for Ca^{2+} and/or a shorter average cross-bridge life cycle.

In order to assess the relative importance of the proposed two main mechanisms of anesthetic-induced negative inotropy, *i.e.*, a reduction in intracellular Ca^{2+} availability and a decrease in myofibrillar responsiveness, the intracellular Ca^{2+} transients of isometric twitch contractions were examined before and during exposure to anesthetic. Right ventricular papillary muscles were microinjected with the Ca^{2+}-regulated protein aequorin.[16] Force and aequorin luminescence were recorded in control conditions and during exposure to anesthetic. In the presence of anesthetic, extracellular $[Ca^{2+}]$ was rapidly raised so that steady state peak force development in high $[Ca^{2+}]_o$ plus anesthetic was equal in amplitude to that in the pre-anesthetic control isometric twitch (figure 4, top). Peak aequorin luminescence was higher in anesthetic plus high $[Ca^{2+}]_o$ than that in control at equal force development. This observation suggests that the volatile anesthetic under consideration causes a desensitization of the contractile apparatus to Ca^{2+}. Results from such "Ca^{2+} back titration" experiments indicate that at concentrations of halothane, enflurane and isoflurane equal or greater than 0.5 MAC there is a decrease in myofibrillar Ca^{2+} responsiveness, an effect that is most pronounced for isoflurane.

In order to provide for a quantitative measure of the possible importance of this mechanism, we derived the presumed time course of Ca^{2+} occupancy of troponin C, of Ca^{2+} occupancy of calmodulin, of Ca^{2+} uptake from the SR, and of Ca^{2+} release from the SR with a computational approach similar to that of Baylor *et al.*[17] This approach utilized published values of cytoplasmic concentrations of troponin, calmodulin, of rate constants of Ca^{2+}-troponin and Ca^{2+}-calmodulin association and dissociation and of SR Ca^{2+} uptake.[18] Myoplasmic Ca^{2+} delivery was equated to Ca^{2+} release from the SR in this model, since transsarcolemmal Ca^{2+} entry contributes to less than 10% of myoplasmic Ca^{2+} delivery.[19] The model accounts for an exponential dependence of force on the Ca^{2+} occupancy of troponin C. The model solves a series of differential equations that relate free myoplasmic $[Ca^{2+}]$, Ca^{2+} occupancy of

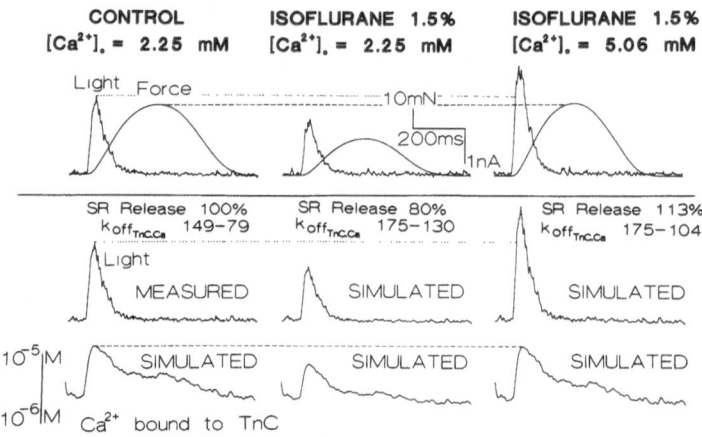

FIGURE 4. Ca^{2+} back titration experiment in aequorin-injected right ventricular papillary muscle. One hundred twenty-eight twitches were averaged to improve the signal-to-noise ratio in the aequorin luminescence signals. Top panel shows the experimental records of force and of aequorin luminescence. Bottom panel shows traces of aequorin luminescence obtained by computational simulation except where marked "measured," and simulated traces of Ca^{2+} occupancy of troponin C. Variables in the simulation were: [troponin] 70 μM, [calmodulin] 24 μM, Ca^{2+}-troponin k_{on} 3.9×10^7, Ca^{2+}-troponin k_{off} 80 in diastole, Ca^{2+}-calmodulin k_{on} 10^8, Ca^{2+}-calmodulin k_{off} 238.

troponin C, Ca^{2+} occupancy of calmodulin, SR Ca^{2+} uptake, and SR Ca^{2+} release for a given set of rate constants and concentrations, and allows one to predict the amplitude and time course of the aequorin luminescence signal that would be obtained if one or more of these variables were assigned a different value. With this approach, starting from experimentally measured force and aequorin luminescence signals, the computational model predicts by least squares fitting what values have to be assigned to SR Ca^{2+} release (myoplasmic Ca^{2+} delivery) and Ca^{2+}-troponin C k_{off} to simulate the aequorin luminescence signal in the presence of anesthetic in elevated $[Ca^{2+}]_o$ at equal developed force as the control. The assumption implicit to this type of analysis is that peak Ca^{2+} occupancy of troponin C is the same in both twitch contractions at equal peak force, so that the myofibrillar Ca^{2+} sensitivity can be assessed from the relationship between myoplasmic free $[Ca^{2+}]_o$ (detected with aequorin) and Ca^{2+} occupancy of troponin C. If the anesthetics under consideration significantly alter mechanisms "downstream" from the binding of Ca^{2+} to troponin C and therefore modify the relationship between Ca^{2+} occupancy of troponin C and force, this approach would not be a valid one. Yet, there is no evidence that this occurs in physiological conditions. The possibility that volatile anesthetics alter mechanisms "downstream" from the Ca^{2+} binding to troponin C cannot be completely excluded, as high concentrations of volatile anesthetics decrease maximal Ca^{2+}-activated force in rat skinned cardiac fibers by 5 to 10%.[5] The simulation provides for a unique solution of values of SR release and Ca^{2+}-troponin C k_{off} that simultaneously satisfies the condition 1) of a best fit (by least squares criterion) to the observed aequorin luminescence in anesthetic plus high $[Ca^{2+}]_o$ (figure 4, right) and 2) of equal Ca^{2+} occupancy of troponin C as in the control (figure 4, bottom). In a further step, the myoplasmic Ca^{2+} delivery is calculated for the experimentally measured aequorin luminescence in the presence of anesthetic in control $[Ca^{2+}]_o$ (2.25 mM) (figure 4, middle); SR release is determined to provide the best least squares fit of the simulated aequorin luminescence to the experimentally measured aequorin luminescence. In this step, Ca^{2+}-troponin C k_{off} is maintained at the value found originally, as anesthetic is present in both conditions. This type of analysis permits a quantitative evaluation of the changes in myoplasmic Ca^{2+} delivery and in myofibrillar Ca^{2+} responsiveness that result from anesthetic action on intact, living cardiac muscle. Results from such analyses indicate that at 1 MAC anesthetic, the intracellular Ca^{2+} availability is decreased by 20 to 40%, depending on the anesthetic. Myofibrillar Ca^{2+} responsiveness, expressed as Ca^{2+}-troponin C k_{off} in this type of analysis, is decreased by 5 to 15% at 1 MAC of anesthetic. These preliminary results and those of previous studies in intact papillary muscle[4] and in skinned cardiac fibers[15] are consistent with the hypothesis that the negative inotropic effects of halothane, enflurane, and isoflurane are mostly a consequence of a reduction of intracellular Ca^{2+} availability and that anesthetic-induced decreases in myofibrillar Ca^{2+} responsiveness play only a minor role. This field of investigation may benefit from analytical computational modeling which also incorporates variables of transsarcolemmal Ca^{2+} entry, Na^+/Ca^{2+} exchange and other contractile regulatory mechanisms in cardiac muscle.

ACKNOWLEDGEMENTS

This work was supported in part by National Institutes of Health grant GM36365.

REFERENCES

1. B. F. Rusy and H. Komai, Anesthetic depression of myocardial contractility: A review of possible mechanisms, *Anesthesiology* 67:745-766 (1987).

2. P. R. Housmans and I. Murat, Comparative effects of halothane, enflurane, and isoflurane at equipotent anesthetic concentrations on isolated ventricular myocardium of the ferret: I. Contractility, *Anesthesiology* 69:451-463 (1988).

3. P. R. Housmans and I. Murat, Comparative effects of halothane, enflurane, and isoflurane at equipotent anesthetic concentrations on isolated ventricular myocardium of the ferret: II. Relaxation, *Anesthesiology* 69:464-471 (1988).

4. P. R. Housmans, Negative inotropy of halogenated anesthetics in ferret ventricular myocardium, *Am J Physiol* 259:H827-H834 (1990).

5. Z. J. Bosnjak and J. P. Kampine, Effects of halothane on transmembrane potentials, Ca^{2+} transients, and papillary muscle tension in the cat, *Am J Physiol* 251:H374-H381 (1986).

6. P. R. Housmans, L. A. Wanek and E. G. Carton, Halothane, enflurane and isoflurane decrease myofibrillar Ca^{2+} responsiveness in intact mammalian ventricular muscle, *Biophys J* 57:554a (1990).

7. M. Katsuoka, K. Kobayashi and S. T. Ohnishi, Volatile anesthetics decrease calcium content of isolated myocytes, *Anesthesiology* 70:954-960 (1989).

8. D. M. Wheeler, R. T. Rice, R. G. Hansford and E. G. Lakatta, The effect of halothane on the free intracellular calcium concentration of isolated rat heart cells, *Anesthesiology* 69:578-583 (1988).

9. C. Lynch III and M. J. Frazer, Depressant effects of volatile anesthetics upon rat and amphibian ventricular myocardium: Insights into anesthetic mechanisms of action, *Anesthesiology* 70:511-522 (1989).

10. J. S. Herland, F. J. Julian and D. G. Stephenson, Halothane increases Ca^{2+} efflux via Ca^{2+} channels of sarcoplasmic reticulum in chemically skinned rat myocardium, *J Physiol* (London) 426:1-18 (1990).

11. T. E. Nelson and T. Sweo, Ca^{2+} uptake and Ca^{2+} release by skeletal muscle sarcoplasmic reticulum: Differing sensitivity to inhalational anesthetics, *Anesthesiology* 69:571-577 (1988).

12. E. S. Casella and T. J. J. Blanck, The effect of halothane on the binding of calcium by cardiac troponin C, *Biophys J* 53:583a (1988).

13. T. Shibata, T. J. J. Blanck, K. Sagawa and W. Hunter, The effect of halothane, enflurane, and isoflurane on the dynamic stiffness of rabbit papillary muscle, *Anesthesiology* 70:496-702 (1989).

14. P. W. Brandt, R. N. Cox, M. Kawai and T. Robinson, Regulation of tension in skinned muscle fibres: Effect of cross-bridge kinetics on apparent Ca^{2+} sensitivity, *J Gen Physiol* 79:997-1016 (1982).

15. I. Murat, R. Ventura-Clapier and G. Vassort, Halothane, enflurane, and isoflurane decrease calcium sensitivity and maximal force in detergent-treated rat cardiac fibers, *Anesthesiology* 69:892-899 (1988).

16. J. R. Blinks, P. H. Mattingly, B. R. Jewell, M. van Leeuwen, G. C. Harrer and D. G. Allen, Practical aspects of the use of aequorin as a calcium indicator: assay, preparation, microinjection, and interpretation of signals, *Meth Enzymol* 57:292-328 (1978).

17. S. M. Baylor, W. K. Chandler and M. W. Marshall, Sarcoplasmic reticulum calcium release in frog skeletal muscle fibres estimated from arsenazo III calcium transients, *J Physiol* (Lond) 344:625-666 (1983).

18. A. Fabiato, Calcium release in skinned cardiac cells: Variations with species, tissues, and development, *Fed Proc* 41:2238-2244 (1982).

19. W. G. Wier, Cytoplasmic [Ca^{2+}] in mammalian ventricle: Dynamic control by cellular processes, *Annu Rev Physiol* 52:467-485 (1990).

Part III — Smooth Muscle

19

Anesthetic Effects on Vascular Smooth Muscle

Sheila M. Muldoon

INTRODUCTION

For many years it was generally believed that volatile anesthetic agents, with the exception of diethyl ether and cyclopropane, had a depressant effect on the vascular system. Because the physiological effects of anesthetic agents were measured in man and whole animal, the precise site or sites of action could not be determined. In 1962 Price and Price,[1] using isolated tissue and standardized experimental conditions, reported that the effect of halothane was primarily at the adrenergic receptor on the vascular smooth muscle (VSM) cell. Other studies[2] concluded that although the response to norepinephrine (NE) in VSM was depressed by halothane, antagonism between NE and halothane was difficult to explain in terms of classical pharmacological kinetics.

In the 1960's and early 1970's, understanding of the neural control of blood vessels and the role of endogenous catecholamines in that control developed rapidly. A number of studies were designed to determine how anesthetic agents alter sympathetic activation of VSM. Muldoon et al.[3] used the canine saphenous vein as a model to show that halothane depressed the response to sympathetic nerve stimulation. Lunn and Rorie[4] showed that halothane inhibited NE release and that the decrease of NE release was prevented when the muscarinic receptors were blocked, suggesting an action by halothane on presynaptic muscarinic receptors.[5] This latter observation was consistent with studies indicating that the depression of ganglionic transmission by halothane was also mediated by muscarinic receptors.[6]

Anesthetic actions appear to be specific for both the type of anesthetic and the vascular bed. For example, in the canine pulmonary artery it has been shown that nitrous oxide significantly increases NE release.[7] In other studies using the anterior tibial artery, isoflurane was reported to decrease NE release more than in the canine saphenous vein.[8]

In 1987 Larach and colleagues[9] examined the interaction between halothane and the α_1 and α_2 adrenoreceptors which exist on vascular smooth muscle. These post-synaptic receptors produce vasoconstriction when stimulated by agonists, although each acts through different activation mechanisms. In the canine saphenous vein, halothane significantly attenuated α_2 but not α_1 adrenoreceptor responsiveness. The impairment of the α_2-mediated constriction implies that halothane interferes with α_2

SHEILA M. MULDOON, Depatment of Anesthesiology, Uniformed Services University of the Health Sciences, 4301 Jones Bridge Road, Bethesda, Maryland 20814-4799.

Mechanisms of Anesthetic Action in Skeletal, Cardiac, and Smooth Muscle
Edited by T.J.J. Blanck and D.M. Wheeler, Plenum Press, New York, 1991

excitation-contraction coupling, which is thought to be dependent upon the influx of extracellular Ca^{2+} into the VSM cell. The observation of unimpaired α_1-mediated constriction implies that α_1-related mechanisms such as intracellular calcium release remain intact in the presence of halothane.

The recognition of the importance of the endothelium in the regulation of vascular function has increased greatly since 1980 when Furchgott[10] reported the obligatory role of the endothelial cells in the relaxation of arterial smooth muscle. Endothelium-derived relaxing factors (EDRF) have been reported to be continuously released under basal conditions; and when the endothelium is activated, additional EDRF release and vasodilatation occurs.[11] The production of EDRF requires an increase in intracellular Ca^{2+}. Following its production and release, EDRF acts on VSM cells to activate soluble guanylate cyclase and increase cyclic GMP concentration. Recently, the discovery was made that one type of EDRF is nitric oxide (NO).[12] NO is synthesized from the amino acid L-arginine.[13] An arginine analogue, N-monomethyl-L-arginine (L-NMMA) has been shown to inhibit the synthesis of NO. This analogue has been shown to increase mean arterial pressure in both animals[14] and man,[15] further establishing the importance of EDRF activity in maintaining vascular tone. While EDRF is of interest to anesthesiologists because of its role in basic cardiovascular physiology and pathophysiology, it has also been shown to have a direct interaction with volatile anesthetic agents. Muldoon et al.[16] reported that halothane decreased EDRF-induced relaxation in a number of blood vessels. Stone and Johns[17] reported that volatile anesthetics cause vasoconstriction through inhibition of basal EDRF production. Blaise et al.[18] reported that in canine coronay arteries isoflurane attenuated contractions induced by prostaglandin $F_2\alpha$, phenylephrine and serotonin (5HT) and that this effect was dependent on an intact endothelium.

CURRENT RESULTS

Eight papers were presented at this symposium which examined the mechanism of action of anesthetics on VSM. These studies were conducted on cells grown in culture and on preparations of coronary and systemic blood vessels. The anesthetic effects were examined under basal and stimulated conditions using several different agonists to stimulate vascular activity. Collectively, these papers show that there are several mechanisms and sites of action, such as the vascular endothelium, involved in the effect of anesthetics on VSM. Several papers discussed the effects of EDRF on vascular smooth muscle. Studies on the direct actions of anesthetics on signal transduction pathways in vascular smooth muscle were also presented.

Roger Johns used a multifaceted approach to determine the relationship between EDRF, its action on vascular smooth muscle tension, and cyclic GMP production (Chapter 20). He demonstrated that, in phenylephrine-contracted pulmonary artery rings, endothelium-dependent vasodilation correlates with an increase in vascular smooth muscle cyclic GMP concentrations. Endothelium-dependent dilators such as bradykinin, mellitin and arachidonic acid did not cause relaxation of endothelium-denuded bioassay rings unless the agonists were first passed through a column containing endothelial cells grown on beads. Endothelium-dependent dilators had no significant effect on cyclic GMP concentration in either endothelial cells or vascular smooth muscle cultured cells alone, but caused a significant increase in cyclic GMP when the two types of cells were co-cultured and exposed to the vasodilators. A selective inhibitor of EDRF production, L-NMMA, reversed endothelium-dependent relaxations and this inhibition could be overcome by an excess of L-arginine. Another analog, nitro-L-arginine, completely inhibited cyclic GMP accumulation in co-cultures stimulated by the endothelium-dependent dilators. The calmodulin antagonist trifluoperazine inhibited bradykinin- and ATP-induced cyclic GMP accumulation in co-cultures, demonstrating the importance of calmodulin in EDRF synthesis. Production and release of EDRF stimulated by endothelium-dependent dilators was inhibited by hypoxia in rabbit pulmonary artery rings, while the action of endothelium-independent dilators was not altered by hypoxia.

Gilbert Blaise also presented data on the interaction between volatile anesthetic agents and EDRF (Chapter 21). He reported that halothane (2.5%) had a biphasic effect on rabbit aortic rings contracted with submaximal doses of phenylephrine. Halothane caused an initial increase in tension which was followed by a slow, continuous relaxation. These changes in tension were not endothelium-dependent. In other experiments designed to further study the interaction between EDRF and volatile anesthetics, the effect of halothane on NO-induced relaxation in the rabbit aorta ring was examined. Halothane significantly decreased the relaxation induced by NO. This experiment supports earlier EDRF studies and demonstrates that volatile anesthetics are not acting on EDRF synthesis. Separate experiments on isolated rabbit hearts perfused at a constant pressure showed that acetylcholine (ACh) increased coronary flow and that the increase was endothelium-dependent. Halothane or isoflurane at 0.35 MAC did not alter the effect of ACh on coronary flow, but at anesthetic concentrations of 0.7 MAC the ACh-induced flow was increased. From this study plus the results from large vessel rings, the author concluded that halothane has a different action on endothelial responses in large arteries and small arterioles.

Flynn and colleagues examined the effect of anesthetics on canine cerebral arteries *in vitro* (Chapter 22). It is well documented that the volatile anesthetics halothane and isoflurane cause cerebral vasodilation and that isoflurane causes less cerebral vasodilation than halothane. However, the mechanism by which volatile anesthetics exert their action on cerebral vascular smooth muscle is unclear. Possible modes of action suggested by Flynn and colleagues include the following: volatile anesthetics may act on the endothelium, stimulating EDRF release or inhibiting production of endothelium-dependent constricting factor (EDCF), or the anesthetics may act directly on vascular smooth muscle at the cellular level by affecting calcium transport. To distinguish between these possibilities, canine middle cerebral artery rings were prepared and suspended in tissue baths containing aerated Krebs-Ringer solution and maintained at 37°C. The results confirm that halothane is a more potent vasodilator of isolated cerebral vessels than isoflurane. Isoflurane's action on cerebral arteries was not endothelium-mediated, and the release of EDRF was unaffected by isoflurane or halothane. To further examine the mechanism of action of volatile anesthetics on cerebral arteries, the authors performed patch clamp experiments on isolated cerebral arterial muscle cells. In these cells isoflurane significantly depressed Ca^{2+} channel currents, demonstrating that these agents have a direct action on VSM.

Vascular endothelial cells have been shown to produce reactive oxygen intermediates (ROIs), and these are important in several physiological and pathological processes. William Freas and colleagues used isolated canine saphenous vein rings as a model of sympathetically innervated vascular smooth muscle to study the interaction between ROIs and anesthetic agents (Chapter 23). ROIs were produced by administration of a photosensitizer, hematoporphyrin derivative (HpD), and application of laser light. HpD pre-treatment and application of laser light consistently induced a contraction in saphenous vein preparations. These contractions have been shown to be ROI-dependent.[19] While ROI-induced contractions are endothelium-independent, ROI generation attenuates EDRF release. When halothane (1-3%) was added to the aerating gas mixture, the contractions induced by HpD and laser were attenuated in a dose-dependent manner. In contrast, isoflurane did not attenuate the contractile response. In an effort to determine the mechanism of halothane's action on these ROI-induced contractions, studies were conducted in the presence and absence of calcium. The results suggest that ROIs increase calcium permeability in vascular smooth muscle cells and that halothane attenuates this increase in permeability.

Christopher Sill reported on the site of anesthetic action on coronary vasculature (Chapter 24). Because anesthetic agents could exert their effect at numerous sites in VSM signal transduction pathways, four independent methodologies were used. Quantitative angiography was used to measure coronary artery diameters and their response to serotonin in the presence of halothane (1%) and isoflurane (1%). In addition to his whole-animal studies, isolated coronary arteries were used to quantify the effect of these anesthetics on changes in tension induced by a variety of agonists. Cultured VSM and endothelial cells were used to measure changes in intracellular Ca^{2+}. Inositol phosphate concentration measurements were made using both cultured and isolated segments of coronary arteries.

The results of the whole animal experiments indicate that halothane and isoflurane dilate coronary vessels preconstricted with serotonin. Pretreatment of these animals with either halothane or isoflurane was also effective in attenuating serotonin-induced constriction. These effects appear to be endothelium-independent.

The effect of isoflurane and halothane on contractions induced by agonists linked to phospholipase C, inositol phosphate formation and Ca^{2+} mobilization in coronary artery rings was studied *in vitro*. Isoflurane and halothane (1-2%) attenuated the contractile responses evoked by serotonin, acetylcholine, and endothelin. However, responses evoked by histamine were not attenuated. Agonists such as serotonin, acetylcholine, histamine and endothelin interact with receptors that stimulate a G protein to activate phospholipase C and hydrolyze membrane phosphoinositide lipids to form inositol phosphate and diacylglycerol (DAG). DAG and Ca^{2+} stimulate protein kinase C (PKC) activity. PKC is a complex family of related intracellular enzymes which are involved in initiation and control of contraction in VSM. To determine if halothane and isoflurane altered PKC regulation of vascular smooth muscle, Sill and colleagues used a phorbol ester analog (PDBU) to bypass the cell surface receptors and stimulate PKC directly. The data suggest that the site of action of volatile anesthetics is not on PKC, and does not involve steps in the signal transduction pathway distal to this enzyme.

Since in the coronary vasculature halothane and isoflurane decreased the response to serotonin, which is one of the agonists dependent on PI activation, Sill examined the effect of these anesthetics on inositol phosphate formation. In both coronary rings and cultured aortic VSM cells, isoflurane and halothane inhibited the inositol phosphate formation stimulated by serotonin and acetylcholine.

The action of anesthetic agents on intracellular calcium was also examined. Contractile agonists such as endothelin, ATP and vasopressin increase cytosolic Ca^{2+} both by promoting release of Ca^{2+} from sarcoplasmic reticulum, and by signalling Ca^{2+} entry *via* channels in the cell surface membrane. The fluorescent, Ca^{2+}-sensitive probe indo-1 was used and the results indicate that the increases in Ca^{2+} evoked by endothelin, ATP and vasopressin were depressed in the presence of 1.25% and 2% halothane. Isoflurane was less effective and less consistent in depressing increases in cytosolic Ca^{2+} than halothane.

In summary, Sill's studies showed that halothane and isoflurane depressed agonist-induced inositol phosphate formation and agonist-induced increases in intracellular Ca^{2+}. Neither anesthetic had a marked effect on contractions evoked by activation of PKC. Since the contractile response to serotonin was markedly depressed but the response to histamine was not, Sill concluded that volatile anesthetics may act at the proximal part of the signal transduction pathway.

Audrey Robinson-White presented a study directed at examining the effect of barbiturates, lidocaine, and halothane on signal transduction pathways in endothelial cells and in a rat basophilic leukemia cell line (Chapter 25). Receptor activation by angiotensin II (AII) and DNP_{24}-BSA stimulates the hydrolysis of inositol phospholipids, which is mediated by a GTP-binding protein (G-protein). This hydrolysis results in the production of two intracellular signals (see figure 1 of Chapter 25). Hydrolysis of phosphatidylinositol 4,5-biphosphate to inositol 1,4,5-triphosphate causes the transient release of Ca^{2+} from intracellular stores and the production of diacylglycerol (DAG). These signals act synergistically to activate PKC, and subsequently evoke a cell response. The studies by Robinson-White indicate that barbiturates inhibit the release of inositol phosphates, in a highly specific manner, possibly by acting on a G-protein (G_p), or on the G-protein-phospholipase C complex. Barbiturates, however, do not alter levels of membrane inositol phospholipids. In contrast, the local anesthetic lidocaine, altered both inositol phosphate release and membrane inositol lipid levels. Lipid levels were altered, whether in the presence or absence of the agonist of inositol phospholipid hydrolysis. Studies with halothane showed an effect on both inositol phosphates and inositol lipids; however, no effect of halothane was observed in the absence of the agonist. The results suggested three separate mechanisms of action for the three classes of anesthetics on this signal transduction pathway.

David Larach and colleagues used isolated, perfused, arrested rat hearts (Langendorff preparation) to examine the effects of halothane and isoflurane on coronary vascular activity (Chapter 26). The isolated hearts were subjected to retrograde aortic perfusion with modified Krebs-Hensleit solution at constant pressure. Each heart was arrested with tetrodotoxin and the left ventricular cavity drained to minimize preload and intra-myocardial coronary compression. The effect of halothane and isoflurane at concentrations between 0.5 and 2 MAC on coronary blood flow was measured at steady state. The results indicate that halothane and isoflurane are equipotent in vasodilating coronary resistance vessels in the isolated, perfused, arrested heart. Coronary blood flow approximately doubled with either agent at 2 MAC. Administrationtion of adenosine (5×10^{-5} M) after anesthetic recovery generated higher coronary flows than did either halothane or isoflurane. However, the increase in coronary flow with adenosine was only slightly greater than with the anesthetic, which suggested that both agents reduce coronary reserve in a dose-dependent manner. In contrast to the results with halothane and isoflurane the authors cite other data which suggests that sevoflurane appears to preserve coronary flow reserve. Preservation of flow reserve by an anesthetic may afford some advantages in a clinical setting but this possibility needs further study.

Robert Merin and colleagues used conscious, chronically-instrumented animals to compare the effects of isoflurane, sevoflurane and the isoflurane analog, desflurane, on coronary blood flow (Chapter 27). He used healthy mongrel dogs, instrumented to measure cardiac output, aortic and left atrial pressures, left ventricular pressure, dp/dt and coronary blood flow. All three anesthetics provided significant increases in coronary blood flow when compared to awake controls. All anesthetics produced dose-related decreases in mean arterial pressure and hence coronary perfusion pressure. There was no significant difference between isoflurane, sevoflurane and desflurane on coronary blood flow or coronary vascular resistance. These effects are proposed to be the result of a direct action of the anesthetics on coronary VSM. Future studies including measurements of myocardial oxygen consumption will provide important information as to whether these anesthetic agents act directly on coronary VSM or act by an indirect effect on myocardial performance.

In summary, in this symposium we have learned that there are several sites of action of anesthetics on vascular smooth muscle. The response of each agent depends on type of vascular bed, vascular activity, type of agonist activation, and concentration of anesthetic investigated. There has yet to be a unifying concept which is capable of clarifying these various results. In most of the relevant preparations, volatile anesthetics appear to interfere with EDRF action. While this action probably does not occur at the level of EDRF synthesis, it is not determined whether EDRF release, transit or effect is compromised. In any case, the result should be a tendency toward vasoconstriction. There also appears to be direct action at the VSM cell by volatile anesthetics, such as inhibition of Ca^{2+} influx. Furthermore, a variety of anesthetics have been found to inhibit inositol phospholipid hydrolysis, an effect likely to result in inhibition of contraction. Since these direct and EDRF-related actions are essentially antagonistic, it perhaps is not surprising that results do vary among different experimental preparations. In addition to defining more precisely the site and nature of anesthetic action in the vasculature, the challenge for the future is to determine if anesthetic action in human blood vessels correlates with observations such as those reported at this symposium.

REFERENCES

1. M. L. Price and H. L. Price, Effects of general anesthetics on contractile responses in rabbit aortic strips, *Anesthesiology* 23:16-20 (1962).
2. S. C. Clark and K. L. MacCannell, Vascular responses to anesthetic agents, *Can Anaesth Soc J* 22:20-33 (1975).
3. S. M. Muldoon, P. A. Vanhoutte, R. R. Lorenz and R. A. Van Dyke, Venomotor changes caused by halothane acting on the sympathetic nerves, *Anesthesiology* 43:41-48 (1975).
4. J. J. Lunn and D. K. Rorie, Halothane-induced changes on the release and disposition of norepinephrine at adrenergic nerve endings in the dog saphenous vein, *Anesthesiology* 61:377 (1984).
5. D. K. Rorie, G. M. Tyce and R. A. MacKenzie, Evidence that halothane inhibits norepinephrine release from sympathetic nerve endings in dog saphenous vein by stimulation of presynaptic inhibitory muscarinic receptors, *Anesth Analg* 63:1059-1064 (1984).
6. Z. J. Bosnjak, J. L. Seagard, A. Wu and J. P. Kampine, The effects of halothane on sympathetic ganglionic transmission, *Anesthesiology* 57:473-479 (1982).
7. D. K. Rorie and G. M. Tyce, Effects of hypoxia on norepinephrine release and metabolism in dog pulmonary artery, *J Appl Physiol* 55:750-758 (1983).
8. L. W. Edwards, W. Freas, J. McGehee and S. M. Muldoon, Mechanism of vasomotor changes with isoflurane, *Anesth Analg* 67:S56 (1988).
9. D. R. Larach, H. G. Schuler, J. A. Derr, M. G. Larach, F. A. Hensley Jr. and R. Zelis, Halothane selectively attenuates α_2-adrenoceptor mediated vasoconstriction, *in vivo* and *in vitro*, *Anesthesiology* 66:781-791 (1987).
10. R. F. Furchgott and J. V. Zawadzki, The obligatory role of endothelial cells in the relaxation of arterial smooth muscle by acetylcholine, *Nature* 288:373-376 (1980).
11. P. M. Vanhoutte, G. M. Rubanyi, V. M. Miller and D. S. Houston, Modulation of vascular smooth muscle contraction by the endothelium, *Annu Rev Physiol* 48:307-320 (1986).
12. R. M. J. Palmer, A. G. Ferrige and S. Moncada, Nitric oxide accounts for the biological activity of endothelium-derived relaxing factor, *Nature* 327:524-526 (1987).
13. R. M. J. Palmer, D. S. Ashton and S. Moncada, Vascular endothelial cells synthesize nitric oxide from L-arginine, *Nature* 333:644-666 (1988).
14. D. D. Rees, R. M. J. Palmer and S. Moncada, Role of endothelium-derived nitric oxide in the regulation of blood pressure, *Proc Natl Acad Sci* (USA) 86:3375-3378 (1989).
15. P. Vallance, J. Collier and S. Moncada, Effects of endothelium-derived nitric oxide on peripheral arteriolar tone in man, *Lancet* 997-1000 (1989).
16. S. M. Muldoon, J. L. Hart, K. A. Bowen and W. Freas, Attenuation of endothelium-mediated vasodilation by halothane, *Anesthesiology* 68:31-37 (1988).
17. D. J. Stone and R. A. Johns, Endothelium dependent effects of halothane enflurane and isoflurane on isolated rat aortic vascular rings, *Anesthesiology* 71:126-132 (1989).

18. G. Blaise, J. G. Sill, M. Nugent, R. A. Van Dyke and P. M. Vanhoutte, Isoflurane causes endothelium-dependent inhibition of contractile responses of canine coronary arteries, *Anesthesiology* 67:513-517 (1987).
19. W. Freas, J. L. Hart, D. Golightly, H. McClure, D. R. Rodgers and S. M. Muldoon, Vascular interactions of calcium and reactive oxygen intermediates produced following photoradiation, *J Cardiovasc Pharmacol* 17:27-35 (1991).

20

Endothelium-Derived Relaxing Factor (EDRF)
Production from L-Arginine

Roger A. Johns, Appavoo Rengasamy

INTRODUCTION

Recent investigations have greatly improved our understanding of the chemical nature of endothelium-derived relaxing factor (EDRF) and the regulation and metabolic pathway of its production. EDRF is a potent but labile relaxing factor with a biologic half-life of between 6.3 and 50 seconds in an oxygenated aqueous medium.[1,2] The production of EDRF from the endothelium requires an increase in intracellular calcium.[3-6] Following its production and release from the endothelial cell, EDRF is transferred to the vascular smooth muscle (VSM) where it activates soluble guanylate cyclase resulting in an increase in smooth muscle cyclic GMP concentration, which correlates with its relaxing action.[7-13] Extremes of both high and low oxygen tension inhibit the production or stability of EDRF.[14-15] Early investigations into the chemical nature of EDRF implicated an unstable, non-prostanoid oxidation product of arachidonic acid or some type of free radical.[1,16-18] A great deal of recent evidence, however, suggests that EDRF is nitric oxide or a similar nitrogen oxide species.[19-29] EDRF can be formed from L-arginine by a pathway involving a calcium-, calmodulin- and NADPH-dependent enzyme.[30-39] EDRF synthesis has now been described in a wide range of cell types in addition to the endothelium, and indeed EDRF may be the second messenger responsible for the activation of guanylate cyclase in most cells containing the enzyme.[34,35,38-43] This manuscript will present data from our laboratory which support these and other pharmacologic characteristics of EDRF.

MATERIALS AND METHODS

Vascular Ring Preparation and Cell Culture Techniques

Male Sprague-Dawley rats or New Zealand White rabbits were killed by exsanguination following induction of anesthesia with ketamine in accordance with our institutional animal care committee guidelines. The thoracic aortae and/or first branch pulmonary arteries were carefully removed and placed in cold modified Kreb's buffer (NaCl 111 mM, KCl 5 mM, NaH_2PO_4 1 mM, $MgCl_2$ 0.5 mM, $NaHCO_3$ 25 mM, $CaCl_2$

ROGER A. JOHNS, APPAVOO RENGASAMY, Department of Anesthesiology, University of Virginia Health Sciences Center, Charlottesville, Virginia 22908.

Mechanisms of Anesthetic Action in Skeletal, Cardiac, and Smooth Muscle
Edited by T.J.J. Blanck and D.M. Wheeler, Plenum Press, New York, 1991

2.5 mM, and dextrose 11.1 mM). The vessels were then cleaned of fat and extraneous tissue and divided into 2.5 mm ring segments which were mounted for isometric tension measurements in 37°C water-jacketed tissue baths continuously gassed with 95% O_2/5% CO_2 unless otherwise indicated. In some vessels the endothelium was deliberately removed by rubbing with a wooden applicator stick or the tip of a watch maker's forceps. Rings were mounted at optimal resting tension as determined by preliminary length-tension experiments, and active tension was induced with a 50% maximum dose of phenylephrine. Prior to experimentation, all rings were tested for the integrity of the endothelium by observing the presence or absence of a relaxation response to methacholine (1×10^{-6} M).

Bovine pulmonary artery endothelial cells and rat aortic vascular smooth muscle cells were isolated, characterized and maintained in culture by methods previously described.[12,44-45]

Endothelial Cell Culture

Endothelial cells were placed into microcarrier culture by seeding 2×10^7 endothelial cells onto 0.6 g of Cytodex 3 (Pharmacia) microcarrier beads (4600 cm^2 of surface area per gram of beads) in 200 ml of M199 medium (GIBCO, Inc.) in a 2 L roller bottle containing 20% fetal calf serum. After a two-hour incubation period with occasional stirring to allow the cells to attach to the beads, the bottles were rolled at 1 RPM. The medium was changed every third day and the bottles gassed with 95% oxygen/5% CO_2.

Column Transfer Experiments

The apparatus used to transfer EDRF from cultured endothelial cells for bioassay by either isometric tension measurement in a denuded vascular ring or cyclic GMP accumulation in cultured VSM is shown in figure 1. A column for transfer of EDRF was prepared by placing endothelial cells grown on microcarrier beads (0.0325 g or 1×10^7 cells) into the barrel of a 20 ml syringe with 100 μm nylon mesh fixed in its tip to prevent the beads from washing out. The cell column was superfused with oxygenated (95% O_2/5% CO_2) Kreb's buffer (3 ml/min) maintained at 37°C using a peristaltic pump. The endothelial cells were stimulated by bolus injection of various drugs into the superfusate immediately proximal to the column, or the bioassay alone was stimulated by infusing drugs below the column of endothelial cells. EDRF released from the column was detected by allowing the effluent to drip: 1) onto an endothelium-denuded rabbit thoracic aortic ring contracted with phenylephrine (1×10^{-7} M) and measuring relaxation; or 2) into wells (2 cm^2) of cultured confluent VSM and measuring cyclic GMP accumulation. The bioassay ring was allowed to equilibrate at 2 g of resting tension for 1.5 hours before experimentation and active tension (60% of maximal) was induced by adding phenylephrine (1×10^{-7} M) to the buffer solution.

Co-Culture Experiments

After reaching confluency, endothelial cells on microcarrier beads were washed with serum-free medium and placed into monolayer culture wells (2 cm^2) of VSM. The number of each cell type was calculated to provide a 1:1 ratio of endothelial cells to smooth muscle cells in each co-culture. Co-cultures were allowed to incubate for four hours at which time the medium was changed to phosphate-buffered saline. Co-cultures were stimulated by a variety of endothelium-dependent (ATP, 1×10^{-4}M;

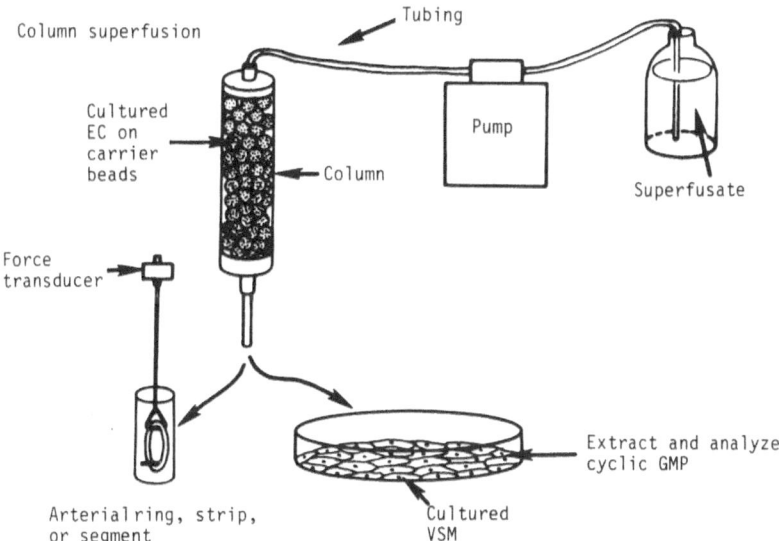

FIGURE 1. EDRF column transfer apparatus. Endothelial cells on cytodex 3 microcarrier beads were placed in a column which was superfused with Kreb's buffer, and the effluent was bioassayed for EDRF on an endothelium denuded vascular ring or by collecting the effluent into wells (2 cm²) of vascular smooth muscle in monolayer and subsequently assaying for cyclic GMP.

bradykinin, 1×10^{-6}M; melittin, 3 μg/ml; A23187, 1×10^{-6} M) or endothelium-independent (sodium nitroprusside, 1×10^{-6} M; isoproterenol, 1×10^{-7} M) vasodilators in the presence or absence of a variety of pharmacologic inhibitors or varying oxygen atmospheres as described below. Forty seconds following the addition of stimulants, the media was aspirated and 500 μl of 0.1 N HCl was added to the culture wells to extract cyclic GMP. Cyclic GMP concentrations were determined by radioimmunoassay as previously described.[44]

Cyclic GMP Extraction from Vascular Rings

The time and dose dependence of methacholine-stimulated, endothelium-dependent accumulation of cyclic GMP in rat aortic vascular rings was determined. Vessels were prepared and equilibrated as described above. Following precontraction with phenylephrine, the rings were stimulated with varying concentrations of methacholine for 30 seconds at which time they were flash frozen by immersion in acetone (cooled in dry ice) and stored at $-90°$C for subsequent cyclic GMP and protein determinations. In time course experiments, methacholine (1×10^{-6} M) was incubated at varying time intervals prior to freezing. Cyclic GMP was extracted by homogenizing the frozen rings in 2 ml iced 1 N HCl and centrifuging at $1000 \times g$ for 10 minutes. The supernatant was lyophilized and redissolved in 250 μl of 0.1 N HCl for cyclic GMP radioimmunoassay, as previously described.[44] The pellet was suspended in 2 ml of 2 N NaOH, incubated at 60°C until dissolved (approximately 1 hour), and assayed for protein by the method of Lowry.

Assay of EDRF Synthase

EDRF synthase was prepared and its activity assayed by modification of the methods of Bredt and Snyder.[35] Bovine cerebellum (10 g) was homogenized in 50 ml

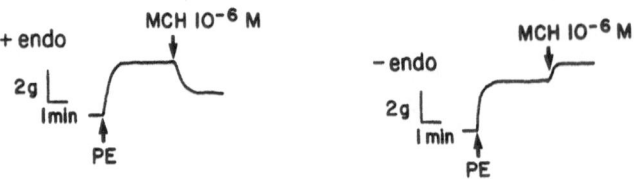

FIGURE 2. A typical isometric force tracing of the effect of methacholine (MCH) on phenylephrine (PE, 1×10^{-6} M) precontracted rabbit first branch pulmonary artery segments with and without an intact endothelium (+endo and −endo, respectively).

buffer containing 0.32 M sucrose, 20 mM HEPES, 0.5 mM EDTA, 1 mM dithiothreitol (DTT) and 5 mg phenylmethylsulfonyl fluoride (PMSF) at pH 7.4. The homogenate was centrifuged at 20,000 × g for 1 hour at 4°C and the supernatant was passed over a 4 ml Dowex AG50WX-8 (Na$^+$ form) column. The EDRF synthase assay mixture (1 ml) contained 850 μl homogenate, 2 mM NADPH, 0.45 mM CaCl$_2$ (1 μM free CaCl$_2$) and 200 μM [^{3}H]-L-arginine (1 μCi) and was incubated at 37°C. To stop the reaction, a 150 μl aliquot was mixed with 2 ml ice-cold 10 mM Hepes buffer, pH 5.5 and 2 mM EDTA. The mixture was applied to a 2 ml Dowex AG50WX-8 column and eluted with 2.5 ml distilled water. [^{3}H]-citrulline was determined by liquid scintillation spectrophotometry and used as a measure of EDRF production.

RESULTS

Figure 2 presents a typical isometric force tracing of a first branch rabbit pulmonary artery preconstricted with phenylephrine and stimulated by methacholine in the presence and absence of endothelium. When the endothelium was intact, methacholine caused a profound relaxation, whereas, in the absence of endothelium, methacholine had a direct vasoconstricting effect. Endothelium-dependent vasodilation correlates with an increase in vascular smooth muscle cyclic GMP. Figure 3A demonstrates the dose-dependence of methacholine-stimulated cyclic GMP accumulation in endothelium-intact rat thoracic aorta rings. Figure 3B presents the cyclic GMP time course in response to stimulation by methacholine (1×10^{-6} M), demonstrating a peak response at 10 s.

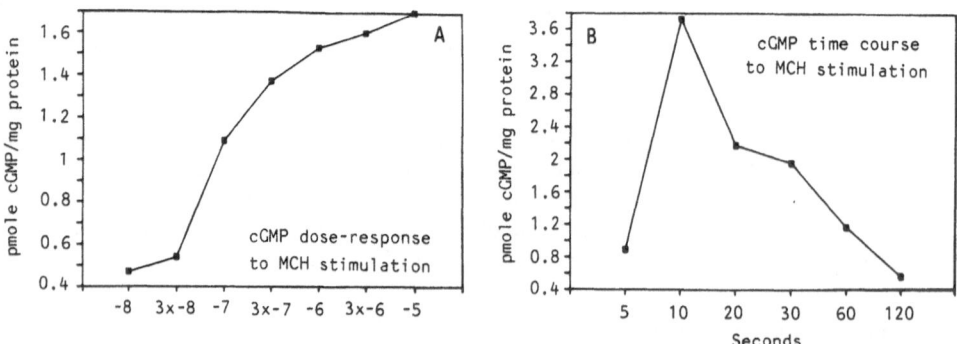

FIGURE 3. Cyclic GMP content of rat thoracic aortic rings stimulated with increasing concentrations of methacholine (MCH) (panel A) or with a fixed concentration of methacholine (1×10^{-6} M) for varying time periods (panel B).

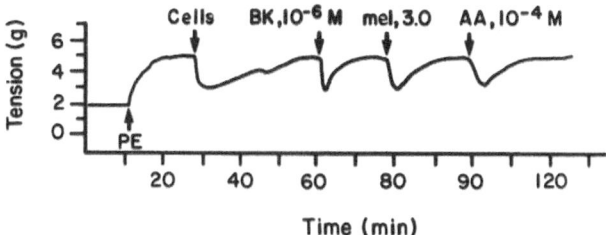

FIGURE 4. Response of rabbit aortic ring without endothelium to EDRF released from cultured bovine pulmonary artery endothelial cells. The bioassay ring was contracted with phenylephrine (1×10^{-7} M) in the superfusate. EDRF agonists were transiently infused (45 s) beginning at the arrows. BK = bradykinin (1×10^{-6} M), mel = melittin (3 μg/ml), AA = arachidonic acid (1×10^{-4} M).

Figure 4 shows the response of a rabbit aortic ring without endothelium to EDRF released from cultured bovine pulmonary endothelial cells. The addition of cells to the column caused relaxation of the bioassay ring which may relate to basal EDRF release. Bolus injections of the endothelium-dependent dilators bradykinin (10^{-6} M), melittin (3 μg/ml) and arachidonic acid (10^{-4} M) caused relaxation of the bioassay ring if they were injected above the endothelial cell column but not if injected below the endothelial cell column.

Co-cultures of endothelial cells on microcarrier beads and monolayers of VSM provide an excellent model for the study of EDRF. Figure 5 demonstrates that the endothelium-dependent dilators melittin (3 μg/ml) and bradykinin (1×10^{-5} M) had no significant effect on cyclic GMP concentration in endothelial or VSM alone but caused a significant increase in cyclic GMP when the two cell types are in co-culture. Sodium nitroprusside, an endothelium-independent vasodilator which also stimulates vascular

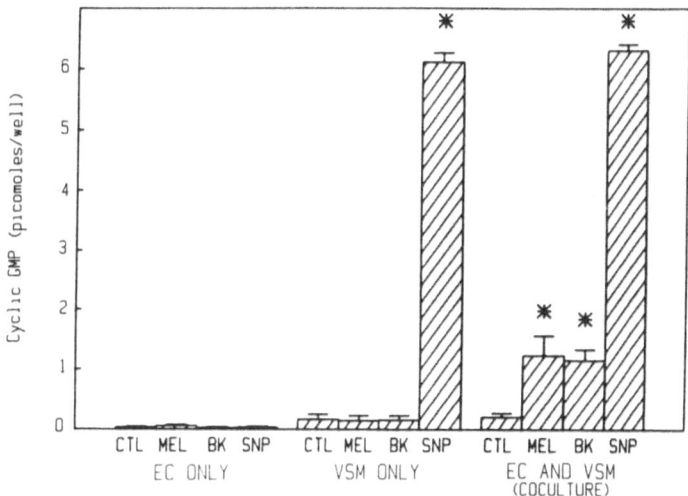

FIGURE 5. The effects of endothelium-dependent and independent dilators on cyclic GMP accumulation in endothelial cells (EC) or VSM alone and in co-cultures of EC and VSM. The cultures were stimulated with control buffer (CTL), melittin (MEL, 3 μg/ml), bradykinin (BK, 1×10^{-5} M), or sodium nitroprusside (SNP, 1×10^{-6} M). Asterisks (*) indicate significant differences from control ($P < 0.01$; n = 8).

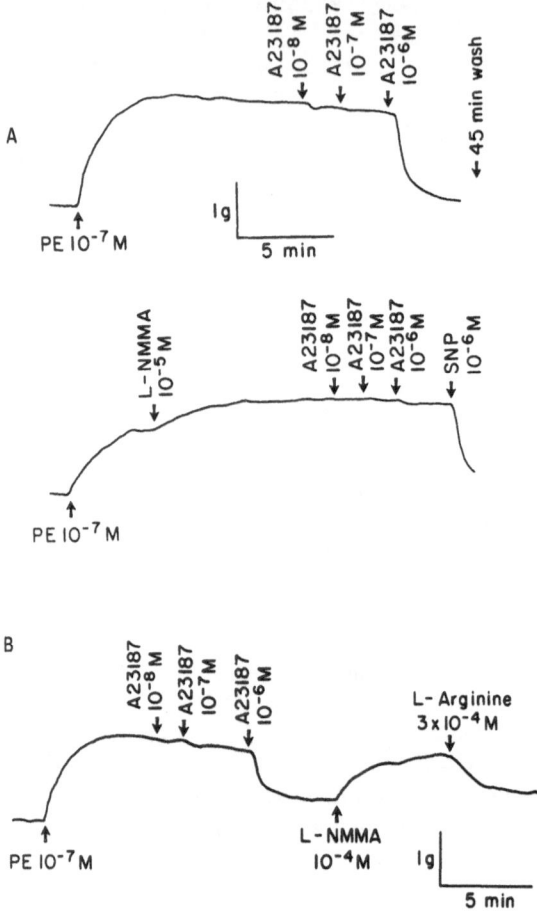

FIGURE 6. A: Upper trace represents endothelium-dependent relaxation of a phenylephrine (PE) precontracted rabbit aortic ring by the calcium ionophore A23187. Following washout, the ring was again contracted with PE (lower trace), and LNMMA (1×10^{-5} M) was added to the bath. A23187 no longer caused vasodilation, but the endothelium-independent dilator sodium nitroprusside (SNP) was still active. B: LNMMA also acutely reversed endothelium-dependent dilation by A23187. This inhibition was overcome by an excess of L-arginine.

smooth muscle cyclic GMP, causes a significant increase in cyclic GMP both in co-culture and in vascular smooth muscle cells alone.

Since the discovery of L-arginine as a precursor for EDRF, several analogs of L-arginine have been found to be specific and selective inhibitors of EDRF production. We examined the ability of one such analog, N^G-monomethyl-L-arginine (LNMMA), to inhibit EDRF production. Figure 6A presents an experiment in which endothelium-dependent relaxation of a rabbit thoracic aorta ring by the calcium ionophore A23187 was prevented by preincubation with LNMMA (1×10^{-4} M). LNMMA (figure 6B) also acutely reversed A23187-induced vascular relaxation. This inhibition was overcome by an excess of L-arginine (3×10^{-4} M). D-arginine, on the other hand, was ineffective in reversing endothelium-dependent inhibition by LNMMA (data not shown). We have confirmed these inhibitory effects of LNMMA and specifically localized its effect to the endothelium using cell culture models.[44]

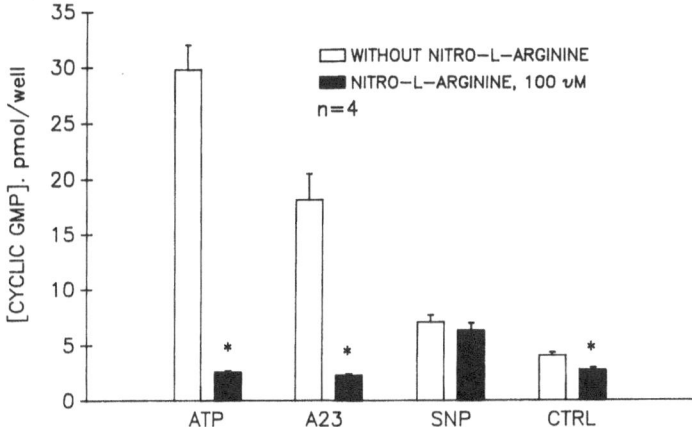

FIGURE 7. The effect of nitro-L-arginine (1×10^{-4} M) on cyclic GMP accumulation in co-cultures stimulated by ATP (1×10^{-4} M), A23187 (A23; 1×10^{-6} M), sodium nitroprusside (SNP, 1×10^{-6} M) or control buffer (CTRL). Asterisks (*) indicate significant differences from control ($P < 0.01$; n = 4).

The inhibitory effect of another L-arginine analog, nitro-L-arginine, was investigated using co-cultures. Nitro-L-arginine (1×10^{-4} M) completely inhibited cyclic GMP accumulation in co-cultures stimulated by the endothelium-dependent dilators ATP (1×10^{-4} M) and A23187 (1×10^{-6} M) (figure 7). Nitro-L-arginine also had an inhibitory effect on basal cyclic GMP concentration, implying basal release of EDRF in co-cultures. Nitro-L-arginine had no significant effect on increases in cyclic GMP induced by the non-endothelium-dependent dilator sodium nitroprusside (1×10^{-6} M).

The enzyme EDRF synthase has been shown to require calmodulin. In figure 8 we demonstrate in co-cultures that the calmodulin antagonist trifluoperazine (1×10^{-4} M) inhibited the cyclic GMP accumulation induced by bradykinin (1×10^{-6} M) and ATP (1×10^{-4} M) but had no effect on the cyclic GMP accumulation induced by non-endothelium-dependent sodium nitroprusside (1×10^{-6} M).

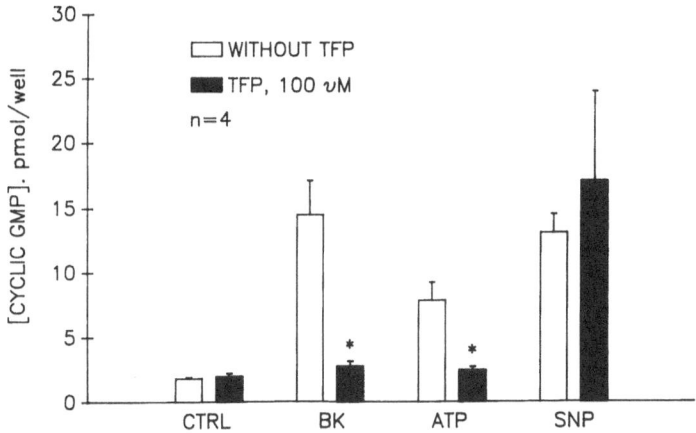

FIGURE 8. Effect of trifluoperazine (TFP, 1×10^{-4} M) on cyclic GMP accumulation in co-cultures stimulated by bradykinin (BK, 1×10^{-6} M), ATP (1×10^{-4} M) sodium nitroprusside (SNP, 1×10^{-6} M) or control buffer (CTRL). Asterisks (*) indicate significant inhibition by TFP ($P < 0.05$; n = 4).

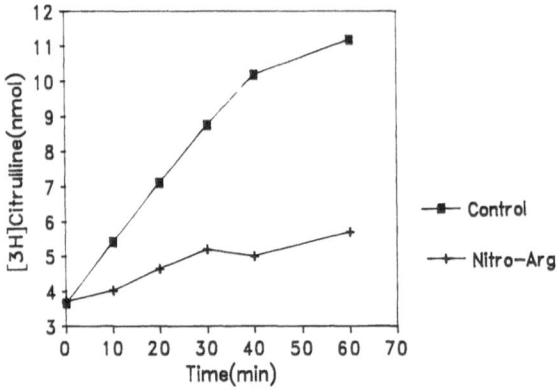

FIGURE 9. Time course of the activity of EDRF synthase from bovine cerebellum and the effect of the EDRF synthase inhibitor, nitro-L-arginine (6×10^{-4} M). [^{3}H]-citrulline was measured as an index of enzyme activity.

The activity of the EDRF synthase enzyme from bovine cerebellum was linear with time and was markedly inhibited by nitro-L-arginine (6×10^{-4} M) (figure 9).

The production or release of EDRF from endothelium also was inhibited by hypoxia. We observed that the endothelium-dependent relaxation of isolated rabbit first branch pulmonary artery by the endothelium-dependent dilators methacholine, ATP, and A23187 was significantly inhibited by moderate hypoxia ($PO_2 = 40$ mm Hg), while the endothelium-independent vasodilation by sodium nitroprusside or isoproterenol was not affected by this level of hypoxia (table 1).

DISCUSSION

EDRF is a potent endogenous vasodilator which is produced and released from endothelial cells and subsequently causes the relaxation of vascular smooth muscle

Table 1. Effect of Hypoxia on the Relaxation of Rabbit Pulmonary Artery

Drug	First Cycle (Normoxia)	Second Cycle (Hypoxia)	Third Cycle (Normoxia)
Methacholine (10^{-6} M)	54.4 ± 6.3	19.4 ± 11.4*	55.6 ± 4.8
A23187 (3×10^{-7} M)	82.8 ± 6.1	27.3 ± 16.6*	71.8 ± 3.1
ATP (3×10^{-5} M)	95.0 ± 5.0	69.8 ± 9.0*	96.5 ± 4.0
ATP (3×10^{-5} M) + BW-A1433U	71.8 ± 6.4	31.0 ± 9.3*	70.0 ± 6.5
SNP (3×10^{-7} M)	97.8 ± 1.1	96.1 ± 1.6	97.7 ± 2.4
Isoproterenol	82.8 ± 1.9	85.7 ± 3.0	85.6 ± 2.4

Data are expressed as the mean percent relaxation ± SEM. The relaxations in response to endothelium-dependent agents (italics) were significantly impaired by hypoxia, while those in response to endothelium-independent agents were not. BW-A1433U is an adenosine receptor antagonist. Asterisks (*) indicate $P < 0.05$. N = 6 to 10.

through the activation of soluble guanylate cyclase and an increase in VSM cyclic GMP (figure 10). Chemically, EDRF is likely to be nitric oxide, a related nitrogen oxide containing compound or a combination of such compounds. EDRF is synthesized by endothelial cells and other cell types from L-arginine by an enzyme dependent on calcium, calmodulin and NADPH (figure 10). It is possible that other or additional enzymatic pathways or cofactors are involved in EDRF production in different cell types. This is particularly true for the macrophage, where EDRF synthesis appears to require tetrahydrobiopterin. In addition, others have described an EDRF synthase activity in particulate cell fractions in addition to the soluble fraction described above.[37]

The action of EDRF parallels that of the nitrovasodilators which act directly upon vascular smooth muscle. EDRF appears to be quite ubiquitous. It is present in all vascular beds, in large and small vessels and in a wide range of species. Its role in human vascular physiology and pathophysiology is just beginning to be investigated. Newly developed specific inhibitors such as LNMMA and nitro-L-arginine are sure to be of benefit in this regard. In addition to its role as a potent endogenous vasodilator, EDRF is an inhibitor of platelet aggregation and adhesion. Several animal and human studies have suggested that its activity or production is impaired in hypertension and atherosclerosis and that its absence, due to endothelial cell damage, may play a role in cerebral and coronary vasospasm. Recent reports that EDRF serves as a second

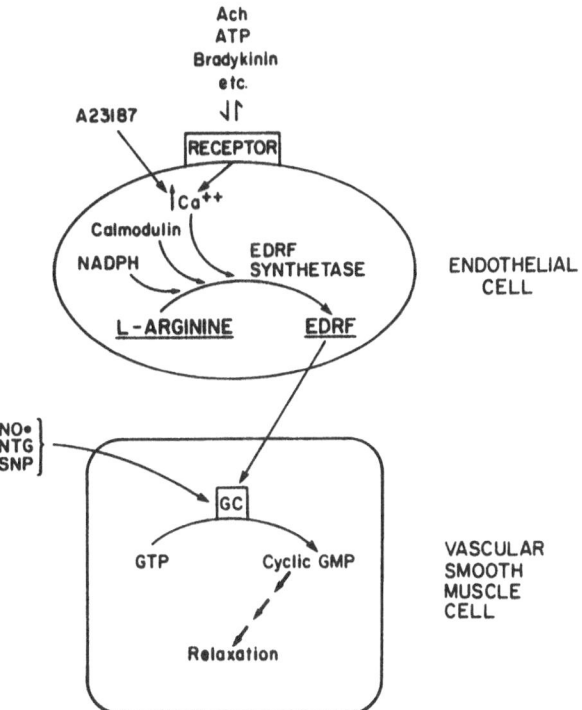

FIGURE 10. Both receptor- and non-receptor-mediated increases in endothelial cell cytosolic calcium lead to the activation of EDRF synthase which produces EDRF from L-arginine. This enzyme also requires calmodulin and NADPH as co-factors. Citrulline is a byproduct of this reaction. EDRF is then released from the endothelium and transferred to vascular smooth muscle (VSM) where it activates soluble guanylate cyclase causing an increase in cyclic GMP which subsequently leads to VSM relaxation. This parallels the mechanism of relaxation by the exogenous agents nitric oxide (NO•), nitroglycerin (NTG) and sodium nitroprusside (SNP).

messenger for guanylate cyclase activation and cyclic GMP production in a variety of cell types, including renal and respiratory epithelium, cerebellar neurons, macrophages and adrenocytes, suggest even broader implications.

EDRF is of interest to the anesthesiologist because of its role in cardiovascular physiology and in pathophysiologic states, but its importance to the anesthesiologist may be even more direct. Studies have shown that the endothelium may play a role in mediating the vascular actions of anesthetics[46] and that anesthetics can inhibit the production or release or action of EDRF.[47] Recent work in our laboratory has demonstrated that the volatile anesthetic agents halothane, enflurane, and isoflurane are capable of inhibiting endothelium-dependent vasodilation of rat thoracic aorta rings induced by both receptor-mediated, endothelium-dependent dilators such as methacholine and receptor-independent, endothelium-dependent vasodilators such as the calcium ionophore A23187 and ionomycin. These anesthetics had no effect on vasodilation induced by the endothelium-independent vasodilator sodium nitroprusside. Attempts at localizing the mechanism of anesthetic inhibition of endothelium-dependent vasodilation are now focused on the effect of anesthetics on receptor-mediated calcium increases in endothelial cells and on the effect of volatile anesthetics on the activity of isolated EDRF synthase.

Similarly, investigations into the mechanisms by which hypoxia can impair endothelium-dependent vasodilation and EDRF production from the endothelium are an active area of research in our laboratory. We have recently observed that hypoxia selectively inhibits endothelium-dependent cyclic GMP accumulation in co-cultures and have definitively localized this effect to the endothelial cell through the use of column transfer experiments. This hypoxic inhibition of EDRF production appears to occur readily in both pulmonary and systemic vessels. Preliminary experiments in our laboratory suggest that hypoxia decreases the activity of EDRF synthase from bovine cerebellum.

The specific chemical identity of EDRF remains a key controversy. While few doubt that the specific agent which activates guanylate cyclase is the nitric oxide free radical (NO), there is evidence that EDRF is a chemically bound form of NO such as a nitrosothiol. Methods used to measure NO and to prepare exogenous NO solutions are nonspecific and subject to misinterpretation.[48] There are reports that EDRF is a more specific relaxant of vascular versus non-vascular smooth muscle compared to nitric oxide and that these two compounds are differentially bound by ion exchange columns.[49-56] Myers et al.[57] recently have reported that EDRF more closely resembles S-nitrosocysteine. Clarification of these issues parallels ongoing investigations into the physiologic and pathophysiologic importance of EDRF.

ACKNOWLEDGEMENTS

Supported in part by NIH grants R29 HL39706 and P01 HL19242 and an Anesthesia Young Investigator/Parker B. Francis Award (RAJ).

REFERENCES

1. T. M. Griffith, D. H. Edwards, M. J. Lewis, A. C. Newby and A. H. Henderson, The nature of endothelium-dependent vascular relaxing factor, *Nature* 308:645-647 (1984).
2. R. J. Gryglewski, R. M. J. Palmer and S. Moncada, Superoxide anion is involved in the breakdown of endothelium-derived vascular relaxing factor, *Nature* 320:454-456 (1986).
3. C. J. Long and T. W. Stone, The release of endothelium-derived relaxing factor is calcium dependent, *Blood Vessels* 22:205-208 (1985).
4. H. A. Singer and M. J. Peach, Calcium- and endothelial-mediated vascular smooth muscle relaxation in rabbit aorta, *Hypertension* 4(Suppl 2):19-25 (1982).

5. N. J. Izzo, A. L. Loeb, R. A. Johns and M. J. Peach, Intracellular calcium flux accompanies the release of endothelium-derived relaxing factor (EDRF) and prostacyclin (PGI_2) from cultured endothelial cells, *Fed Proc* 45:198 (1986).

6. A. L. Loeb, N. J. Izzo, R. M. Johnson, J. C. Garrison and M. J. Peach, Endothelium-derived relaxing factor release associated with increased endothelial cell inositol trisphosphate and intracellular calcium, *Am J Cardiol* 62:366-406 (1988).

7. S. Holzmann, Endothelium-induced relaxation by acetylcholine associated with larger rises in cyclic GMP in coronary arterial strips, *J Cyclic Nucleotide Res* 8:409-419 (1982).

8. R. Rapoport and F. Murad, Endothelium-dependent and nitrovasodilator-induced relaxation of vascular smooth muscle, role of cyclic GMP, *J Cyclic Nucleotide Prot Phos Res* 9:281-296 (1983).

9. R. Furchgott and D. Jothianandan, Relation of cyclic GMP levels to endothelium-dependent relaxation by acetylcholine in rabbit aorta, *Fed Proc* 42:619 (1983).

10. L. J. Ignarro, T. M. Burke, K. S. Wood, M. S. Wolin and P. J. Kadowitz, Association between cyclic GMP accumulation and acetylcholine-elicited relaxation of bovine intrapulmonary artery, *J Pharmacol Exp Ther* 228:682-690 (1984).

11. R. M. Rapoport, M. B. Draznin and F. Murad, Mechanisms of adenosine triphosphate, thrombin and trypsin-induced relaxation of rat thoracic aorta, *Circ Res* 55:468-479 (1984).

12. R. A. Johns and M. J. Peach, Parabromophenacyl bromide inhibits endothelium-dependent arterial relaxation and cyclic GMP accumulation by effects produced exclusively in the smooth muscle, *J Pharmacol Exp Ther* 244:859-865 (1988).

13. A. L. Loeb, R. A. Johns, P. Milner and M. J. Peach, Studies on endothelium-derived relaxing factor from cultured cells, *Hypertension* 9(Suppl III):186-192 (1987).

14. R. F. Furchgott and J. V. Zawadzki, The obligatory role of endothelial cells in the relaxation of arterial smooth muscle by acetylcholine, *Nature* 288:373-376 (1980).

15. J. G. DeMey and P. M. Vanhoutte, Anoxia and endothelium-dependent reactivity of the canine femoral artery, *J Physiol* (Lond) 335:65-74 (1983).

16. H. A. Singer and M. J. Peach, Endothelium-dependent relaxation of rabbit aorta: I. Relaxation stimulated by arachidonic acid, *J Pharmacol Exp Ther* 226:790-795 (1983).

17. R. F. Furchgott, Role of endothelium in responses of vascular smooth muscle, *Circ Res* 53:557-573 (1983).

18. R. F. Furchgott, The role of endothelium in the responses of vascular smooth muscle to drugs, *Annu Rev Pharmacol Toxicol* 24:175-197 (1984).

19. J. L. Amezcua, G. J. Dusting, R. M. Palmer and S. Moncada, Acetylcholine induces vasodilation in the rabbit isolated heart through the release of nitric oxide, the endogenous nitrovasodilator, *Br J Pharmacol* 95:830-834 (1988).

20. M. Kelm, M. Feelisch, R. Spahr, H.-M. Piper, E. Noack and J. Schrader, Quantitative and kinetic characterization of nitric oxide in EDRF released from cultured endothelial cells, *Biochem Biophys Res Comm* 154:236-244 (1988).

21. R. M. Palmer, A. G. Ferrige and S. Moncada, Nitric oxide release accounts for the biological activity of endothelium-derived relaxing factor, *Nature* 327:524-526 (1987).

22. R. M. Palmer, D. S. Ashton and S. Moncada, Vascular endothelial cells synthesize nitric oxide from L-arginine, *Nature* 333:664-666 (1988).

23. L. J. Ignarro, G. M. Buga, K. S. Wood, R. E. Byrns and G. Chaudhuri, Endothelium-derived relaxing factor produced and released from artery and vein is nitric oxide, *Proc Natl Acad Sci* (USA) 84:9265-9269 (1987).

24. L. J. Ignarro, R. E. Byrns, G. M. Buga and K. S. Wood, Endothelium-derived relaxing factor from pulmonary artery and vein possesses pharmacologic and chemical properties identical to those of nitric oxide radical, *Circ Res* 61:866-879 (1987).

25. L. J. Ignarro, G. M. Buga, R. E. Byrns, K. S. Wood and G. Chaudhuri, Endothelium-derived relaxing factor and nitric oxide possess identical pharmacologic properties as relaxants of bovine arterial and venous smooth muscle, *J Pharmacol Exp Ther* 246:218-226 (1988).

26. L. J. Ignarro, R. E. Byrns, G. M. Buga, K. S. Wood and G. Chaudhuri, Pharmacological evidence that endothelium-derived relaxing factor is nitric oxide: use of pyrogallol and superoxide dismutase to study endothelium-dependent and nitric oxide-elicited vascular smooth muscle relaxation, *J Pharmacol Exp Ther* 244:181-189 (1988).

27. L. J. Ignarro, M. E. Gold, G. M. Buga, R. E. Byrns, K. S. Wood, G. Chaudhuri and G. Frank, Basic polyamino acids rich in arginine, lysine, or ornithine cause both enhancement of and refractoriness to formation of endothelium-derived nitric oxide in pulmonary artery and vein, *Circ Res* 64:315-329 (1989).

28. L. J. Ignarro, Endothelium-derived nitric oxide: actions and properties. *FASEB J* 3:31-36 (1989).

29. S. Moncada, M. W. Radomski and R. M. Palmer, Endothelium-derived relaxing factor: Identification as nitric oxide and role in the control of vascular tone and platelet function, *Biochem Pharmacol* 37:2495-2501 (1988).

30. B. Mayer and E. Bohme, Ca^{++}-dependent formation of an L-arginine derived activator of soluble guanylate cyclase in bovine lung, *FEBS Lett* 256:211-214 (1989).

31. B. Mayer, K. Schmidt, P. Humbert and E. Bohme, Biosynthesis of endothelium-derived relaxing factor: A cytosolic enzyme in porcine aortic endothelial cells Ca^{++}-dependently converts L-arginine into an activator of soluble guanylal cyclase, *Biochem Biophys Res Comm* 164:678-685 (1989).

32. M. A. Marletta, P. S. Yoon, R. Iyengar, C. D. Leas and J. S. Wishnok, Macrophage oxidation of L-arginine to nitrite and nitrate: Nitric oxide is an intermediate, *Biochemistry* 27:8706-8711 (1988).

33. D. J. Stuehr, N. S. Kwon, S. S. Gross, B. A. Thiel, R. Levi and C. F. Nathan, Synthesis of nitrogen oxides from L-arginine by macrophage cytosol: Requirement for inducible and constituitive components, *Biochem Biophys Res Comm* 161:420-426 (1989).

34. D. S. Bredt and S. H. Snyder, Isolation of nitric oxide synthetase, a calmodulin-requiring enzyme, *Proc Natl Acad Sci* (USA) 87:682-685 (1990).

35. D. S. Bredt and S. H. Snyder, Nitric oxide mediates glutamate-linked enhancement of cGMP levels in the cerebellum, *Proc Natl Acad Sci* (USA) 86:9030-9033 (1989).

36. A. Mulsch, E. Bassenge and R. Busse, Nitric oxide synthesis in endothelial cytosol: Evidence for a calcium-dependent and a calcium-independent mechanism, *Naunyn-Schmiedeberg's Arch Pharmacol* 340:767-770 (1989).

37. K. M. Boje and H. L. Fung, Endothelial nitric oxide generating enzyme(s) in the bovine aorta: Subcellular localization and metabolic characterization, *J Pharmacol Exp Ther* 253:20-26 (1990).

38. R. M. Palmer, D. D. Rees, D. S. Ashton and S. Moncada, L-arginine is the physiological precursor for the formation of nitric oxide in endothelium-dependent relaxation, *Biochem Biophys Res Comm* 153:1251-1256 (1988).

39. R. M. J. Palmer and S. Moncada, A novel citrulline-forming enzyme implicated in the formation of nitric oxide by vascular endothelial cells, *Biochem Biophys Res Comm* 158:348-352 (1989).

40. U. Forstermann, K. Ishii, L. D. Gorsky and F. Murad, The cytosol of N1E-115 neuroblastoma cells synthesizes an EDRF-like substance that relaxes rabbit aorta, *Naunyn-Schmiedeberg's Arch Pharmacol* 340:771-774 (1989).

41. H. Schroder and K. Schror, Cyclic GMP stimulation by vasopressin in LLC-PK_1 kidney epithelial cells is L-arginine-dependent, *Naunyn-Schmiedeberg's Arch Pharmacol* 340:475-477 (1989).

42. R. G. Knowles, M. Palacios, R. M. J. Palmer and S. Moncada, Formation of nitric oxide from L-arginine in the central nervous system: Mechanism for stimulation of the soluble guanylate cyclase, *Proc Natl Acad Sci* (USA) 86:5159-5162 (1989).

43. M. Palacios, R. G. Knowles, R. M. J. Palmer and S. Moncada, Nitric oxide from L-arginine stimulates the soluble guanylate cyclase in adrenal glands, *Biochem Biophys Res Comm* 165:802-809 (1989).

44. R. A. Johns, M. J. Peach, J. M. Linden and A. Tichotsky, N^G-monomethyl-L-arginine causes specific, dose-dependent inhibition of cyclic GMP accumulation in cocultures of bovine pulmonary endothelium and rat vascular smooth muscle through an action specific to the endothelium, *Circ Res* 67:979-985 (1990).

45. R. A. Johns, N. J. Izzo, P. J. Milner, A. L. Loeb and M. J. Peach, Use of cultured cells to study the relationship between arachidonic acid and endothelium-derived relaxing factor, *Am J Med Sci* 295:287-292 (1988).

46. D. J. Stone and R. A. Johns, Endothelium-dependent effects of halothane, enflurane, and isoflurane on isolated rat aortic vascular rings, *Anesthesiology* 71:126-132 (1989).

47. S. M. Muldoon, J. L. Hart, K. A. Bowen and W. Freas, Attenuation of endothelium-mediated vasodilation by halothane, *Anesthesiology* 68:31-37 (1988).

48. W. R. Tracey, J. L. Linden, M. J. Peach and R. A. Johns, Comparison of spectrophotometric and biological assays for nitric oxide and EDRF: Nonspecificity of the diazotization reaction for nitric oxide and failure to detect EDRF, *J Pharmacol Exp Ther* 252:922-928 (1990).

49. K. Shikano, C. J. Long, E. H. Ohlstein and B. A. Berkowitz, Comparative pharmacology of endothelium derived relaxing factor and nitric oxide, *J Pharmacol Exp Ther* 247:873-881 (1988).

50. K. Shikano, E. H. Ohlstein and B. A. Berkowitz, Differential selectivity of endothelium-derived relaxing factor and nitric oxide in smooth muscle, *Br J Pharmacol* 92:483-485 (1987).

51. C. J. Long, K. Shikano and B. A. Berkowitz, Anion exchange resins discriminate between nitric oxide and EDRF, *Eur J Pharmacol* 142:317-318 (1987).

52. C. J. Long and B. A. Berkowitz, What is the relationship between the endothelium-derived relaxant factor and nitric oxide? *Life Sci* 45:1-14 (1989).

53. J. L. Beny and P. C. Brunet, Neither nitric oxide nor nitroglycerin accounts for all the characteristics of endothelially mediated vasodilatation of pig coronary arteries, *Blood Vessels* 25:308-311 (1988).

54. G. J. Dusting, M. A. Read and A. G. Stewart, Endothelium-derived relaxing factor released from cultured cells: Differentiation from nitric oxide, *Clin Exp Pharmacol Physiol* 15:83-92 (1988).

55. U. Hoeffner, C. Boulanger and P. M. Vanhoutte, Proximal and distal dog coronary arteries respond differently to basal EDRF but not to NO, *Am J Physiol* 256:H828-H831 (1989).

56. P. R. Myers, R. Guerra Jr and D. G. Harrison, Release of NO and EDRF from cultured bovine aortic endothelial cells, *Am J Physiol* 256:H1030-H1037 (1989).

57. P. R. Myers, R. L. Minor, R. Guerra, J. N. Bates and D. G. Harrison, Vasorelaxant properties of the endothelium-derived relaxing factor more closely resemble S-nitrosocysteine than nitric oxide, *Nature* 345:161-163 (1990).

21

Effect of Volatile Anesthetic Agents on Endothelium-Dependent Relaxation

Gilbert A. Blaise

INTRODUCTION AND REVIEW

Volatile Anesthetic Actions on Vascular Smooth Muscle

The vasodilatory mechanism of volatile anesthetic agents is complex and not yet fully understood. It is known that the endothelium controls smooth muscle tone as well as its response to most vasoactive substances. The endothelium not only metabolizes vasoactive agents, but it secretes several vasodilators (prostacyclin, EDRF, hyperpolarizing factor) and vasoconstrictors (endothelin, EDCF).[1] In the last few years, an effect of volatile anesthetic agents on endothelium-dependent vasodilation has been suspected. Blaise et al.[2] have shown that isoflurane's attenuation of the contractile response of isolated canine coronary artery rings to 5-hydroxytryptamine and prostaglandin F2α was dependent on the presence of the endothelium. As this attenuation was only slightly reduced by pretreatment of the vascular rings with indomethacin, it was postulated that isoflurane's effect was mediated through EDRF release or action. Muldoon et al.[3] have shown that halothane reduces the response of precontracted rabbit and canine vascular rings to acetylcholine and bradykinin. They suggest that halothane inhibits the release or the action of EDRF. Our findings were similar and showed that, contrary to isoflurane, halothane attenuates the serotoninergic response of isolated canine coronary artery rings in denuded vessels (without endothelium), but not in intact vessels.[4] These preliminary findings suggest a positive interaction of isoflurane and a negative interaction of halothane on endothelium-dependent dilation. Recently, Stone et al.[5] have demonstrated that volatile agents have a biphasic, dose-dependent effect on tension of isolated, preconstricted rat aorta. At lower doses, the tension increased; at higher doses, the isometric force decreased. The increase in tension induced by the volatile agents occured only in vessels with endothelium and was clearly potentiated by indomethacin. These data also suggest that these agents affect endothelium function and prostaglandin synthesis.

GILBERT A. BLAISE, Department d'Anesthesie, Universite de Montreal, Hopital Notre Dame, Montreal, Quebec H2L 4K8.

Mechanisms of Anesthetic Action in Skeletal, Cardiac, and Smooth Muscle
Edited by T.J.J. Blanck and D.M. Wheeler, Plenum Press, New York, 1991

The EDRF/NO System

In 1980, Furchgott and Zawadski[6] demonstrated the obligatory role of endothelium in ACh-induced relaxation of rabbit aorta rings contracted with norepinephrine. They concluded that the endothelial cells release a vasodilator called EDRF, which is different from prostacyclin. It was later shown that EDRF is probably composed of several substances.[7,8] One is nitric oxide (NO)[9,10] or a substance containing NO;[11] another is a hyperpolarizing factor[12,13,14] which relaxes smooth muscles by interacting with the voltage-operated calcium channel (VOC).

NO is derived from the guanido N-terminal of the L-arginine molecule, and the activity of the NO-forming enzyme (NO synthetase) is dependent on the presence of free Ca^{2+} and NADPH.[15] An increase in the intracellular free calcium concentration is an important signal in endothelial cells involved in the production of EDRF. Both intracellular and extracellular sources of calcium are involved in this signal[16] and a sustained elevation of intracellular Ca^{2+}, due to transmembrane Ca^{2+} influx, appears to be a prerequisite.[17] The molecular mechanisms whereby Ca^{2+} enters the cytosol and the nature of the calcium channel involved in calcium entry into endothelial cells are not well established. No L-type, voltage-operated channel has been found in endothelial cells; and therefore, the dihydropyridine calcium antagonists and agonists do not interfere with EDRF release.[18]

Calcium influx in endothelial cells is inhibited by membrane depolarization and is augmented by agonists such as bradykinin and ATP.[17,18] Some agonists activate a receptor-operated calcium channel (ROC).[19] Intracellular second messengers, such as inositol triphosphate or inositol tetraphosphate, are also released following membrane receptor activation and serve to release intracellular calcium as well as open a calcium channel in the sarcolemma.[20] G-proteins (G_i type) are the links between the activation of some receptors (5-HT and alpha-2 receptors) and the intracellular response.[21] It is not known whether other types of G-proteins, or in fact any G-proteins, are involved in the mediation of the effects of the other agonists. Bradykinin,[17,22,23] ACh[24] and ATP[17,22,25] also activate Ca^{2+}-dependent K^+ channels in endothelial cells. This results in a hyperpolarization and an increase in calcium entry. The activation of the K^+ channel is potentiated by K^+ channel activators such as chromkalin and pinacidil.[17,22]

The hyperpolarizing factor hyperpolarizes smooth muscle and probably endothelial cells; the nature of the hyperpolarizing factor as well as its mechanism of action are not definitively established. Komori et al. have shown that the hyperpolarizing factor is different from NO.[13] However, Tare et al. have proven that NO itself can hyperpolarize smooth muscle cells[14]. The hyperpolarization mechanism is also a matter of controversy. Feletou et al.[12] and Komori et al.[13] have demonstrated that hyperpolarization is prevented by ouabain and that Na^+-K^+ exchange is involved. Chen et al.[26] and Brunet et al.[27] have indicated that hyperpolarization is mediated through the activation of K^+ channels and not through Na^+-K^+ exchange.

The effect of hyperpolarization on smooth muscle and on endothelial cells is different: hyperpolarization of smooth muscle cells reduces Ca^{2+} entry and induces relaxation, hyperpolarization of endothelial cells increases the driving force for Ca^{2+} entry, increases intracellular free Ca^{2+} and increases EDRF synthesis or release.[17] These two complementary mechanisms potentiate smooth muscle relaxation. Hyperpolarization of endothelial cells, either through the direct activation of the K^+ channel by agonist receptor interaction or by the hyperpolarizing factor, is an important amplification mechanism for EDRF release.

Possible Sites of Halothane's Effect on EDRF Activity

Halothane may interfere with calcium entry in endothelial cells. It has been shown before that halothane is a weak antagonist of the Ca^{2+} VOC in cardiac[28] and vascular smooth muscle cells[29] and an antagonist of the ROC in pancreatic cells.[30] There is no VOC on the endothelial cells[31] and the effect of halothane on endothelial ROC is not known.

The role of membrane potential as an amplification mechanism for EDRF release or action has been suggested by several authors.[17,22] Halothane interferes with membrane potential and depending on the type of tissue can increase or decrease K^+ conductance. Scharff and Foder[32] have shown that halothane inhibits hyperpolarization and potassium channels in human red blood cells. However, previous publications have shown that halothane activates the Ca^{2+}-sensitive K^+ channel and hyperpolarizes neurological tissue.[33,34] The Ca^{2+}-dependent K^+ channel in human red cells has the same electrical characteristics as the K^+ channel in endothelial cells; however, the effect of halothane on the endothelial Ca^{2+}-dependent K^+ channel is not known.

G-proteins are a complex of proteins that mediate the intracellular response to some agonists and modulate adenylate cyclase activity.[35] A G_i-protein mediated the effect of halothane on electrically induced contraction of guinea pig ileum.[36] A G_i-protein is one of the intermediates in EDRF release induced by 5-hydroxytryptamine and the alpha-2 agonists.[21] Fluoride, which is an activator of the G-proteins, is an endothelium-dependent vasodilator.[37] Its vasodilatory effect is inhibited by pertussis toxin. These data suggest that G-proteins are involved in the endothelial response to some agonists and that halothane can interfere with the cellular response mediated by Gi-proteins.

Halothane activates adenylate cyclase and increases cyclic AMP in vascular and non-vascular smooth muscle.[38,39] Halothane's effect on cyclic AMP levels in endothelial cells is not known and the effect of halothane on guanylate cyclase has not been measured. There is a negative interaction between cyclic AMP levels in endothelial cells and intracellular free Ca^{2+} and EDRF synthesis.[40]

CURRENT STUDIES

Effect on Basal Tension and on Tension of Contracted Rings

Neither halothane nor isoflurane have any effect on the basal tension of isolated canine coronary arteries, canine femoral arteries, or rabbit and rat aorta rings. Our data show that these agents can modify the tension only of contracted rings. Halothane (2.5%) has a biphasic effect when added to the mixture gassing rabbit aorta rings contracted with submaximal concentrations of phenylephrine. An initial increase in tension is followed by a slow, continued relaxation. After 40 min of treatment with the volatile agent, the tension was the same in control rings (the ones not treated with halothane) and in treated rings with and without endothelium (figure 1).

Effect of Halothane on Vasodilatation Induced by Nitric Oxide

Nitric oxide was prepared for the experiments as follows. A gas bulb fitted with a silicone rubber injection septum was filled with nitric oxide from a cylinder (Union

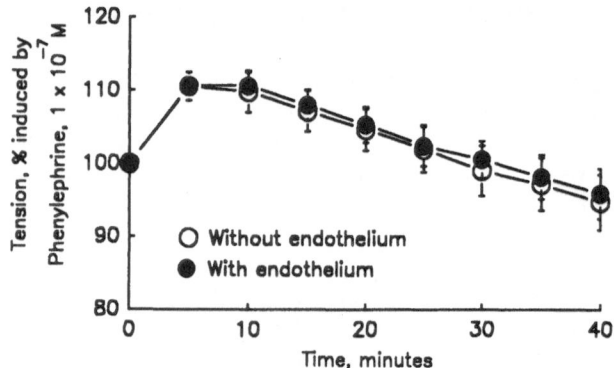

FIGURE 1. Effect of 2.5% halothane on the tension of isolated intact and denuded rabbit aorta rings precontracted with phenylephrine (10^{-7} M). Halothane induces a biphasic response: an initial increase in tension (10%) is followed by a slow, continuous decrease in tension which reaches the initial level after 30 minutes. There is no difference in the time course between rings with and without endothelium. Each point represents the mean ± SEM of 6 different rings which came from 6 animals.

Carbide, LINDE). Appropriate volumes (100 and 1000 microliters) were removed with a glass syringe and injected into other bulb filled with 100 ml of distilled water, which had been gassed with helium for approximately 3 hours, giving stock solutions of nitric oxide of 4×10^{-5} and 4×10^{-4} M.[7,9]

Non-cumulative concentration response curves to nitric oxide were determined for isolated, denuded rabbit aorta rings contracted with 10^{-7} M phenylephrine. Half of the rings were treated with halothane (2.5%), the other half were untreated (controls). Halothane significantly attenuated the relaxation induced by nitric oxide (figure 2).

Effects of Volatile Anesthetics on Coronary Arterioles

The effects of volatile anesthetic agents on endothelium-dependent vasodilation of coronary arterioles was investigated using isolated working rabbit hearts, perfused at a constant pressure with an oxygenated Krebs-Ringer solution. ACh, added to the perfusate (final concentration 10^{-6} M) increased the coronary flow. This increase was

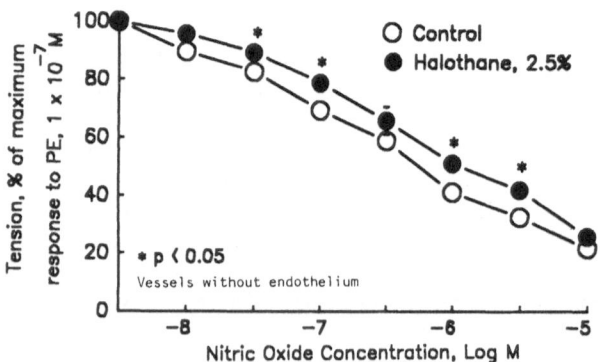

FIGURE 2. Response of isolated rabbit aorta rings without endothelium to increasing concentrations of NO from 10^{-8} M to 10^{-5} M. Solid circles represent rings in the presence of halothane 2.5%. Data are expressed as mean ± SEM; n = 6 rings, each originating from a different animal. Asterisks (*) indicate $P < 0.05$ by paired t-test.

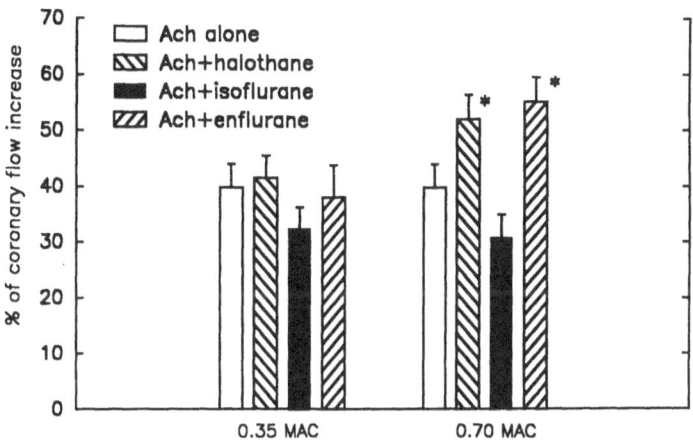

FIGURE 3. Effect of two concentrations of volatile anesthetics on ACh-induced coronary arteriolar dilatation in isolated perfused rabbit hearts (Langendorff preparation). The volatile agents have no effect at 0.35 rabbit MAC. Halothane and enflurane, at 0.70 MAC, potentiate the coronary dilation induced by ACh. Data are presented as mean ± SEM with n = 7. Asterisks (*) indicate P < 0.05 for the comparison of coronary flow in the presence and absence of each volatile agent (ACh alone *vs.* ACh plus agent).

endothelium-dependent. In this preparation, the volatile agents have a weak effect on the coronary flow.[41] However, after 10 min of stabilization despite a pretreatment with the volatile anesthetics, ACh added to the perfusate increased the coronary flow by the same magnitude in both control hearts as in hearts treated with the anesthetics (figure 3).

Conclusion

The data suggest that halothane interferes with the endothelium-dependent relaxation of large arteries (coronary, femoral, aorta). This effect is dose dependent. However, the volatile anesthetics do not reduce endothelium-dependent dilation of coronary arterioles. The response to NO is attenuated by halothane. Halothane's effect on NO synthesis, release, or NO half-life is unknown. It is also unknown if the halothane-induced attenuation of endothelium-dependent relaxation is due to its effect on EDRF/NO, hyperpolarizing factor or prostaglandin.

FUTURE RESEARCH

The effect of the volatile agents on the changes in intracellular free Ca^{2+} induced in endothelial cells by several agonists can be measured by loading the cells with fluorescent Ca^{2+} indicators. The effect of these agents on membrane potential and on K^+ current activated by the endothelial agonists can be measured by patch clamp techniques.[42] The effect of these anesthetics on the release, half life and the action of EDRF will be differentiated using a bioassay system.[43] The effluent of an intact perfused vascular segment superfuses an isolated vascular ring without endothelium. The vascular ring suspended between 2 stirrups is precontracted and

EDRF released from the intact segment induces a relaxation of the vascular ring. Anesthetics can be added to the perfusate either before or after the intact segment. By measuring their effects on endothelium-dependent relaxation, it will be possible to separate their effects on EDRF metabolism and EDRF action.

REFERENCES

1. J. R. Vane, E. E. Anggard, R. M. Botting, Mechanisms of disease: Regulatory functions of the vascular endothelium, *N Eng J Med* 323:27-36 (1990).
2. G. Blaise, J. C. Sill, M. Nugent, R. A. Van Dyke, P. M. Vanhoutte, Isoflurane causes endothelium-dependent inhibition of contractile responses of canine coronary arteries, *Anesthesiology* 67:513-517 (1987).
3. S. M. Muldoon, J. L. Hart, K. A. Bowen, W. Freas, Attenuation of endothelium-mediated vasodilation by halothane, *Anesthesiology* 68:31-37 (1987).
4. G. Blaise, S. G. Lenis, D. Girard, C. Hollman, R. Meloche, Halothane and enflurane attenuate the K^+ induced contractile response of isolated canine coronary artery rings, *Can J Anesth* 35:S72 (1988).
5. D. J. Stone and R. A. Johns, Endothelium-dependent effects of halothane, enflurane and isoflurane on isolated rat aortic rings, *Anesthesiology* 71:126-130 (1989).
6. R. F. Furchgold and J. V. Zawadski, The obligatory role of endothelial cells in the relaxation of arterial smooth muscle by acetylcholine, *Nature* 288:373-376 (1980).
7. C. Boulanger, H. Hendrickson, R. R. Lorenz, P. M. Vanhoutte, Release of different relaxing factors by cultured porcine endothelial cells, *Circ Res* 64:1070-1078 (1989).
8. U. Hoeffner, M. Feletou, N. A. Flavahan, P. M. Vanhoutte, Canine arteries release two different endothelium-derived relaxing factors, *Am J Physiol* 257:H330-H333 (1989).
9. R. M. J. Palmer, A. G. Ferrige, S. Moncada, Nitric oxide release accounts for the biological activity of endothelium-derived relaxing factor, *Nature* 327:524-526 (1987).
10. L. J. Ignarro, G. M. Buga, K. S. Wood, R. E. Byrns, G. Chadhur, Endothelium-derived relaxing factor produced and released from artery and vein is nitric oxide, *Proc Natl Acad Sci USA* 84:9265-9269 (1987).
11. P. R. Myers, R. L. Minor Jr., R. Guerra Jr., J. N. Bates, D. G. Harrison, Vasorelaxant properties of the endothelium-derived relaxing factor more closely resemble S-nitrocysteine than nitric oxide, *Nature* 345:161-163 (1990).
12. M. Feletou and P. M. Vanhoutte, Endothelium-dependent hyperpolarization of canine coronary smooth muscle, *Br J Pharmacol* 93:515-524 (1988).
13. K. Komori, R. R. Lorenz, P. M. Vanhoutte, Nitric oxide, ACh, and electrical and mechanical properties of canine arterial smooth muscle, *Am J Physiol* 255:H207-H212 (1988).
14. M. Tare, H. C. Parkington, H. A. Coleman, T. O. Neild, G. J. Dusting, Hyperpolarization and relaxation of arterial smooth muscle caused by nitric oxide derived from the endothelium, *Nature* 346:69-71 (1990).
15. K. M. Boje and H.-L. Fung, Endothelial nitric oxide generating enzyme(s) in the bovine aorta: Subcellular location and metabolic characterization, *J Pharmacol Exp Ther* 253:20-26 (1990).
16. M. J. Peach, H. A. Singer, N. J. Izzo, A..L. Loeb, Role of calcium in endothelium-dependent relaxation of arterial smooth muscle, *Am J Card* 59:35A-45A (1987).
17. A. Lückhoff and R. Busse, Calcium influx into endothelial cells and formation of endothelium-derived relaxing factor is controlled by the membrane potential, *Pflügers Arch* 416:305-311 (1990).
18. P. M. Vanhoutte, Vascular endothelium and Ca^{2+} antagonists, *J Cardiovasc Pharmacol* 12 (suppl.6):S21-S28 (1988).
19. A. Johns, T. W. Lategan, N. J. Lodge, U. S. Ryan, C. Van Breeman, D. J. Adams, Calcium entry through receptor-operated channels in bovine pulmonary artery endothelial cells, *Tissue and Cell* 19:733-745 (1987).
20. N. G. Morgan, Inositol phospholipid turnover in cell calcium homeostasis, *in:* "Cell Signaling," N. G. Morgan, ed., The Guilford Press, New York (1989) pp. 91-117.
21. N. A. Flavahan, H. Shimokawa, P. M. Vanhoutte, Pertussis toxin inhibits endothelium-dependent relaxations to certain agonists in porcine coronary arteries, *J Physiol* (Lond) 408:549-560 (1989).
22. A. Lückhoff and R. Busse, Activators fo potassium channels enhance calcium influx into endothelial cells as a consequence of potassium currents, *Naunyn-Schmiedeberg's Arch Pharmacol* 342:94-99 (1990).

23. R. Sauvé, M. Chahine, J. Tremblay, P. Hamet, Single channel analysis of the electrical response of cells to bradykinin stimulation: Contribution of Ca^{++}-dependent K^+ channel, *J Hypertension* 8 (suppl. 7):F193-F201 (1990).

24. T. Sakai, Acetylcholine induces Ca^{++}-dependent K^+ current in rabbit endothelial cells. *J Pharmacol* 53:234-246 (1990).

25. R. Sauvé, L. Parent, C. Simoneau, G. Roy, External ATP triggers a biphasic activation process of a calcium-dependent K^+ channel in cultured bovine endothelial cells, *Pflügers Arch* 412:469-481 (1988).

26. G. Chen, H. Hashitani, H. Suzuki, Endothelium-dependent relaxation and hyperpolarization of canine coronary artery smooth muscles in relation to the electrogenic Na-K pump, *Br J Pharmacol* 98:950-956 (1989).

27. P. C. Brunet and J.-L. Beny, Substance P and bradykinin hyperpolarize pig coronary artery endothelial cells in primary culture, *Blood Vessels* 26:228-234 (1989).

28. C. Lynch III, S. Vogel, N. Sperelakis, Halothane depression of myocardial slow action potentials, *Anesthesiology* 55:360-368 (1981).

29. G. Blaise, J. Hughes, J. C. Sill, J. Buluran, G. Caille, Attenuation of contraction of isolated canine coronary arteries by enflurane and halothane *Can J Anaesth* 38:111-115 (1990).

30. N. Yashima, A. Wada, F. Izumi, Halothane inhibits the cholinergic-receptor-mediated influx of calcium in primary culture of bovine adrenal medulla cells, *Anesthesiology* 64:466-472 (1986).

31. R. L. Jayakody, C. T. Kappagod, M. P. Senaratn, N. Sreehara, Absence of effect of calcium antagonists on endothelium-dependent relaxation of rabbit aorta, *Br J Pharmacol* 91:155-164 (1987).

32. O. Scharff, B. Foder, Halothane inhibits hyperpolarization and potassium channels in human red blood cells, *Eur J Pharmacol* 159:165-173 (1989).

33. R. A. Nicoll and D. V. Madison, General anesthetics hyperpolarize neurons in the vertebrate central nervous system, *Science* 217:1055-1057 (1982).

34. K. Krnjevic, Cellular and synaptic effects of general anesthetics, *in*: "Molecular and Cellular Mechanisms of Anesthetics," S.H. Roth and K.W. Miller, eds., Plenum Medical Book Co., New York (1986).

35. N. G. Morgan, Guanine nucleotide binding proteins as signal transducers, *in:* "Cell Signaling," N. G. Morgan, ed., The Guilford Press, New York (1989) pp. 35-57.

36. M. M. Puig, H. Turndorf, W. Warner, Effect of pertussis toxin on the interaction of azepoxide and halothane, *J Pharmacol Exp Ther* 252:1156-1159 (1990).

37. N. A. Flavahan and P. M. Vanhoutte, Pertussis toxin inhibits endothelium-dependent relaxations evoked by fluoride, *Eur J Pharmacol* 178:121-124 (1990).

38. L. Triner, Y. Vulliemoz, M. Verosky, Action of halothane on adenylate cyclase, *Mol Pharmacol* 13:976-979 (1977).

39. K. J. Bernstein, M. Verosky, L. Triner, Effect of halothane on rat liver adenylate cyclase: Role of cytosolic components, *Anesth Analg* 64:531-537 (1987).

40. A. Lückhoff, A. Mulsh, R. Busse, cAMP attenuates autacoid release from endothelial cells: Relation to internal calcium, *Am J Physiol* 258:H960-966 (1990).

41. M. Tanguay, G. Blaise, L. Dumont, G. Beique, C. Hollman, Beneficial effects of volatile anesthetics on the decrease of coronary flow and myocardial contractility induced by oxygen-derived free radicals in isolated rabbit hearts, *J Cardiovasc Pharmacol* (in press).

42. R. Sauvé, Le patch clamp: une nouvelle façon de voir des canaux ioniques, *Med Sci* 9:536-545 (1987).

43. G. M. Rubanyi, R. R. Lorenz, P. M. Vanhoutte, Bioassay of endothelium-derived relaxing factors: Inactivation by catecholamines, *Am J Physiol* 249:H95-H101 (1985).

22

Cerebral Vascular Responses to Anesthetics

Noel Flynn, Nediljka Buljubasic, Zeljko J. Bosnjak and John P. Kampine

INTRODUCTION

The volatile anesthetics, halothane and isoflurane, cause cerebral vasodilation, especially at concentrations required to induce deep planes of anesthesia. Furthermore, most,[1,2,3] but not all,[4,5] studies in animals have demonstrated that isoflurane causes less cerebral vasodilation than halothane. However, low concentrations of either volatile anesthetic do not appear to affect the cerebral blood flow (CBF). Manohar and Parks[6] demonstrated that CBF in swine models was not significantly altered at 1 MAC isoflurane. Similar observations were made by Cucchiara et al.[1] in dogs. Murphy et al.[7] reported no change in CBF in humans receiving 0.6 MAC halothane or 1.1 MAC isoflurane.

The mechanisms by which volatile anesthetics exert their action on vascular smooth muscle is unclear. In the cerebral arteries the volatile anesthetics may act via the endothelium by stimulating endothelium-derived relaxing factor (EDRF), or inhibiting endothelium-derived constricting factor (EDCF) release, or they may act directly at the cellular level affecting calcium transport. EDRF, first described by Furchgott and Zawadzki,[8] is a diffusible substance(s) which relaxes vascular smooth muscle. Recent work suggests that it is nitric oxide,[9] although there may be more than one EDRF.[10,11] Its role and interaction with volatile anesthetics are unclear. Muldoon et al.[12] suggested that halothane attenuates EDRF-mediated responses in canine carotid arteries. These findings were supported by Stone and Johns[13] who proposed that the initial vasoconstriction associated with isoflurane and enflurane in rat aorta may be due to inhibition of basal EDRF release. Blaise et al.[14] demonstrated that isoflurane-induced vasodilation occurring in canine coronary arteries is endothelium-mediated. These studies indicate that volatile agents have variable effects on the endothelium in different vascular beds.

Finally, recent studies from our laboratory[15] and others[16] have shown that isoflurane reduces whole cell calcium current in canine and rat cardiac cells. This action could also be the mechanism by which isoflurane exerts its vasodilating effect in cerebral arterial muscle cells. Therefore, in order to examine the volatile anesthetic's effect on canine cerebral arteries, a series of studies were designed to determine if: 1) the effects of isoflurane and halothane on cerebral arteries were

NOEL FLYNN, NEDILJKA BULJUBASIC, ZELJKO J. BOSNJAK, JOHN P. KAMPINE, Department of Anesthesiology, Medical College of Wisconsin, Milwaukee, Wisconsin 53226.

Mechanisms of Anesthetic Action in Skeletal, Cardiac, and Smooth Muscle
Edited by T.J.J. Blanck and D.M. Wheeler, Plenum Press, New York, 1991

similar; 2) isoflurane's effect on cerebral arteries is endothelium-mediated; 3) the volatile anesthetics affect endothelium-mediated responses in the cerebral arteries; 4) isoflurane decreases the calcium channel current in isolated cerebral arterial cells.

METHODS

All experimental procedures and protocols used in these investigations were reviewed and approved by the Animal Care Committee at the Medical College of Wisconsin and were in accordance with NIH guidelines. Adult mongrel dogs of either sex were anesthetized with halothane, and their brains were carefully dissected and removed. The middle cerebral arteries were gently dissected from the brain and placed under a microscope in a Sylgard-coated dish containing cold modified Krebs solution. Fat and other non-vascular tissue were carefully removed without touching the luminal surfaces or stretching the vessels, which were then cut into 2.5 mm segments.

The endothelium was gently removed from some rings by rubbing the vessel lumen with a finely shaved wooden pick. Rings with and without endothelium were always taken from adjacent segments of the same vessels. The rings were suspended in 15 ml, water-jacketed, temperature-controlled (37°C) tissue baths, containing modified Krebs solution which was continuously bubbled with 93.5% O_2 and 6.5% CO_2 to maintain pH within the range 7.38-7.42 and PCO_2 35-40 mmHg. Following gradual stretching to an optimal resting tension (750 mg) previously determined by tension-length measurements, the vessels were allowed to equilibrate for 90 min. The induced tensions were measured with a Grass FT103 isometric force transducer attached to a Grass Model 7 polygraph.

An ED_{50} (0.2 μM) dose of serotonin (5HT) was used to obtain prolonged, stable vasoconstriction of the cerebral vessel rings. 5HT has been implicated as a neurotransmitter regulating CBF[17] and has been shown to initiate contractions of the canine basilar artery.[18] The presence or absence of functional endothelium was determined by the vessel ring's response to the Ca^{2+} ionophore A23187 (0.2 μM). A23187 causes relaxation of isolated arteries that is completely dependent on the endothelium,[19,20] and the relaxation is not inhibited by the cyclooxygenase inhibitors indomethacin or flurbiprofen. To determine if cerebral arterial relaxation by isoflurane is related to stimulation of EDRF release, we measured the effects of increasing concentrations of isoflurane in intact vessel rings prior to and following treatment with 300 μM of N-G-monomethyl-L-arginine (LnMMA). LnMMA, an analog of L-arginine, inhibits the synthesis of nitric oxide in a dose-dependent manner.[21] Rees et al.[22] have shown that LnMMA (300 μM) inhibited endothelium-dependent relaxation induced by A23187 without affecting the endothelium-independent relaxations induced by glyceryl trinitrate or sodium nitroprusside. In our study, following the addition of LnMMA (300 μM) to the tissue baths, the vessels rings did not respond to a concentration of the Ca^{2+} ionophore A23187 which was 30 times the usual dose required to stimulate EDRF release. A further study was undertaken to determine if EDRF or factors other than EDRF had any role in the anesthetic's effect on cerebral arteries. We measured the effects of increasing concentrations of isoflurane on intact and endothelium-stripped vessel rings. Furthermore, we examined the effects of volatile anesthetics on endothelium-mediated responses by constricting the vessels with 5HT (5HT also stimulates EDRF release)[23] in the presence of halothane and isoflurane. Since it has been postulated[13] that the vasodilator action of isoflurane is mediated by PGI_2, a further subset of experiments with isoflurane and halothane was performed on vessel rings pretreated with indomethacin (10 μM).

Both halothane and isoflurane were bubbled through the tissue baths using Drager vaporizers (Drager Werk, Lubeck, Germany). Concentrations of both agents in the baths were determined by gas chromatography. Their content in the gas was measured by mass spectrometry. The millimolar concentrations of both anesthetic agents in the baths remained consistently proportional to their measured concentrations in the gas throughout the entire study. The millimolar concentrations were then converted to equivalent partial pressures for the Krebs solution and expressed as percentage of the volatile agent in the gas phase.[24] A ratio of 1 to 1.5 for halothane to isoflurane was used to calculate equianesthetic doses (figure 2). We chose 0.86% halothane and 1.28% isoflurane to represent 1 MAC in dogs.

In order to measure the effects of isoflurane (1.2 mM) on Ca^{2+} channel currents, dispersed single cerebral arterial muscle cells were placed in a perfusion chamber (22°C) on the stage of an inverted microscope. At 500× magnification, a hydraulic micromanipulator was used to position heat-polished patch pipettes with tip resistance of 1-5 MΩ on the membranes of single arterial cells. High resistance seals (2-20 GΩ) were formed, after which the pipette patch was removed by negative pressure to give electrical access to the whole cell. Whole-cell Ca^{2+} currents were elicited every 5 seconds by 200 msec depolarizing pulses generated by a computerized system (Axon Instruments). The Ca^{2+} channel current was activated by gradually increasing (in 10 mV increments) the membrane potentials from -60 to $+20$ mV. Isoflurane was added to the inflow perfusate and steady-state recordings were performed after an interval of two minutes.

The contractile responses of the middle cerebral arteries to 5HT were expressed as a percentage of the constriction induced by KCl (40 mM) in each preparation. The relaxation produced by halothane and isoflurane was expressed as a percentage of the 5HT-induced constriction. The responses of the vessel rings to each concentration of each anesthetic agent were compared by repeated measures analysis of variance. Voltage- and isoflurane-dependent changes in Ca^{2+} currents were analyzed by 2-way ANOVA and Student's t-test. Data were expressed as mean ± SEM, and $P < 0.05$ was considered statistically significant.

RESULTS

Isolated Vessel Rings

Both halothane and isoflurane induced significant relaxation of canine cerebral arteries (figure 1). These relaxations were dose-dependent although the dose-response relationship between anesthetics differed. The response to halothane began to plateau at 0.75 MAC, whereas the response to isoflurane remained almost linear over the range of concentrations used (figure 2). Since LnMMA was shown to attenuate the EDRF-mediated relaxation caused by the Ca^{2+} ionophore A23187 in cerebral vessels,[22] we repeated the above experiments in the presence of LnMMA. The results of these experiments show that LnMMA does not alter the response of cerebral arteries to isoflurane (figure 3). Furthermore, stripping the vessels of their endothelium (which abolished the relaxation response to the Ca^{2+} ionophore A23187) did not affect the degree of relaxation induced by isoflurane (figure 4). Pretreatment of the cerebral vessel rings with indomethacin did not alter their response to isoflurane or halothane. The relaxation phase of the biphasic response (figure 5) to 5HT was abolished by the removal of the endothelium and by pretreatment with LnMMA. Also, the 5HT-induced constriction of the cerebral vessel rings was significantly greater in the endothelium-stripped vessels compared to the intact vessels (figure 6). These results

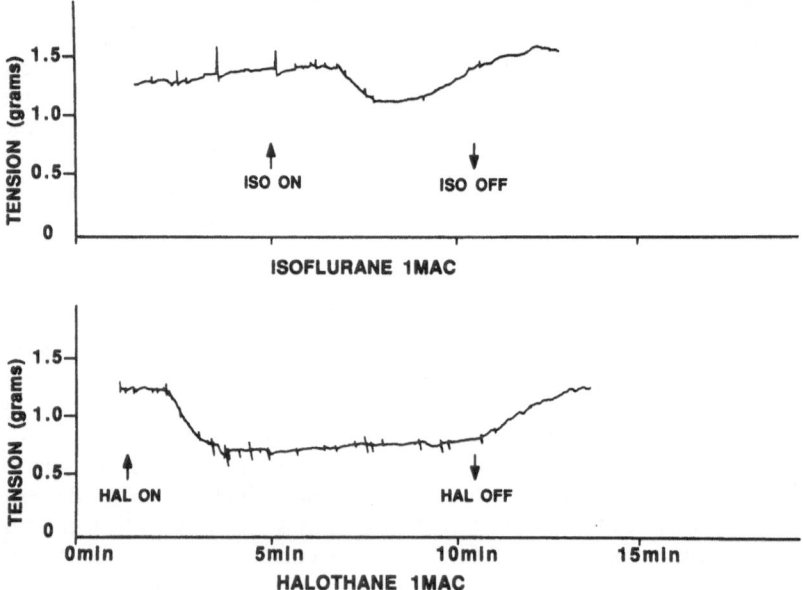

FIGURE 1. Isometric tension recordings of canine middle cerebral artery rings preconstricted with 5HT (0.2 μM) and exposed to isoflurane or halothane.

confirm that a vasodilator component of the 5HT response is dependent on the endothelium and is mediated by EDRF. Isoflurane (1.0 mM) significantly reduced the response to 5HT (to 70% of control); however, it did not affect the biphasic action of 5HT. Halothane at high concentrations (1.2 mM) also caused a significant reduction in the 5HT response (to 52% of control), and the EDRF component of 5HT action was absent in 75% of the vessels (figure 7).

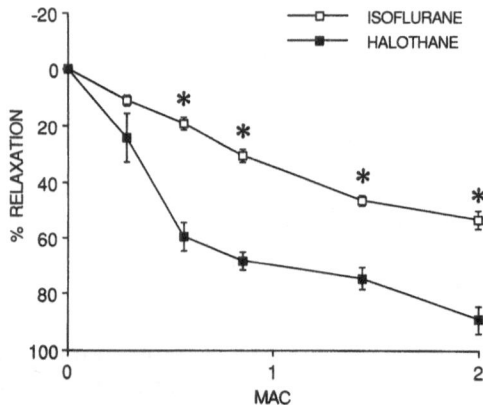

FIGURE 2. Comparison of equipotent concentrations of halothane and isoflurane. The anesthetic-induced relaxations were expressed as a percentage of the 5HT-induced tensions prior to addition of anesthetic (mean ± SEM). Halothane caused greater depression of contractions compared to isoflurane (P < 0.05).

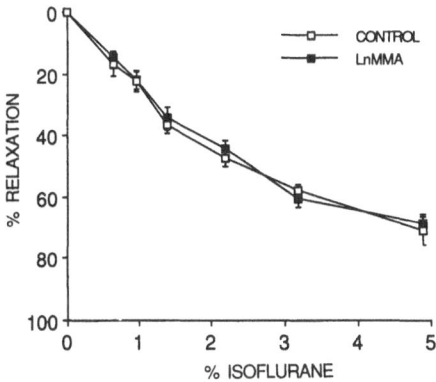

FIGURE 3. Isoflurane's effect on vessel rings prior to and following treatment with LnMMA. Relaxation was expressed as a percentage of 5HT-induced constriction. The attenuation of EDRF did not significantly alter the cerebral vessel's response to increasing concentrations of isoflurane.

Calcium Current

The effect of isoflurane on peak Ca^{2+} channel current in a single cerebral arterial cell is shown in figure 8. Isoflurane (1.2 mM) depressed the current amplitude elicited by a depolarizing step from -60 to $+20$ mV to 36.5% of control. This depressant effect of isoflurane can be observed in figure 9, where peak Ca^{2+} current is plotted as a function of membrane potential to analyze the effect of isoflurane on the current-voltage relationship for Ca^{2+} activation when progressive test pulses were applied from a constant holding potential of -60 mV. Exposure to isoflurane depressed peak Ca^{2+} current amplitudes at most of the voltages studied without shifting the current-voltage relationship. Similar results were obtained in four additional preparations.

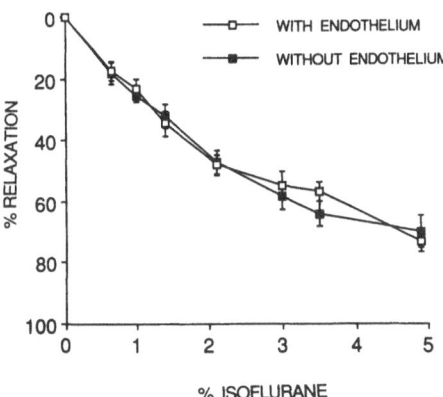

FIGURE 4. The effect of isoflurane on middle cerebral artery segments preconstricted with 5HT. This graph demonstrates that isoflurane's effect on the cerebral vasculature is not endothelium-dependent (P < 0.05). The relaxation response to increasing concentrations of isoflurane was expressed as a percentage of 5HT-induced tension prior to the addition of isoflurane.

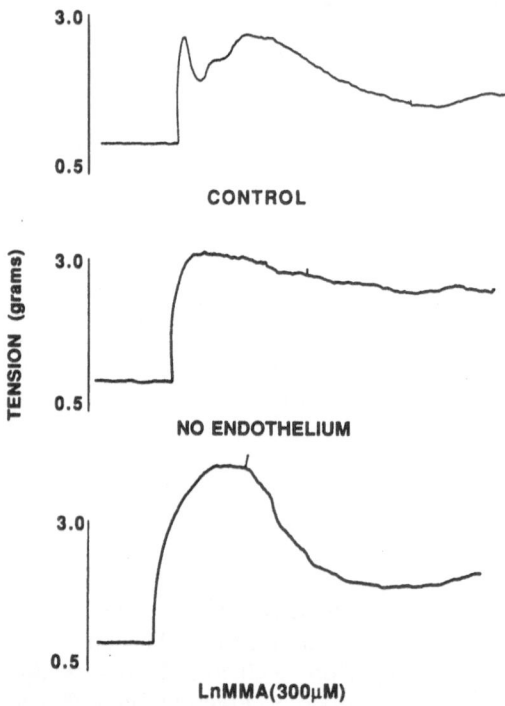

FIGURE 5. 5HT (0.2 μM) constriction of the middle cerebral arteries in a control, following removal of endothelium and in the presence of LnMMA. The dilator component of the 5HT constriction is no longer apparent in the presence of LnMMA or in endothelium-denuded vessel rings.

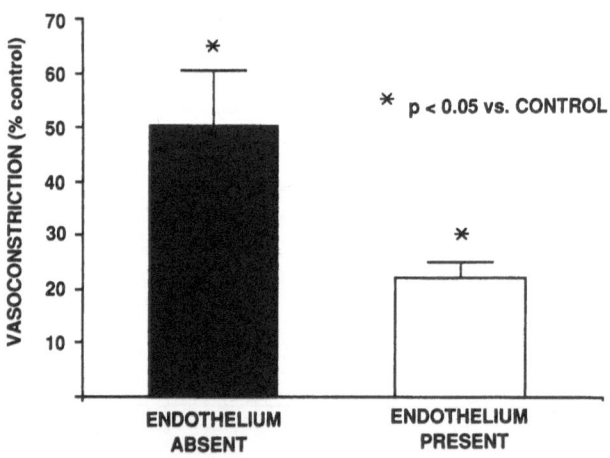

FIGURE 6. Removal of the endothelium significantly increases the response of vessel rings to 5HT ($P < 0.05$). The constrictions were expressed as a percentage of the vessel's response to 40 mM KCl (mean ± SEM).

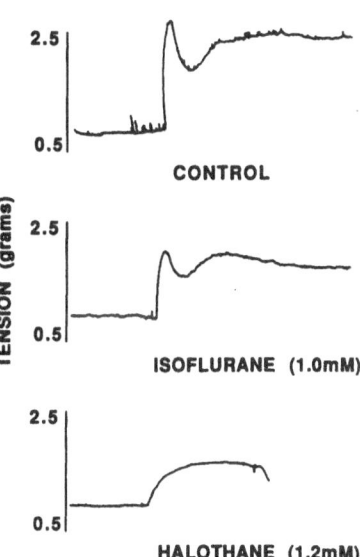

FIGURE 7. Isometric tension recordings of cerebral vessels constricted with 0.2 μM 5HT in the absence of anesthetic and in the presence of isoflurane or halothane. The biphasic response to 5HT shows initial constriction followed by a dilatation and return to final constriction. Halothane appears to have blunted this response.

CONTROL

ISOFLURANE (1.0mM)

HALOTHANE (1.2mM)

DISCUSSION

When comparing the effects of the two volatile anesthetics at equipotent doses on cerebral vessel rings, halothane is a more powerful cerebral vasodilator than isoflurane. Halothane causes almost maximal cerebral vasodilation at clinically low concentrations (*i.e.*, 0.5 to 1 MAC). The addition of LnMMA or removal of the endothelium did not alter the arterial response to isoflurane. Furthermore, the EDRF component of 5HT's action on cerebral vessels was abolished by stripping the vessels of endothelium or pretreating the vessels with LnMMA. Isoflurane did not affect the EDRF response, although this response appeared to be blunted in most of the vessels by exposure to 1.2 mM halothane. Finally, our preliminary investigations indicate that isoflurane has a direct depressant effect on Ca^{2+} current in the cerebral arterial muscle cell.

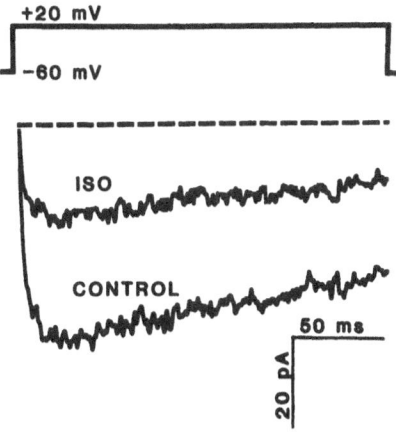

FIGURE 8. Whole cell calcium channel current in a middle cerebral arterial cell in response to stepwise depolarizing pulses (200 msec) from −60 mV to +20 mV. 1.2 mM isoflurane (ISO) reduced peak Ca^{2+} current amplitude to 36.5% of control in this cell.

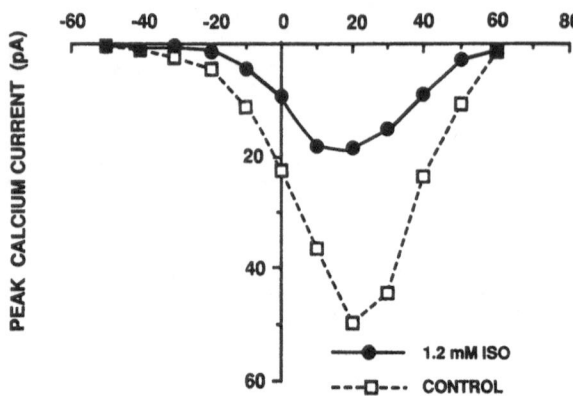

FIGURE 9. Peak current-voltage relationship for Ca^{2+} channel activation obtained before (control) and during exposure to 1.2 mM isoflurane (ISO) in the same arterial cell. Currents were generated from a holding potential of −60 mV to the command potentials indicated on the abscissa. Isoflurane depressed peak Ca^{2+} channel current at most of the voltages studied.

Previous *in vivo* human and animal studies[1,6,7] showed insignificant changes in CBF at 0.5 to 1 MAC halothane or 1 MAC isoflurane. In contrast, our *in vitro* study showed significant relaxation of the vessel rings occurring at MAC values as low as 0.3 for both halothane and isoflurane. These differences could be due to the fact that in our study (unlike *in vivo*) the vessels were preconstricted prior to anesthetic exposure. Also, *in vivo*, cerebral autoregulation is maintained at lower MAC concentrations of halothane and isoflurane.[25] In addition, our results demonstrated that both halothane and isoflurane produced dose-dependent relaxation in cerebral vessel rings. Although other *in vivo* studies have demonstrated a vasoconstrictor response to low concentrations of volatile agents followed by a vasodilator response to higher concentrations in some vascular beds,[12,13] vasodilatation was the only response to halothane and isoflurane in the cerebral rings. These apparently contradictory findings confirm that there may be regional variation in vascular response to volatile anesthetics.[26,27]

Attenuation of the EDRF response with LnMMA did not alter the cerebral vessel response to isoflurane. These findings suggest that EDRF, specifically nitric oxide, does not have a role in isoflurane-induced vasodilation. Furthermore, stripping the vessels of endothelium, while maintaining vascular smooth muscle integrity, again did not alter the cerebral vascular response to isoflurane. Thus, isoflurane acts as an endothelium-independent vasodilator on canine cerebral vascular smooth muscle.

In the cerebral arteries, isoflurane (1.0 mM) neither inhibits nor stimulates the release of EDRF. Although halothane at 1.2 mM appears to have suppressed the EDRF component of 5HT's action, it is likely that this was due to functional antagonism, as a result of the high concentrations of halothane (approximately 4 MAC) used on the cerebral vessels. Further studies are required to determine the effects of lower concentrations of halothane on EDRF release.

Although isoflurane depresses calcium channel current, this might not be its only site of action in the cerebral arterial muscle cell. Indeed, studies in cardiac tissue suggest that isoflurane may affect calcium movement or availability at the cell membrane, sarcoplasmic reticulum and contractile proteins.[28-30]

In conclusion, both halothane and isoflurane cause a relaxation of middle cerebral arteries which is dose-dependent. At equianesthetic concentrations, halothane is a more potent cerebral vasodilator than isoflurane. Isoflurane acts directly on the cerebral arterial muscle cells independently of endothelium and at least one of the mechanisms of this action is depression of the calcium channel current in these smooth muscle cells.

REFERENCES

1. R. F. Cucchiara, R. A. Theye and J. D. Michenfelder, The effects of isoflurane on canine cerebral metabolism and blood flow, *Anesthesiology* 40:571-574 (1974).
2. R. Y. Chen, F. C. Fan, R. D. Carlin, G. B. Schuessler and S. Chien, Comparison of regional cerebral blood flow during isoflurane and halothane induced hypotension, *Anesthesiology* 61:A21 (1984).
3. J. C. Drummond, M. M. Todd and M. S. Scheller, A comparison of the intrinsic cerebral vasodilating potencies of halothane and isoflurane in the New Zealand white rabbit, *Anesthesiology* 61:A364 (1984).
4. E. H. Stulken Jr, J. H. Milde, J. D. Michenfelder and J. H. Tinker, The non-linear responses of cerebral metabolism to low concentrations of halothane, enflurane, isoflurane, and thiopental, *Anesthesiology* 46:28-34 (1977).
5. T. D. Hansen, D. S. Warner, M. M. Todd, L. J. Vust and D. C. Trawick, Distribution of cerebral blood flow during halothane versus isoflurane anesthesia in rats, *Anesthesiology* 69:332-337 (1988).
6. M. Monahar and C. Parks, Regional distribution of brain and myocardial perfusion in swine while awake and during 1.0 and 1.5 MAC isoflurane anaesthesia produced without or with 50% nitrous oxide, *Cardiovasc Res* 18:344-353 (1984).
7. F. L. Murphy Jr, E. M. Kennel, R. E. Johnstone, P. L. Lief, D. R. Jobes, B. M. Tompkins, B. B. Gutsche, M. G. Behar and J. Wallman, The effect of enflurane, isoflurane and halothane on cerebral blood flow and metabolism in man, *in:* "Abstracts of Scientific Papers: Annual Meeting of the ASA" (1974) pp. 61-62.
8. R. F. Furchgott and J. V. Zawadzki, The obligatory role of endothelial cells in the relaxation of arterial smooth muscle by acetylcholine, *Nature* 288:373-376 (1980).
9. R. F. Furchgott, Studies on relaxation of rabbit aorta by sodium nitrite: The basis for the proposal that the acid-activatable inhibitory factor from bovine retractor penis is inorganic nitrite and the endothelium-derived relaxing factor is nitric oxide, *in:* "Vasodilatation: Vascular Smooth Muscle, Peptides, Autonomic Nerves, and Endothelium," P. M. Vanhoutte, ed., Raven Press, New York (1988) pp. 401-414.
10. L. J. Ignarro, R. E. Byrns and K. S. Wood, Biochemical and pharmacological properties of endothelium-derived relaxing factor and its similarity to nitric oxide radical, *in:* "Vasodilatation: Vascular Smooth Muscle, Peptides, Autonomic Nerves, and Endothelium," P. M. Vanhoutte, ed., Raven Press, New York (1988) pp. 427-435.
11. H. A. Kontos, E. P. Wei, J. T. Povlishock and C. W. Christman, Oxygen radicals mediate the cerebral arteriolar dilation from arachidonate and bradykinin in cats, *Circ Res* 55:295-303 (1984).
12. S. M. Muldoon, J. L. Hart, K. A. Bowen and W. Freas, Attenuation of endothelium mediated vasodilation by halothane, *Anesthesiology* 68:31-37 (1988).
13. D. J. Stone and R. A. Johns, Endothelium-dependent effects of halothane, enflurane, and isoflurane on isolated rat aortic vascular rings, *Anesthesiology* 71:126-132 (1989).
14. G. Blaise, J. C. Sill, M. Nugent, R. A. Van Dyke and P. M. Vanhoutte, Isoflurane causes endothelium-dependent inhibition of contracile responses of canine coronary arteries, *Anesthesiology* 67:513-517 (1987).
15. Z. J. Bosnjak, F. D. Supan and N. J. Rusch, The effects of halothane, enflurane and isoflurane on calcium current in isolated canine ventricular cells, *Anesthesiology* 74:340-345 (1991).
16. Y. Ikemoto, A. Yatani, H. Arimura and J. Yoshitake, Reduction of slow inward current of isolated rat ventricular cells by thiamylal and halothane, *Acta Anaesthesiol Scand* 29:583-586 (1985).
17. E. J. Marco, G. Balfagen, M. Solaices, C. F. Sanchez-Ferrar and J. Marin, Serotonergic innervation of cat cerebral arteries, *Brain Res* 338:137-139 (1985).

18. E. Muller-Schweinitzer and G. Engel, Evidence for mediation by $5HT_2$ receptors of 5-hydroxytryptamine-induced contraction of canine basilar artery, *Naunyn Schmeidebergs Arch Pharmacol* 324:287-292 (1983).

19. R. F. Furchgott, P. D. Cherry and J. V. Zawadzki, Endothelium dependent relaxation of arteries by acetylcholine, bradykinin and other agents, *in:* "Vascular Neuroeffector Mechanisms: 4th International Symposium," J. Bevan *et al.*, eds., Raven Press, New York (1983) pp 37-43.

20. J. G. De Mey, M. Claeys and P. M. Vanhoutte, Endothelium-dependent inhibitory effects of acetylcholine adenosine triphosphate, thrombin and arachidonic acid in the canine femoral artery, *J Pharmacol Exp Therap* 222:166-173 (1982).

21. R. M. Palmer, D. D. Rees, D. S. Ashton and S. Moncada, L-Arginine is the physiological precursor for the formation of nitric oxide in endothelium-dependent relaxation, *Biochem Biophys Res Commun* 153:1251-1256 (1988).

22. D. D. Rees, R. M. Palmer, H. F. Hodson and S. Moncada, A specific inhibitor of nitric oxide formation from L-arginine attenuates endothelium-dependent relaxation, *Br J Pharmacol* 96:418-424 (1989).

23. D. S. Houston and P. M. Vanhoutte, Comparison of serotonergic receptor subtypes on the smooth muscle and endothelium of the canine coronary artery, *J Pharmacol Exp Therap* 244:1-10 (1988).

24. M. J. Halsey, Physicochemical properties of inhalational anaesthetics, *in:* "General Anesthesia," Vol 1, T. C. Gray, J. E. Utting and J. F. Nunn, eds., Butterworths, London (1980) pp. 45-65.

25. J. C. Drummond, M. M. Todd and H. M. Shapiro, Cerebral blood flow autoregulation in the cat during anesthesia with halothane and isoflurane, *Anesthesiology* 59:A305 (1983).

26. R. F. Hickey and E. I. Eger, Circulatory pharmacology of inhaled anesthetics, *in:* "Anesthesia," R. D. Miller, ed., Churchill Livingstone, New York (1986) pp 649-666.

27. S. F. Vatner and N. T. Smith, Effects of halothane on left ventricular function and distribution of regional blood flow in dogs and primates, *Circ Res* 34:155-167 (1974).

28. C. Lynch III, Are volatile anesthetics really calcium entry blockers? *Anesthesiology* 61:644-646 (1984).

29. H. L. Price and S. T. Ohnishi, Effects of anesthetics on the heart, *Fed Proc* 39:1575-1579 (1980).

30. J. Y. Su and J. G. Bell, Intracellular mechanism of action of isoflurane and halothane on striated muscle of the rabbit, *Anesth Analg* 65:457-462 (1986).

23

Interactions of Volatile Anesthetics and Reactive Oxygen Intermediates on Vascular Smooth Muscle

William Freas, Rocio LLave, Jayne Hart, Diane Golightly, John Nagel and Sheila Muldoon

INTRODUCTION

Reactive oxygen intermediates (ROIs, also referred to as oxygen-derived free radicals or reactive oxygen metabolites) have been implicated in several physiological and pathological processes.[1-3] In particular, they have been reported to be associated with several vascular abnormalities, including those that occur during hypertension,[4] reperfusion injury,[5] transplant rejection,[3] inflammation, premature aging,[6] radiation injury, diabetes[7] and endotoxic shock.[8] An increase in P_aO_2, especially after hypoxia, is a major stimulus for production of ROIs.[9]

Reactive oxygen intermediates were produced in this study by a technique presently utilized in photodynamic therapy (PDT). PDT is presently undergoing phase III trials for the treatment of certain human neoplasms. For this treatment a patient is injected with a photosensitizer, generally hematoporphyrin derivative (HpD), and then 48 to 72 hours later, after this photosensitizer has cleared from healthy tissue, tumors are illuminated with red laser light. The interaction of light and photosensitizer produces reactive oxygen intermediates which are toxic to the tumor itself. With reduced laser intensity and lower photosensitizer concentration, this technique can be used to produce a range of non-lethal levels of reactive oxygen intermediates both *in vivo* and *in vitro*. In this study PDT was utilized to examine the interaction of halothane and reactive oxygen intermediates on the vascular contractility of isolated blood vessels.

Enhanced oxygen radical production has been associated with a number of vascular abnormalities ranging from loss of endothelial activity to total vascular collapse. Therefore, understanding the interactions of volatile anesthetics and the expression of this vascular damage may be important in the anesthetic management of patients in conditions where elevated production of oxygen radicals occurs. Volatile anesthetics, especially halothane, have been shown to have a beneficial effect in minimizing ischemic and reperfusion-related damage.[10-12] In the isolated heart preparation, halothane has been shown to inhibit calcium accumulation following

WILLIAM FREAS, ROCIO LLAVE, JAYNE HART, DIANE GOLIGHTLY, JOHN NAGEL, SHEILA MULDOON, Department of Anesthesiology, Uniformed Services University of the Health Sciences, Bethesda, Maryland 20814.

Mechanisms of Anesthetic Action in Skeletal, Cardiac, and Smooth Muscle
Edited by T.J.J. Blanck and D.M. Wheeler, Plenum Press, New York, 1991

247

myocardial ischemia,[13] which has been associated with the extent of myocardial damage following reperfusion injury. This present study was designed to examine directly the effects of volatile anesthetics on vascular function and calcium permeability during enhanced production of oxygen radicals.

METHODS

Saphenous veins were obtained from anesthetized mongrel dogs (pentobarbital sodium, 30 mg/kg iv) and stored in Krebs-Ringer solution. The veins were then dissected free of adherent connective tissue and cut into rings (5 mm in length). Two triangular shaped stainless steel wire holders were inserted in the lumen of the rings and the preparation was suspended in organ baths containing aerated (95% O_2, 5% CO_2) Krebs-Ringer solution (25 ml, 37°C). One wire was secured to the bottom of the bath while the other was attached to an isometric force transducer (Grass FT-O3) connected to a chart recorder for continuous monitoring of tension. Rings were stretched to an initial tension of 2 grams for 15 min and then optimal passive tension was determined by the response to electrical stimulation (10 V, 5-10 Hz, 2 msec, for a duration of 10 sec) from two platinum plates on either side of the vessel connected to a square wave signal splitter and a Grass stimulator (Model S-44).

Reactive oxygen intermediates were produced by pretreating the rings for 30 min with photosensitizer (3 µg/ml final bath concentration of hematoporphyrin derivative, HpD) prepared as previously described.[14] The tissue was then washed to remove extracellular HpD and equilibrated for at least 5 min. Light was supplied by a helium-neon laser (632.8 nm, Melles Groit, with an intensity of 1.5 to 3 mW). Halothane was delivered from a calibrated vaporizer to produce concentrations of 1 to 3% in the aerating gas (95% O2, 5%CO2) as monitored by an infrared halothane analyzer (Sensor Medics, Model LB-2, Anaheim, CA). Bath halothane concentrations were also determined in 100 µl aliquots of the bath Krebs-Ringer solution following extraction in heptane and gas chromatographic analysis as previously described.[15]

Analysis of data was performed using a blocked two-way analysis of variance. After determining if there were significant treatment effects, ranked *a posteriori* comparisons, using a Studentized range test, were performed to determine which means were significantly different. Data are presented as mean ± SEM. A probability level of less than 0.05 was considered significant.

RESULTS

Reactive oxygen intermediates produced during laser illumination of photosensitizer-pretreated canine saphenous veins significantly contracted vascular smooth muscle (figure 1A). When halothane (3%) was added to the carrier gas before laser illumination, it significantly attenuated the ROI-induced contraction. The vascular rings contracted when the halothane was turned off regardless of whether the laser illumination was on (figure 1B) or off (figure 1C). When halothane was turned on during the laser illumination, the tissue significantly relaxed and remained relaxed until the halothane was turned off (figure 1D). This relaxant property of halothane was evident even if halothane was applied 10 min following termination of laser illumination (figure 1E).

A comparison of the effects of 3% halothane and 3% isoflurane was made on these ROI-induced contractions. While 3% halothane significantly attenuated the ROI-induced contractions, 3% isoflurane did not (figure 2). When the anesthetic was turned off, the vascular smooth muscle treated with halothane contracted while the preparation treated with isoflurane relaxed.

Lower concentrations of halothane were examined and concentrations as low as 1% halothane significantly attenuated tension development (figure 3). With all halothane concentrations, the tension increased when the anesthetic delivery was stopped.

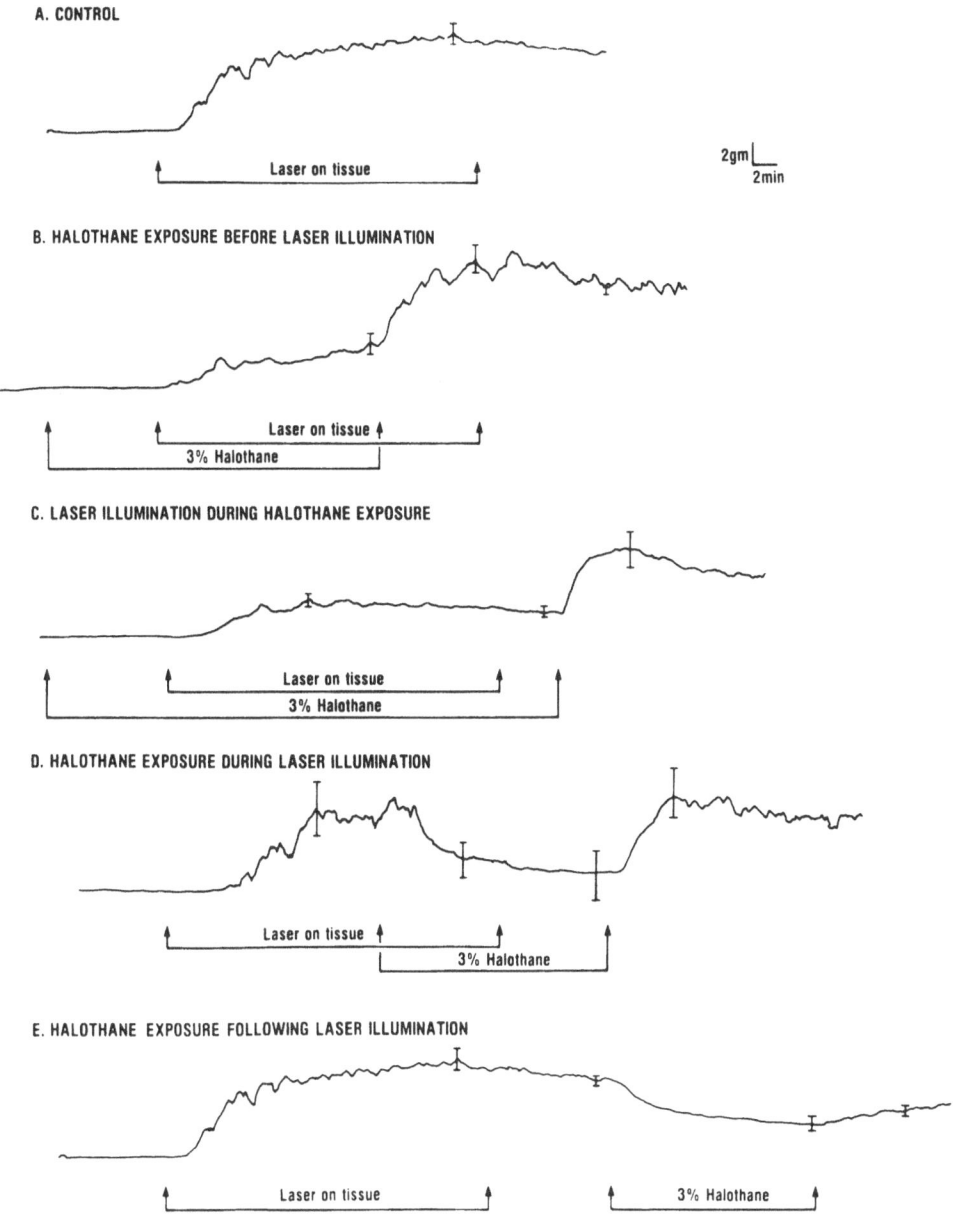

FIGURE 1. Representative tracings of the effects of halothane on tension generation during production of ROIs as a result of laser illumination of the photosensitizer HpD. Halothane was applied to the carrier gas at various times to demonstrate the temporal relationship between exposure to halothane (3%) and tension during ROI generation (bars represent means ± SEM, n = 5).

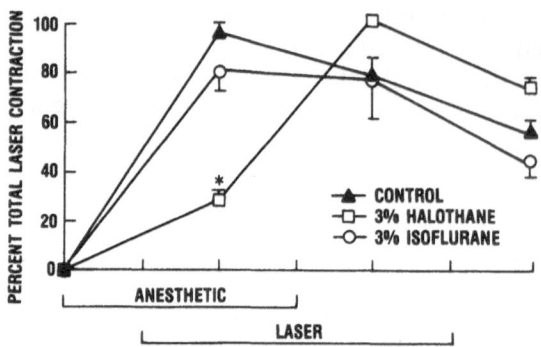

FIGURE 2. Comparison of the effects of halothane (3%) and isoflurane (3%) on canine saphenous vein contractions (n = 5). Results are expressed as percent maximal laser contraction; 100% = 9.1 ± 1.4, 9.5 ± 2.3 and 11.2 ± 2.4 g for control, halothane and isoflurane, respectively. Asterisk (*) indicates significant difference compared with control contraction in absence of anesthetic.

In an effort to determine if this effect of halothane on ROI-induced contractions was due to an effect on membrane permeability, the responses of ROI-exposed saphenous veins to changes in extracellular calcium were examined. ROI-induced contractions promptly relaxed in the presence of calcium-free solutions. After 4 washes with calcium-free solutions over 30 min, calcium was returned to the bathing solution. The tissue started to contract when the level of calcium reached 0.25 mM

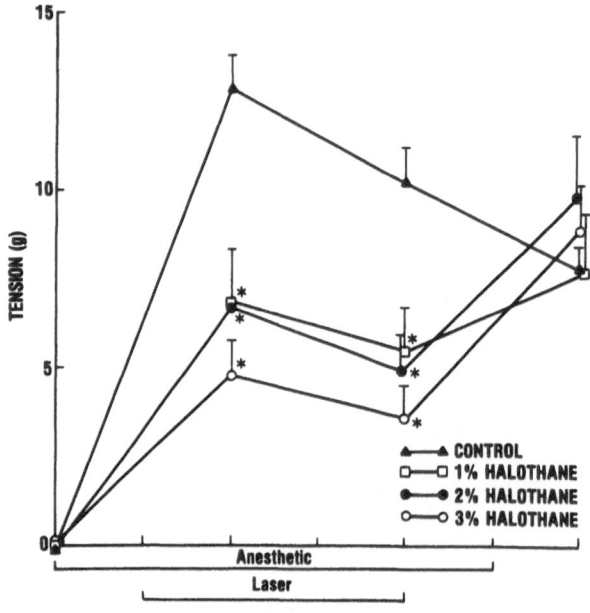

FIGURE 3. Effect of different concentrations of halothane (1, 2, and 3%) on tension development during generation of ROIs by laser illumination of an HpD pretreated canine saphenous vein (n = 5). Note the rapid increase in tension when the anesthetic was turned off. Asterisk (*) indicates significant difference compared with control contraction in absence of halothane.

and contracted to 80% of its original ROI tension when the calcium level reached 2.5 mM (figure 4). Additional increases in bath calcium did not further increase tension. If control vein rings were washed with standard Krebs solution over this period of time, their tension fell to 55% of the initial peak ROI-induced tension. When calcium was incrementally added to the bathing solutions containing these control vascular rings in Krebs solution, the increased extracellular calcium did not significantly alter tension development. If vascular rings did not receive both laser illumination and photosensitizer pretreatment, the tissue remained at its initial resting tension. When these tissues were exposed to a calcium-free solution and subsequently submitted to increases in bath calcium concentration, their tension remained unchanged.

The increase in tension following a return of calcium to canine saphenous vein rings that had previously been contracted by ROIs and then washed in calcium-free solution for 30 min was examined in the presence of two concentrations of halothane. Halothane (3%) significantly attenuated development of tension in vascular rings upon return of calcium to the organ bath containing these precontracted saphenous vein rings (figure 5). Distilled halothane, which did not contain the preservative thymol, produced the same response as the commercial preparations of halothane.

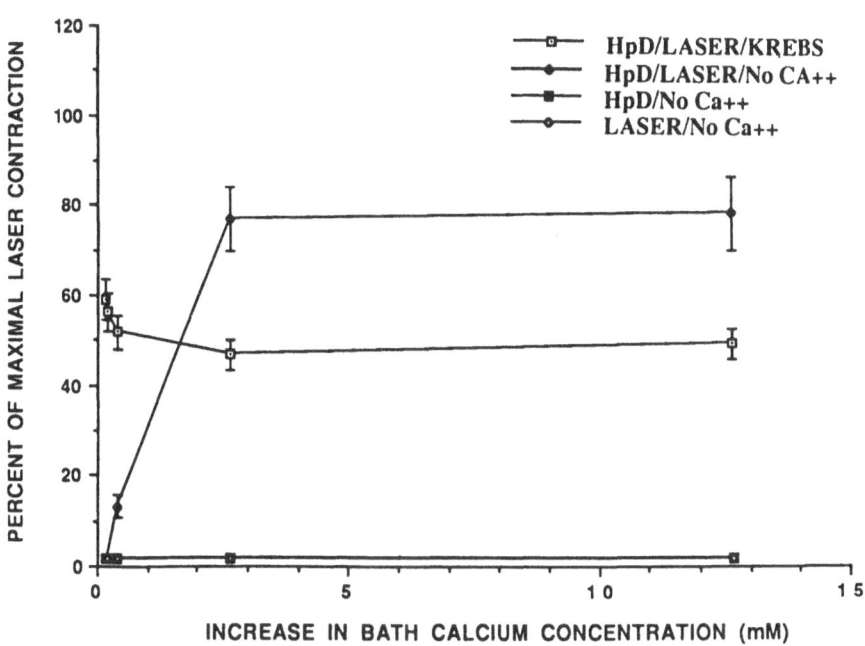

FIGURE 4. Effect of incremental increases in bath calcium concentrations on basal muscle tension of the canine saphenous vein following ROI generation. Two treatment groups received both HpD pretreatment and laser illumination. One group was washed four times with calcium free solution for thirty minutes (solid diamonds) the other group was kept in Krebs Ringer solution for 30 min (open squares). The other two groups (solid squares and open diamonds) received either HpD or laser illumination, not both. They were also washed four times with calcium free Krebs solution and when calcium was returned to the bath they did not contract. Results expressed as percent maximal laser contraction, 100% = 10.8 ± 1.1 and 10.2 ± 1.2 g for laser/Krebs and laser/no Ca^{2+}, respectively.

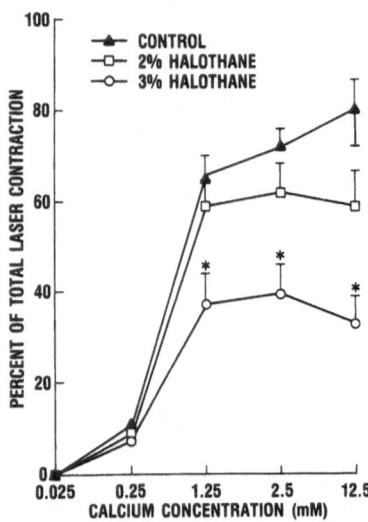

FIGURE 5. Effect of halothane on the contractile response due to increased bath calcium concentrations thirty minutes following laser illumination and photosensitizer pretreatment (n = 5). Asterisk (*) indicates significant difference compared with control contraction in absence of anesthetic. Values expressed as percent maximal laser contraction, 100% = 13.1 ± 1.5, 13.9 ± 1.3 and 13.6 ± 1.0 g for control, halothane 2% and 3%, respectively.

DISCUSSION

ROIs are continually being produced and scavenged by normal cells. During photodynamic therapy, as well as during several pathological conditions, production of these reactive oxygen intermediates exceeds the cells' ability to scavenge them. Under these conditions oxygen radicals may exert toxic effects. This study demonstrates that halothane can attenuate and delay vascular contractions due to the generation of reactive oxygen intermediates following laser illumination of isolated canine saphenous veins.

ROI-induced contractions produced by illumination of the photosensitizer HpD are long lasting and are primarily dependent on changes of the vascular smooth muscle membrane permeability to calcium.[16] These contractions may in part account for the increased peripheral resistance that occurs in vascular conditions associated with elevated production of ROIs, such as reperfusion injury.

The volatile anesthetic halothane can act either as a free radical generator, or as a free radical scavenger. Halothane can either accept electrons as it undergoes reductive metabolism, acting as a scavenger,[17] or it can generate radicals when it undergoes oxidative metabolism,[18] which reportedly accounts for some of its hepatotoxicity. This toxicity is observed following administration of halothane, but not with isoflurane or enflurane, which do not generate free radicals.[19] However, anesthetics do not need to be metabolized to interfere with biochemical processes which involve the transfer of electrons. The presence of a strong electronegative field due to the halogenation of these volatile anesthetics can interfere with electron transport reactions. While halothane, but not isoflurane, inhibited ROI-induced contractions in this study, it is unknown if the effect was due to halothane itself or one of its metabolites.

These experiments demonstrated that halothane sharply attenuated the constrictor response as a result of a vascular ROI production. However, this action was only transitory. When ROI production was terminated and exposure to halothane stopped, the vasoconstriction returned (figure 1). Therefore, the *in vitro* effect of halothane occurred only during halothane delivery. Additional studies must be performed to determine if this effect is due to a direct action of halothane on vascular smooth muscle, or an indirect action as a result of altering endothelial cell or adrenergic nerve terminal activity.

There was a dose-response relationship to halothane (between 1 and 3%). Isoflurane, however (3% in the carrier gas) was not effective in producing the contractile effect following ROI generation.

In order to determine the mechanisms of halothane's interaction with vascular tissue, an experiment was designed to separate out radical scavenger effects of the anesthetic from membrane stabilizing effects. This was performed by relaxing ROI-induced contractions for 30 min in a calcium-free solution. When calcium was reintroduced to the bath the vascular rings contracted for a second time, in the absence of ROI generation.[16] Of these two contractions, the initial ROI contraction could be blocked by oxygen scavengers and reduced in calcium-free solutions. The second contraction, which occurred upon return of calcium to the bathing solution, could be eliminated by calcium-free solutions, significantly attenuated by calcium channel blockers, and not affected by radical scavengers.[16] Therefore, these two sequential contractions of the same preparation by different stimuli can be used to differentiate the effect of oxygen radical scavengers from membrane permeability stabilizers. Halothane significantly attenuated the contraction following laser illumination of HpD-pretreated vascular tissue. However, it also attenuated the second contraction when calcium was added back 30 min after the termination of illumination, when ROIs were no longer produced. Therefore, halothane's action was unlike that of scavengers and anti-oxidants (catalase and ascorbate) and more like that of calcium channel blockers. These results suggest that the action of halothane on ROI-induced vascular contractions is due primarily to its action on the membrane permeability to calcium.

Whether volatile anesthetics have a beneficial effect during pathological conditions where ROI production is elevated is controversial. Several investigators using various animal models have reported a beneficial effect of halothane on cardiac[10,20] and renal tissue[21] following ischemia. While this action is believed to be due to reduced metabolic demand, the anesthetic-ROI interactions in these conditions have not been examined. Some post-ischemic dysfunction has been associated with the influx of calcium and enhanced membrane permeability.[12-13] The results of this *in vitro* study demonstrate that halothane attenuated ROI-induced vasoconstriction of isolated canine saphenous veins. Should halothane have the same effect *in vivo* following ROI generation, then enhanced perfusion of the previously ischemic area may occur.

In addition to an effect on ROI production, anesthetics could also interfere with the expression of the ROIs, once produced. ROIs have a very short half life;[6] however, once produced they will chemically react with cellular components resulting in an irreversible effect which may last long after ROI production is terminated. The second objective of this study was to determine if halothane alters the expression of ROI-induced damage. Understanding and minimizing the expression of ROI-induced damage could have far more clinical relevance than preventing the production of ROIs. While oxygen radical scavengers can prevent formation of oxygen radicals, and therefore prevent their damaging effects, they are not effective unless the tissue is pretreated with scavengers. Free radical scavengers have been shown to be beneficial in experimental animals by reducing infarct size associated with myocardial ischemia and reperfusion.[22-24] The major problem with this proposed therapy is that it requires pretreatment of the tissue with oxygen radical scavengers. In some cases this would not be practical, since one could not always predict the occurrence of the pathological changes in oxygen radical production. If for example during reperfusion these free radicals act primarily as vasoconstrictors, then once the damage is done, vasodilators may be more effective than scavengers in maintaining tissue perfusion and reducing the chance of additional tissue damage. The vasodilatory effect of anesthetics on vascular smooth muscle contractions following radical generation may play a beneficial role in minimizing surrounding tissue damage during reperfusion.

These *in vitro* ROI-induced contractions may be a suitable model to study the mechanisms of action of ROIs on vascular activity and may be of value in understanding vascular abnormalities that occur in certain clinical and pathological conditions.

In summary, ROIs produced by illumination of the photosensitizer HpD resulted in a sustained contraction of vascular smooth muscle. This contraction was attenuated by halothane. This action of halothane appears to be the result of a decrease in the membrane calcium permeability which was increased following generation of ROIs.

ACKNOWLEDGEMENTS

The authors wish to thank Booker Swindall, Drs. D. Eliades, J. Dabney and D. E. Dobbins for their assistance in obtaining canine blood vessels. Our gratitude also goes to Robert Jones for minor equipment fabrication. This work was supported in part by USUHS Grant R08006 and SDIO Grant GM8027.

REFERENCES

1. J. Feher, G. Csomos, A. Vereckei, "Free Radical Reactions in Medicine," Springer-Verlag, Berlin (1985).
2. G. M. Rubanyi, Vascular effects of oxygen derived free radicals. *Free Radical Biol Med* 4:107-120 (1988).
3. D. A. Parks, G. B. Bulkley, and D. N. Granger, Role of oxygen free radicals in shock, ischemia and organ preservation, *Surgery* 94:428-431 (1983).
4. E. P. Wei, H. A. Kontos, C. W. Christman, D. S. Dewitt and J. T. Povlishock, Superoxide generation and reversal of acetylcholine induced cerebral arteriolar dilation after acute hypertension, *Circ Res* 57:781-787 (1985).
5. R. J. Korthius, J. K. Smith and D. L. Carden, Hypoxic reperfusion attenuates postischemic microvascular injury, *Am J Physiol* 256:H315-H319 (1989).
6. W. A. Pryor, Oxy-radicals and related species: Their formation, lifetimes, and reactions, *Ann Rev Physiol* 48:657-667 (1986).
7. P. K. Ganguly, K. S. Dhalla, I. R. Innes, R. E. Beamish, N.J. Dhalla, Altered norepinephrine turnover and metabolism in diabetic cardiomyopathy, *Circ Res* 59:684-693 (1986).
8. K. McKechnie, B. Furman, and J. Parratt, Modification by oxygen free radical scavengers of the metabolic and cardiovascular effects of endotoxin infusion in conscious rats, *Circ Shock* 19:429-439 (1986).
9. J. M. McCord, Oxygen-derived free radicals in postischemic tissue, *N Engl J Med* 312:159-163 (1985).
10. B. A. Macleod, P. Augereau, and M. J. A. Walker, Effects of halothane anesthesia compared with fentanyl anesthesia and no anesthesia during coronary ligation in rats, *Anesthesiology* 58:44-52 (1983).
11. J. Dolman and D. V. Godin, Myocardial ischaemic/reperfusion injury in the anaesthetized rabbit: Comparative effects of halothane and isoflurane, *Can Anaesth Soc J* 33:443-452 (1986).
12. D. C. Warltier, M. H. Al-Wathiqui, J. P. Kampine, W. T. Schmeling, Recovery of contractile function of stunned myocardium in chronically instrumented dogs is enhanced by halothane or isoflurane, *Anesthesiology* 69:552-565 (1988).

13. S. Hoka, Z. J. Bosnjak, J. P. Kampine, Halothane inhibits calcium accumulation following myocardial ischemia and calcium paradox in guinea pig hearts, *Anesthesiology* 67:197-202 (1987).

14. W. Freas, J. L. Hart, D. Golightly, H. McClure, and S. M. Muldoon, Contractile properties of isolated vascular smooth muscle after photoradiation, *Am J Physiol* 256:H655-H664 (1989).

15. S. M. Muldoon, P. M. Vanhoutte, R. R. Lorenz and R.A. Van Dyke, Venomotor changes caused by halothane acting on the sympathetic nerves, *Anesthesiology* 43:41-48 (1975).

16. W. Freas, J. L. Hart, D. Golightly, H. McClure, D. Rodgers, and S. M. Muldoon, Vascular interactions of calcium and reactive oxygen intermediates following photoradiation, *J Cardiovasc Pharmacol* 17:27-35 (1991).

17. R. A. Van Dyke and C. L. Wood, Binding of radioactivity from ^{14}C-labeled halothane in isolated perfused rat livers, *Anesthesiology* 38:328-332 (1973).

18. V. L. Kubic and M. W. Anders, Mechanism of the microsomal reduction of carbon tetrachloride and halothane, *Chem Biol Interact* 34:201-207 (1981).

19. J. L. Plummer, A. L. J. Beckwith, F. N. Bastin, J. F. Adams, M. J. Cousins, P. Hall, Free radical formation in vivo and hepatotoxicity due to anesthesia with halothane, *Anesthesiology* 57:160-166 (1982).

20. T. L. Jang, B. A. Macleod and M. J. A. Walker, Effects of halogenated hydrocarbon anesthetics on responses to ligation of a coronary artery in chronically prepared rats, *Anesthesiology* 59:309-315 (1983).

21. M. J. Rice, J. A. Hjelmhaug, J. H. Southard, The effect of halothane, isoflurane, and verapamil on ischemic-isolated rabbit renal tubules, *Anesthesiology* 71:738-743 (1989).

22. P. J. Simpson, J. K. Mickelson and B. R. Lucchesi, Radical scavengers in myocardial ischemia, *Fed Proc* 46:2413-2421 (1987).

23. S. W. Werns, M. J. Shea, E. W. Driscoll, C. Cohen, G.D. Abrams, B. Pitt, B. R. Lucchesi, The independent effects of oxygen radical scavengers on canine infarct size: Reduction by superoxide dismutase but not catalase, *Circ Res* 56:895-898 (1985).

24. M. L. Myers, R. Bolli, R. F. Lekich, C. J. Hartley, R. Roberts, Enhancement of recovery of myocardial function by oxygen free-radical scavengers after reversible regional ischemia, *Circulation* 72:915-921 (1985).

24

Isoflurane-, Halothane- and Agonist-Evoked Responses in Pig Coronary Arteries and Vascular Smooth Muscle Cells

J. Christopher Sill, M. Ozhan, R. Nelson, C. Uhl

INTRODUCTION

The actions of volatile anesthetics are not restricted simply to depression of consciousness but extend beyond the nervous system to include the heart and circulation. In the circulation, isoflurane and halothane cause vasodilatation; although a part of this effect results simply from decreased nervous system activity, the anesthetics undoubtedly also have direct actions on blood vessels themselves. The nature and magnitude of the anesthetics' vascular effects vary depending upon animal species, vascular bed and type of vessel. In the coronary circulation of intact pigs and in isolated coronary arteries removed from pig, dog and human hearts, isoflurane and halothane have both been shown to attenuate agonist-induced contractile responses.[1-4] The purpose of the current experiments has been to investigate the effects and mechanisms of action of volatile anesthetics in coronary arteries and in cultured vascular cells.

The non-anesthetic effects of isoflurane and halothane on coronary artery constriction are not without possible benefit to patients with coronary artery disease who undergo surgery. In the 1980's a shift occurred in concepts concerning the genesis of myocardial ischemia.[5] It had previously been assumed that diseased coronary arteries were narrowed, rigid tubes of fixed intraluminal diameter. Coronary stenoses were thought to produce a permanently fixed degree of vessel narrowing. However, these views have been abandoned, and it is now generally accepted that even atherosclerotic coronary arteries can constrict and dilate. Episodic increases in the vasomotor tone of epicardial coronary arteries is now recognized as being responsible not only for variant angina but also for angina in patients with typical atherosclerotic coronary artery disease.[5] Myocardial ischemia may be caused by dynamic reduction in coronary blood flow or by an excessive increase in myocardial demand or by both. In the setting of myocardial ischemia, anesthetics may have a dual role in protecting the heart, both by depressing myocardial demand and by preserving blood flow to the heart. However, it must be remembered that currently there is no direct evidence to

J. CHRISTOPHER SILL, M. OZHAN, R. NELSON, C. UHL, Department of Anesthesiology, Mayo Clinic and Mayo Foundation, Rochester, Minnesota 55905.

Mechanisms of Anesthetic Action in Skeletal, Cardiac, and Smooth Muscle
Edited by T.J.J. Blanck and D.M. Wheeler, Plenum Press, New York, 1991

suggest that volatile anesthetics inhibit coronary artery constriction in patients with coronary artery disease.

Evidence obtained from *in vivo* experiments in pigs and from studies of isolated dog, pig and human coronary arteries suggests that volatile anesthetics attenuate contractions of coronary arteries evoked by a number of agonists.[1-4] The mechanisms by which inhalational anesthetics decrease tone is not known. The site of anesthetic action could lie anywhere along the pathway that signals contraction, beginning with receptor function and ending with the effects of Ca^{2+}, protein kinase C or other signals on cell function. Energetics of actin-myosin interaction may also be a possible site of anesthetic effect. However, as volatile anesthetics are lipophilic, an action on structures and events within membranes might be the most likely location of their effects.

A major signal transduction pathway involves inositol lipids.[6] The actions of inositol lipids are of enormous importance to the function of cells. Stimulation of many types of receptors located on the outside surface of cells, especially those associated with vascular smooth muscle contraction and with endothelial cell function, result in hydrolysis of membrane-bound inositol lipids. The membrane-bound lipid-phosphatidylinositol-4,5-biphosphate (PIP_2) is hydrolysed when an agonist interacts with its receptor and activates a G-protein which then stimulates the enzyme phospholipase C. The action of phospholipase C on inositol lipids results in the formation of a least two second messengers—inositol 1,4,5 triphosphate (IP_3) and diacylglycerol (DAG). The latter acts by stimulating protein kinase C whereas IP_3 diffuses into the cytosol to release Ca^{2+} from intracellular stores thought to be represented by the sarcoplasmic reticulum. IP_3 may be tetra-phosphorylated to $I(1345)P_4$ which may have a role, in conjunction with IP_3, in activating Ca^{2+} entry from outside the cell directly into the sarcoplasmic reticulum to replenish stores. The effects of PIP_2 hydrolysis and second messenger formation have important consequences not only for short term functions such as contraction and secretion, but also for longer term events such as gene expression and protein synthesis. It is possible that lipid soluble volatile anesthetics, by an action on cell membranes, may alter signal transduction processes and in this way exert some of their effects.

In cardiac muscle the volatile anesthetics have marked effects on Ca^{2+} movement,[7] suggesting that Ca^{2+} flux may also be a site of anesthetic action in vascular smooth muscle and in endothelial cells. Ca^{2+} is one of the simplest second messengers, entering the cytosol through membrane channels where it ultimately initiates events such as contraction in smooth muscle cells or EDRF secretion in endothelial cells. The Ca^{2+} signal is highly complex and may be localized to discrete regions of the cell, may migrate through the cytosol in the form of a wave, and concentrations of Ca^{2+} within the cell may oscillate.[8] Current evidence suggests that volatile anesthetics have important effects upon intracellular Ca^{2+} transients, but whether or not they occur as a consequence of anesthetic action on the pathways that signal Ca^{2+} release and Ca^{2+} entry or upon Ca^{2+} channels themselves is not known.

In contrast to the volatile anesthetics, nitrous oxide, which is also a versatile and safe inhalational anesthetic, has been regarded as having little effect on the circulation. However, in intact dogs nitrous oxide has been shown to evoke epicardial coronary artery constriction. This effect is puzzling as in cardiac muscle, nitrous oxide causes a degree of depression of contractile function. Subsequently, it was observed that in intact pigs, nitrous oxide's constrictor effect occurs only in vessels with intact endothelium. This evidence implies that the site of nitrous oxide action is not coronary artery smooth muscle, but instead the anesthetic may induce constriction via an endothelium-dependent mechanism.

CURRENT STUDIES

Experiments conducted in the investigators' laboratory were performed using four differing and independent methods. Quantitative angiography was used in anesthetized pigs to permit measurement of coronary artery diameters. In a second type of experiment, isolated coronary arteries were studied *ex-vivo* by suspending rings in aerated Krebs-Ringer-filled organ chambers and measuring generation of isometric force. Thirdly, cultured vascular smooth muscle and endothelial cells were used to investigate changes in intracellular Ca^{2+} by loading the cells with the fluorescent indicator indo-1 and imaging them in a flow cytometer. Inositol phosphate concentration measurements were made using both cultured cells and isolated segments of coronary arteries.

Coronary Artery Constriction

The purpose of the experiments was to determine if clinically useful concentrations of isoflurane and halothane attenuated epicardial coronary artery constriction evoked by serotonin in intact pigs. Serotonin is an endogenous vasoconstrictor that interacts with Ca^{2+}-mobilizing 5HT receptors on vascular smooth muscle. It has been implicated in the genesis of angina in humans. Computerized quantitative coronary angiography was used to determine the dilator effects of 0.5, 1.25 and 2% isoflurane and halothane on left anterior descending coronary arteries preconstricted by constant infusion of serotonin. The effects of pretreatment with 1.0% isoflurane or 1.0% halothane on constriction evoked by bolus infusion of serotonin were also studied. The pigs were anesthetized intravenously with fentanyl and ketamine. Serotonin was infused directly into the coronary arteries via a very small catheter placed in the vessel. The coronary endothelium was gently removed in some animals using a small angioplasty catheter.

Coronary angiograms were obtained during the injection of 7 ml of meglomine diatrizoale (Renografin 76) via the guide catheter; cassette-type x-ray film (Kodak T-Mat L Diagnostic) was exposed at 110 kV, 300 mA for 6 ms (Picker GX 850 and General Electric 300) gated to mid-diastole using an R-wave-triggered time delayed switch (Mayo Engineering). The opacified edges of the coronary artery lumens were manually traced and digitized using a computer (PDP 11/34, Digital Equipment). The program calculated luminal diameter at 1-mm intervals. Three scans of each vessel were performed and the results averaged. Angiograms, with their identification numbers obscured, were shuffled prior to analysis to prevent observer bias.

Results demonstrated that isoflurane and halothane both attenuate coronary artery constriction in intact pigs (figure 1). Coronary artery dilation occurred in vessels both with or without endothelium indicating that the effect was not endothelium dependent. Anesthetic concentrations as low as 0.5% had a small but significant effect in relaxing constriction. Pretreatment of the pigs with isoflurane 1.0% or halothane 1.0% was also effective in attenuating constriction induced by intracoronary bolus infusions of serotonin (figure 2). This study provides the first documentation that isoflurane and halothane dilate constricted coronary arteries in intact animals.

Isolated Coronary Arteries and Contractile Agonists

The purpose of the experiments was to determine if isoflurane and halothane attenuated contractions evoked by a range of agonists or, alternatively, if the anesthetics' effects are limited to specific agonists. Four contractile agonists—serotonin, acetylcholine, histamine and endothelin—were studied. All are known to interact with

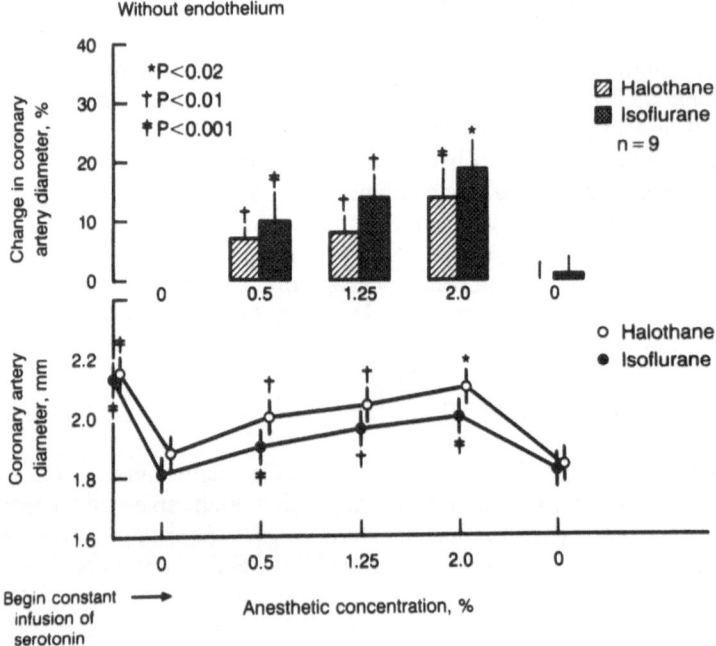

FIGURE 1. Responses of left anterior descending coronary arteries in intact pigs. Serotonin was infused *via* an intracoronary catheter at 3 μg/min in order to induce coronary artery constriction. Isoflurane or halothane was then administered in three concentrations (0.5%, 1.25%, 2.0%). Responses are presented as actual vessel diameters (mm) or as percent change from vessel diameter prior to anesthetic administration. Statistically significant data points are indicated. Analysis of variance demonstrated significant anesthetic effect. (Percent change: isoflurane P < 0.001, halothane P < 0.001; actual diameter: isoflurane P < 0.001, halothane P < 0.001.) Values are mean ± SEM.

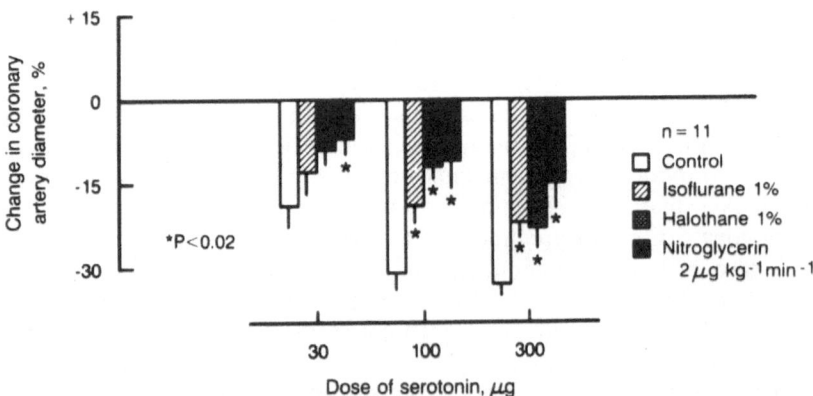

FIGURE 2. Constriction of left anterior descending coronary arteries in intact pigs evoked by intracoronary bolus infusions of serotonin (30, 100, and 300 μg). Percent change in coronary artery diameters are shown. Experiments were performed during control conditions (no volatile anesthetics) and during administration of isoflurane 1.0% and halothane 1.0%. In addition, in order to demonstrate the effects of a known coronary dilator, experiments were also performed during the intravenous infusion of nitroglycerin. Values are mean ± SEM.

receptors linked to phospholipase C activity, inositol phosphate formation and Ca^{2+} mobilization. All have been associated to varying degrees with coronary constriction in humans.

Rings of the left anterior descending and circumflex coronary arteries were obtained from the hearts of pigs anesthetized with ketamine and pentobarbital. Each heart provided eight rings. The endothelium was deliberately removed in some rings by gently rubbing the luminal surface with a wooden implement.

The rings were suspended in organ chambers filled with modified Krebs-Ringer bicarbonate solution at 37°C and aerated with 95% O_2-5% CO_2. Each ring was suspended by two stainless steel clips passed through the lumen. One clip was anchored to the bottom of the organ chamber; the other was connected to a strain gauge (Grass FT03) for the measurement of isometric force. Rings were placed at the optimal point of their length-tension relation by progressively stretching them until the contraction to KCl (20 mM), imposed at each level of distension, was maximal. In all experiments, the presence or absence of endothelium was confirmed by the response to bradykinin (10^{-6} M) of rings contracted with K^+ (20 mM). Contractions to K^+ 40 mM were obtained and were used as a reference to quantitate contractile responses obtained during the experiments.

Isoflurane or halothane (1% or 2%) was delivered to half of the organ chambers. The other chambers were not treated with an anesthetic and the rings served as controls. Sequentially increasing doses of the agonists were added to each organ chamber and isometric force was measured. In additional experiments stable contractions were established and the relaxant effects of 0.5 to 2.5% concentrations of anesthetics were investigated. The results indicate that isoflurane and halothane attenuate pig coronary artery contractions evoked by serotonin, acetylcholine, endothelin and histamine (figures 3 and 4). Halothane had a more marked and consistent effect that isoflurane. However, surprisingly and particularly important, the magnitude of the anesthetic's depressant action varied depending upon the contractile agonist. Contractions evoked by serotonin were readily depressed by both anesthetics (figure 3). In contrast, isoflurane and halothane had little effect on contractions evoked by histamine (figure 5). These differences may provide clues to the mechanism

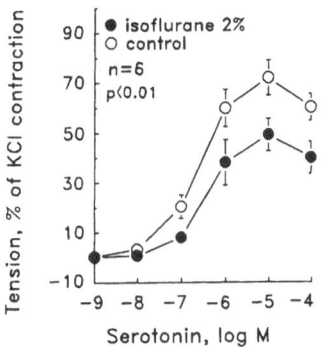 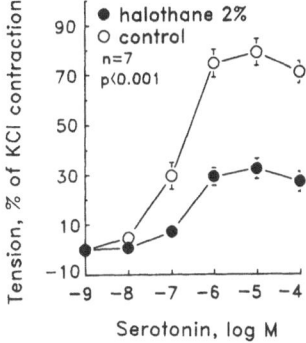

FIGURE 3. Contractile responses evoked by cumulative concentrations of serotonin in pig coronary artery rings without endothelium. Concentration-response relationships are shown for control rings and for rings treated with isoflurane 2% (left) or halothane 2% (right). Contractions are expressed as a percent of a response to 40 mM KCl. Values are mean ± SEM. Statistical significance of comparisons of integrated areas under the dose-response curves are shown.

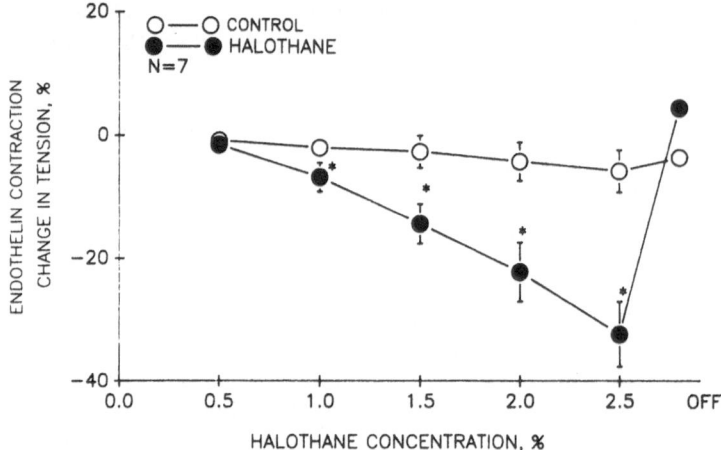

FIGURE 4. Relaxations induced by halothane 0.5% to 2.5% in pig coronary arteries precontracted with endothelin 10^{-8} M. Time controls of precontracted rings not exposed to halothane are also shown. All rings lacked endothelium. Values are mean ± SEM. Asterisks denote significance (Student's t test for paired observation; P < 0.01).

of anesthetic action. As both serotonin and histamine activate phospholipase C, the data may imply that the effect of the anesthetics is predominantly in the proximal part of the signal transduction pathway. The anesthetics may act in part on the receptor or G-protein as the pathway distal to the G-protein is common to both agonists.

Isolated Coronary Arteries and Protein Kinase C Activation

Agonists such as serotonin, acetylcholine, histamine and endothelin interact with receptors that stimulate a G-protein to activate phospholipase C and hydrolyze

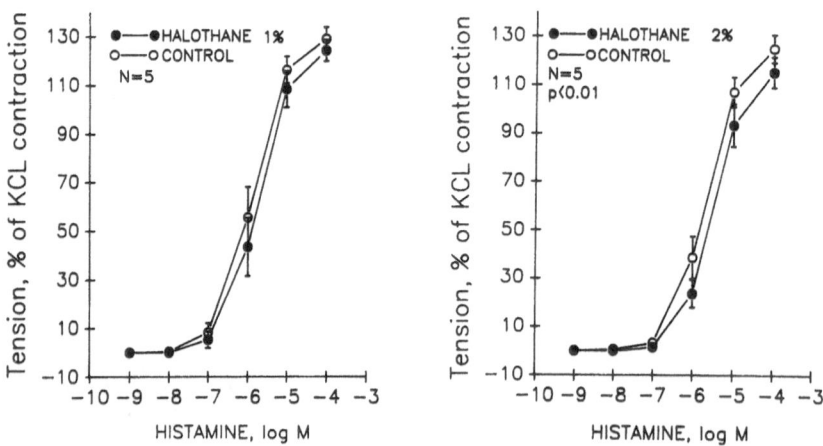

FIGURE 5. Contractile responses evoked by cumulative concentrations of histamine in pig coronary arteries without endothelium in the presence of cimetidine 10^{-5} M. Concentration-response curves are shown for control rings and for rings treated with halothane 1% (left) or halothane 2% (right). Contractions are expressed as a percent of a response to 40 mM KCl. Values are mean ± SEM. Statistical significance of comparisons of integrated areas under the dose-response curves is shown.

membrane phosphoinositide lipids to form inositol phosphate and diacylglycerol.[6] Diacylglycerol stimulates protein kinase C activity. Protein kinase C is not a single entity, but is a complex family of related intracellular enzymes that in vascular smooth muscle are involved in initiation and control of contraction.[9,10] Most protein kinase C isoforms are active at resting levels of intracellular Ca^{2+}, although they may exhibit a varying sensitivity to Ca^{2+}, including lack of Ca^{2+} requirement for activity. In addition, certain isoforms may be stimulated by messenger systems not necessarily linked to inositol lipid turnover and certain isoforms may even be insensitive to DAG.

It is possible, experimentally, to bypass the cell surface receptor linked to IP_3 and DAG formation and instead stimulate protein kinase C directly by using a phorbol ester. The purpose of the experiment reported here was to determine if isoflurane and halothane attenuated contractions evoked by activation of the protein kinase C pathway by stimulating coronary rings with PDBu—a phorbol ester. Results show that neither 2% isoflurane nor 2% halothane attenuated contractions evoked by sequentially increasing concentrations of PDBu (figure 6). However, 2, 2.5 and 3% concentrations of the anesthetics attenuated sustained contractions evoked by PDBu, but the magnitude of this effect was very small. The data suggest that inhibition of protein kinase C activity may contribute only minimally to the anesthetic's vasorelaxant actions.

Inositol Phosphates, Agonists and Anesthetics

When contractile agonists interact with receptors on smooth muscle cells, a G-protein transmits the signal to phospholipase C which hydrolyzes PIP_2 to form inositol phosphates and DAG.[6] The purpose of the experiments was to determine if 1% and 2% concentrations of the anesthetics inhibited the formation of inositol phosphates evoked by serotonin and by acetylcholine. Both cultured vascular smooth muscle cells and coronary artery segments both used in the experiments.

Left anterior descending and circumflex coronary arteries were dissected from the hearts of pigs anesthetized with ketamine and pentobarbital, were freed of fat and connective tissue and then cut into segments. The endothelium was deliberately removed. Segments were incubated for 90 minutes with 10 μCi/ml of [^{3}H]-myoinositol

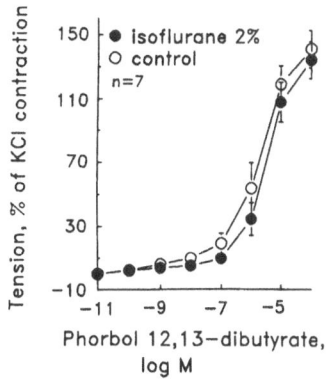 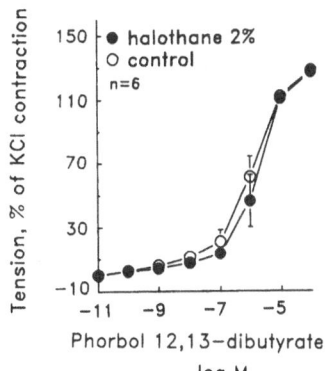

FIGURE 6. Contractile responses evoked by cumulative concentrations of phorbol 12,13-dibutyrate in pig coronary artery rings without endothelium. Concentration-response relationships are shown for control rings and for rings treated with isoflurane 2% (left) or halothane 2% (right). Contractions are expressed as a percent of a response to 40 mM KCl. Values are mean ± SEM. No statistical differences were observed between responses of control rings and rings treated with anesthetics.

(18.2 Ci/mM; Amersham). Tissue was aerated with 95% O_2-5% CO_2 at 37°C in the presence or absence of isoflurane 1% or halothane 1%. Tissues were transferred to borosilicate glass tubes filled with 0.3 ml Krebs buffer solution containing 10 mM LiCl (to prevent inositol phosphate breakdown) and incubated for ten minutes while aeration was continued. Acetylcholine 10^{-4} M was added to evoke inositol phosphate responses and the reaction was stopped at 20 minutes by addition of 0.12 ml 0.22 N HCl and 2.7 ml chloroform:methanol. Phases were separated by adding 0.9 ml chloroform and 0.9 ml water. Aliquots of 2 to 3 ml of the upper phase were diluted 8-fold with water and applied to columns containing approximately 0.5 ml of Dowex-1 in the formate form. The columns were washed with 60 mM Na formate plus 5 mM borax and discarded. [^{3}H]-inositol was then washed through with 5 ml 1.2 M NH_4 formate + 100 mM formic acid solution and this solution collected. Radioactivity was then measured using a scintillation counter.

The rat aortic smooth muscle cell line A_7r_5 was obtained from the American Type Culture Collection (Bethesda, MD) and was used to investigate inositol phosphate responses to serotonin in the presence and absence of 1% and 2% isoflurane. Cells were cultured in a 95% O_2-5% CO_2 incubated in Dulbecco's modified Eagles medium containing 10% fetal calf serum. At confluence the cells were trypsinized and subcultured in 16 mm dishes. Experiments were carried out using confluent monolayers of cells between the 7th and 14th day after plating. Cells were incubated with [^{3}H] inositol (6 μCi/ml) overnight in inositol-free medium (199 Gibco). Following wash with PBS they were lifted off the plates with trypsin, resuspended in Ca-, Mg-free PBS containing 30 mM LiCl. They were incubated in borosilicate glass test tubes for various times (0 to 30 min) in the absence (time controls) and presence of serotonin 10^{-4} M. Analysis of inositol phosphate content was performed using ion exchange chromatography with Dowex columns described above.

The results of both cultured cell and coronary artery experiments demonstrate that agonist induced increases in inositol phosphates are attenuated by isoflurane and by halothane (figures 7 and 8). The effects were not great but may be of sufficient magnitude to account, at least in part, for the anesthetics effects on agonist induced

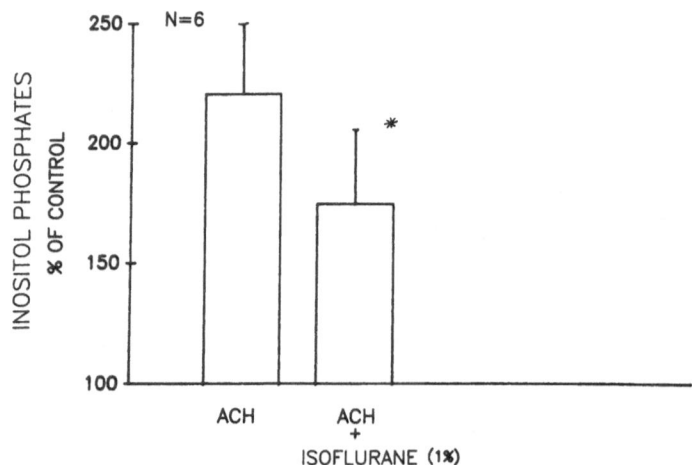

FIGURE 7. Increase in inositol phosphate concentrations evoked by acetylcholine 10^{-4} M in pig coronary artery segments without endothelium. Responses in the presence and absence of isoflurane 1% are shown. Responses are expressed as a percent of inositol phosphate concentrations in segments not treated with acetylcholine. Values are mean ± SEM. Asterisk denotes significance (Student's t test for paired observation; $P < 0.05$).

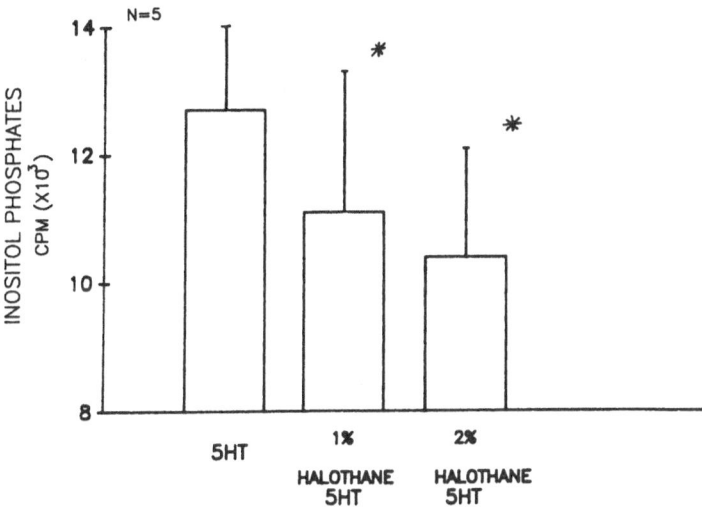

FIGURE 8. Inositol phosphate concentrations in cultured vascular smooth muscle cells (A7r5). Inositol phosphate concentration responses were evoked with serotonin (5HT) 10^{-4} M in the presence and absence of halothane 1% and 2%. Values are mean ± SEM. Asterisks denote significance (Student's t test for paired observation; $P < 0.05$).

contractions. However, whether or not an anesthetic effect on the inositol phosphate cycle is the primary mechanism for attenuation of contractions is not known.

Intracellular Ca^{2+}, Agonists and Anesthetics

An increase in cytosolic Ca^{2+} concentration is a signal that initiates vascular smooth muscle contraction.[11] Contractile agonists such as endothelin, ATP and vasopressin increase cytosolic Ca^{2+} both by promoting release from the sarcoplasmic reticulum and by signalling Ca^{2+} entry via channels in the cell surface membrane. The purpose of the experiments was to determine the effects of isoflurane and halothane on agonist induced changes in intracellular Ca^{2+} concentrations in cultured vascular smooth muscle cells.

The fluorescent, Ca^{2+}-sensitive probe indo-1 was used in the current experiments. Indo-1 was chosen because it causes little Ca^{2+} buffering, has good selectivity against Mg^{2+} and can readily be adapted to measurement of Ca^{2+} concentrations in single cells.[12] In addition, indo-1 has been shown to lack non-specific interactions with halothane.[13] A flow cytometric technique was used.

In the current experiments A10, A7r5 and BC3H-1 vascular smooth muscle cell lines were cultured. Indo-1 AM was loaded by adding the indicator (final concentration 4 μM) and pluronic-127 to cells in culture flasks for 30 minutes. The cells were detached with trypsin, centrifuged and suspended in test tubes containing culture medium. They were then centrifuged and incubated for 30 minutes in culture medium containing isoflurane or halothane in concentrations of 0.5, 1.25 and 2.0% or in culture medium without anesthetics (controls). Anesthetic solutions were prepared by aerating culture medium with isoflurane or halothane delivered from a vaporizer. An analyzer was used to adjust anesthetic concentrations, and samples of the culture medium were assayed for the anesthetics using gas chromatography. Some experiments were performed in the absence of extracellular Ca^{2+}. Cells were passed through a flow cytometer (FACStar plus, Becton-Dickinson) at 400-600 cells/second, where they were excited at 351 nm by ultraviolet light emitted from an argon laser. Agonists (vasopressin, endothelin or ATP) were added, causing an increase in intracellular Ca^{2+} which binds to indo-1 and in doing so shifts the spectrum of light emitted by the indicator. Emitted light was sampled at 490 nm and 390 nm and the ratio of fluorescence at the two wavelengths was used to indicate the concentration of intracellular Ca^{2+}. This measurement is independent of each cell's indo-1 content and is independent of cell size.

Results of the experiments demonstrate that increases in cytosolic Ca^{2+} evoked by ATP, endothelin and vasopressin are depressed by 1.25% and by 2.0% halothane (figures 9 and 10). Halothane also depressed Ca^{2+} release from the sarcoplasmic reticulum (figure 11). Isoflurane was less effective and less consistent in depressing increases in cytosolic Ca^{2+}. Although responses to vasopressin were attenuated (figure 12), isoflurane had less effect on responses evoked by endothelin and by ATP.

CONCLUSIONS

Current work demonstrates that isoflurane and halothane attenuate coronary artery constriction. *In vivo* experiments performed using intact pigs have shown that clinically relevant concentrations of both anesthetics attenuate coronary artery constriction evoked by serotonin. *In vitro* experiments performed using isolated vessels have also shown that clinically relevant concentrations of isoflurane and halothane attenuate contractions evoked by a range of agonists including serotonin, endothelin, acetylcholine and histamine. The magnitude of the anesthetic effect varies depending upon the agonist, the anesthetics being more effective against contractions evoked by

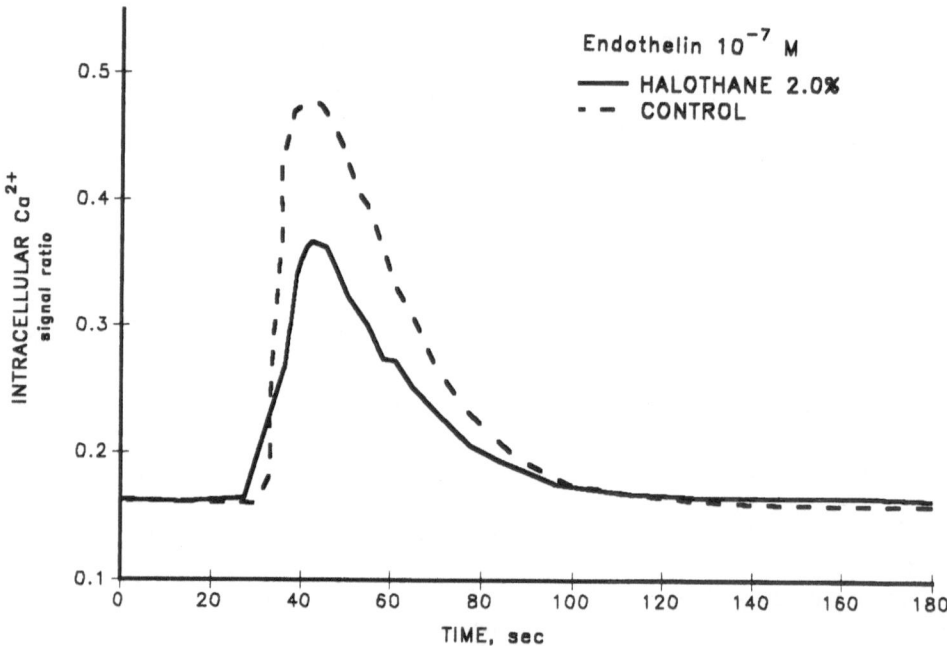

FIGURE 9. Cytosolic Ca^{2+} concentration measured with indo-1 in cultured vascular smooth muscle cells (A10) stimulated with endothelin 10^{-7} M. Indo-1 fluorescence is expressed as the ratio of light emitted at 490 nm and 390 nm. Responses in the presence and absence of halothane 2.0% are shown.

266

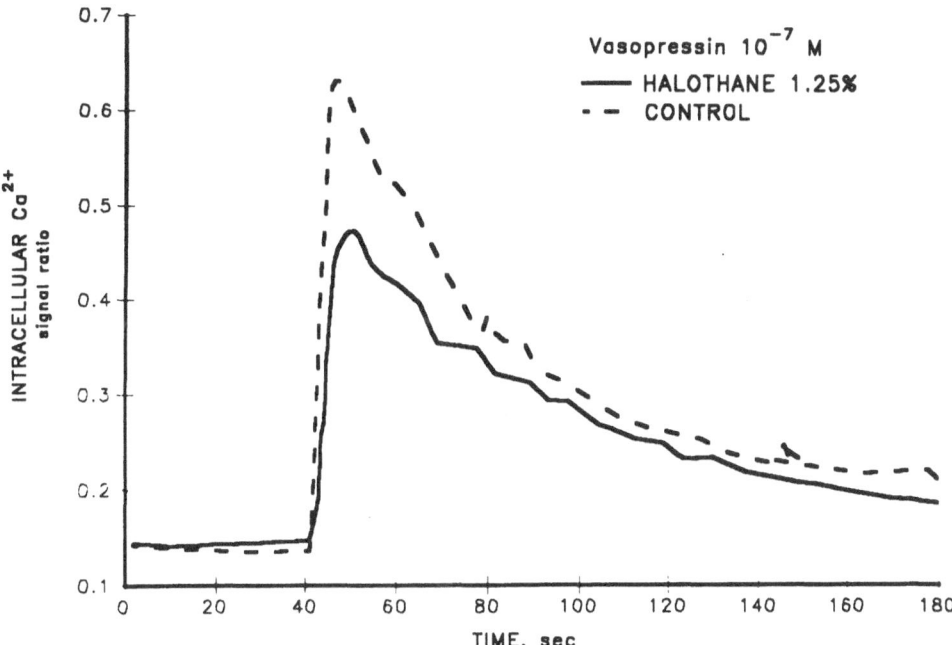

FIGURE 10. Cytosolic Ca^{2+} concentration measured with indo-1 in cultured vascular smooth muscle cells (A7r5) stimulated with vasopressin 10^{-7} M. Indo-1 fluorescence is expressed as the ratio of light emitted at 490 nm and 390 nm. Responses in the presence and absence of halothane 1.25% are shown.

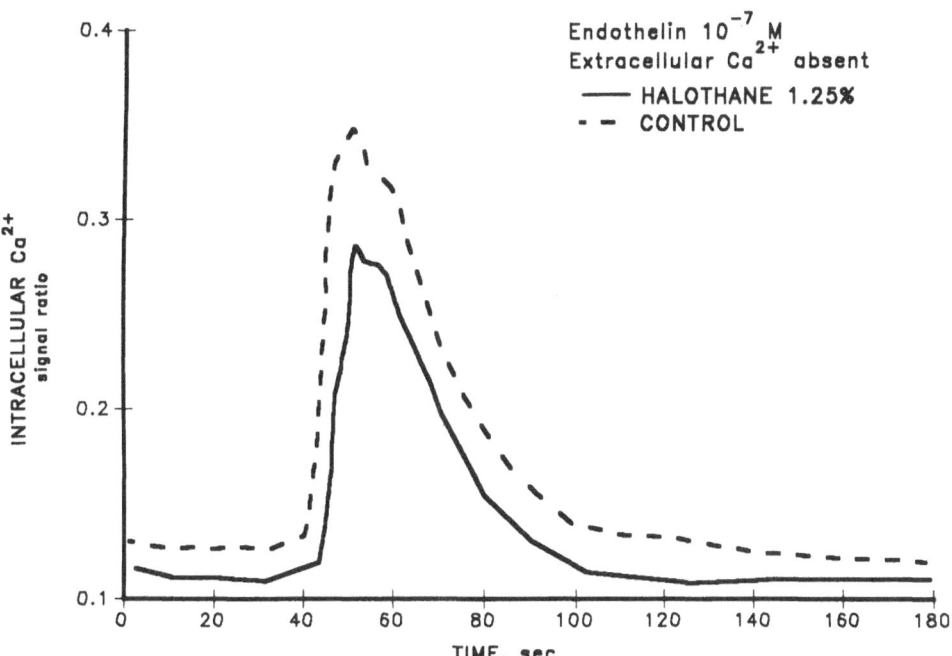

FIGURE 11. Cytosolic Ca^{2+} concentration measured with indo-1 in cultured vascular smooth muscle cells (A10) stimulated with endothelin 10^{-7} M. Extracellular Ca^{2+} was absent. Indo-1 fluorescence is expressed as the ratio of light emitted at 490 nm and 390 nm. Responses in the presence and absence of halothane 1.25% are shown.

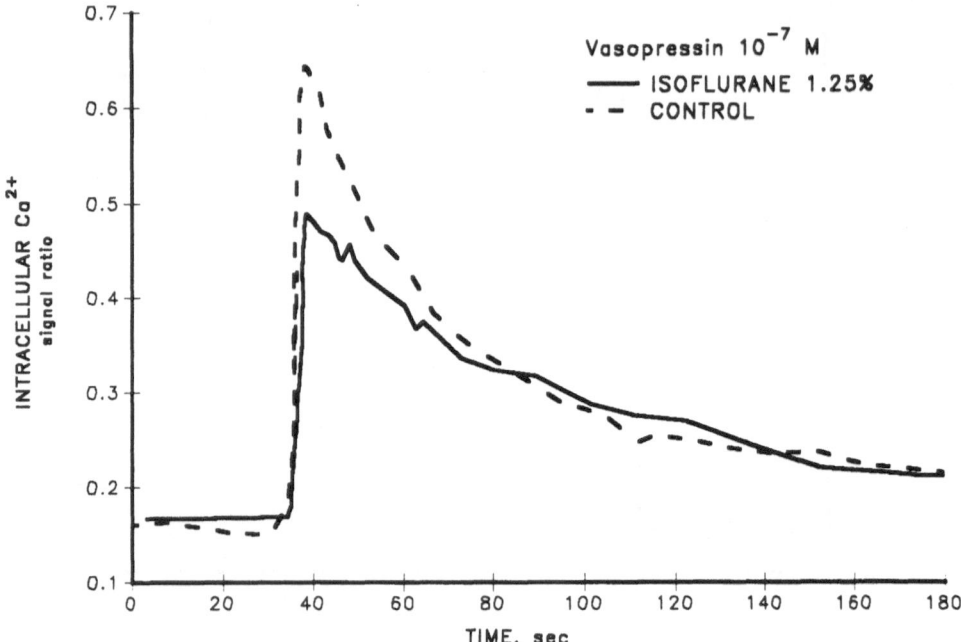

FIGURE 12. Cytosolic Ca^{2+} concentration measured with indo-1 in cultured vascular smooth muscle cells (A7r5) stimulated with vasopressin 10^{-7} M. Indo-1 fluorescence is expressed as the ratio of light emitted at 490 nm and 390 nm. Responses in the presence and absence of isoflurane 1.25% are shown.

serotonin and acetylcholine rather than those evoked by histamine. Current work also shows that clinically relevant concentrations of isoflurane and halothane depress agonist-induced inositol phosphate formation and agonist-induced increases in cytosolic Ca^{2+} concentration.

The mechanism of volatile anesthetic effects remains uncertain. However, results from current experiments do offer some clues. In vascular tissue, isoflurane and halothane depress agonist-induced inositol phosphate formation. Both anesthetics depress agonist-induced increases in cytosolic Ca^{2+}. Current work also indicates that neither anesthetic has a marked effect on contractions evoked by protein kinase C activity. Perhaps most revealing is a comparison between the contractile effects of serotonin and histamine. Both agonists are linked to PLC activity, inositol phosphate formation and Ca^{2+} mobilization. However, contractions evoked by serotonin are readily attenuated by halothane and isoflurane. Neither anesthetic had a marked effect on contractions evoked by histamine. The data may suggest that the anesthetics may act in part upon the proximal part of the signal transduction pathway.

REFERENCES

1. T. M. Witzeling, J. C. Sill, J. M. Hughes, G. A. Blaise, M. Nugent and D. K. Rorie, Isoflurane and halothane attenuate coronary artery constriction evoked by serotonin in isolated porcine vessels and in intact pigs, *Anesthesiology* 73:100 (1990).
2. B. A. Bollen, J. H. Tinker and K. Hermsmeyer, Halothane relaxes previously constricted isolated porcine coronary artery segments more than isoflurane, *Anesthesiology* 66:748 (1987).
3. G. Blaise, J. C. Sill, M. Nugent, R. A. Van Dyke and P. M. Vanhoutte, Isoflurane causes endothelium-dependent inhibition of contractile responses of canine coronary arteries, *Anesthesiology* 67:513 (1987).

4. E. Villeneuve, G. A. Blaise, D. Girard, J. Maucotel and J. Buluran, The effect of halothane on human coronary artery rings vasomotion (abstract), *Anesthesiology* 69:A27 (1988).
5. A. Maseri, G. Davies, D. Hackett and J. C. Kaski, Coronary artery spasm and vasoconstriction: The case for a distinction, *Circulation* 81:1983 (1990).
6. M. J. Berridge and R. F. Irvine, Inositol phosphates and cell signalling, *Nature* 341:197 (1989).
7. B. F. Rusy and H. Komai, Anesthetic depression of myocardial contractility: A review of possible mechanisms, *Anesthesiology* 67:745 (1987).
8. M. J. Berridge, Temporal aspects of calcium signalling, *in*: "Advances in Second Messenger and Phosphoprotein Research: The Biology and Medicine of Signal Transduction," Y. Nishizuka, M. Endo and C. Tanaka, ed., Raven Press, New York (1990).
9. H. Rasmussen, Y. Takuwa and S. Park, Protein kinase C in the regulation of smooth muscle contraction, *FASEB J* 1:177 (1987).
10. Y. Nishizuka, The molecular heterogeneity of protein kinase C and its implications for cellular regulation, *Nature* 334:661 (1988).
11. A. P. Somlyo, J. W. Walker, Y. E. Goldman, D. R. Trentham, S. Kobayashi, T. Kitazawa and A. V. Somlyo, Inositol triphosphate, calcium and muscle contraction, *Phil Trans R Soc Lond* 320:399 (1988).
12. G. Grynkiewicz, M. Poenie and R. Y. Tsien, A new generation of Ca^{2+} indicators with greatly improved fluorescence properties, *J Biol Chem* 260:3440 (1985).
13. A. Klip, G. B. Mills, B. A. Britt and M. E. Elliott, Halothane-dependent release of intracellular Ca^{2+} in blood cells in malignant hyperthermia, *Am J Physiol* 258:C495 (1990).

25

Mechanisms of Action of Anesthetics on Inositol Phospholipid Hydrolysis in Vascular Endothelial Cells and Rat Basophilic Leukemia Cells in Tissue Culture

A. Robinson-White

INTRODUCTION

The different classes of anesthetics (*i.e.*, barbiturates, local anesthetics, and volatile anesthetics) have been shown to exert separate patterns of effects on the vasculature.[1-7] Theories as to the mechanism(s) of these effects suggest that anesthetics act either by the specific alteration of membrane proteins, or in a non-specific manner on membrane proteins, through the perturbation of membrane lipids.[8-10] Although no suggestion as to the degree of specificity of the anesthetics on membrane proteins has been offered, they are thought to act on cellular biochemical processes by alteration of ion fluxes (in particular Ca^{2+})[1,11-13] in the vascular smooth muscle cell.[1] Our recent studies have shown that anesthetics can alter one Ca^{2+}-mediated cellular pathway in vascular endothelial cells, the phosphatidylinositol pathway (inositol phospholipid hydrolysis).[14]

Inositol phospholipid hydrolysis (PI hydrolysis, figure 1) involves a family of receptors, that when activated, stimulate the breakdown of cell membrane inositol phospholipids, through the stimulation of the phosphodiesterase, phospholipase C (PLC).[15-16] Phospholipid breakdown is regulated by the intervention of a G_p-type GTP-binding protein (G-protein) located on the catalytic unit of PLC.[17-18] The resulting hydrolysis generates two intracellular signals, the release of Ca^{2+} from the endoplasmic reticulum by inositol 1,4,5-triphosphate, and the activation of protein kinase C.[19] The interaction of these signals initiates a cell response (*e.g.*, secretion).

We have proposed that the vascular response to anesthetics may be due to an indirect effect on the smooth muscle cell by an alteration of PI hydrolysis in the endothelial cell.[14] The endothelial cell is known to modulate vascular tone by the release of factors that relax[20] or constrict[21-22] the underlying smooth muscle, and are the first cells exposed to circulating hormones and drugs. Thus, changes in endothelial cell function may induce functional changes in the smooth muscle cell. We have, therefore, sought to determine the following: 1) do the anesthetics alter PI hydrolysis in endothelial cells; 2) is the effect the same or different for each anesthetic type; 3) are these alterations due to specific or non-specific effects on proteins of the inositol

A. ROBINSON-WHITE, Department of Anesthesiology, Uniformed Services University of the Health Sciences, Bethesda, Maryland 20814.

Mechanisms of Anesthetic Action in Skeletal, Cardiac, and Smooth Muscle
Edited by T.J.J. Blanck and D.M. Wheeler, Plenum Press, New York, 1991

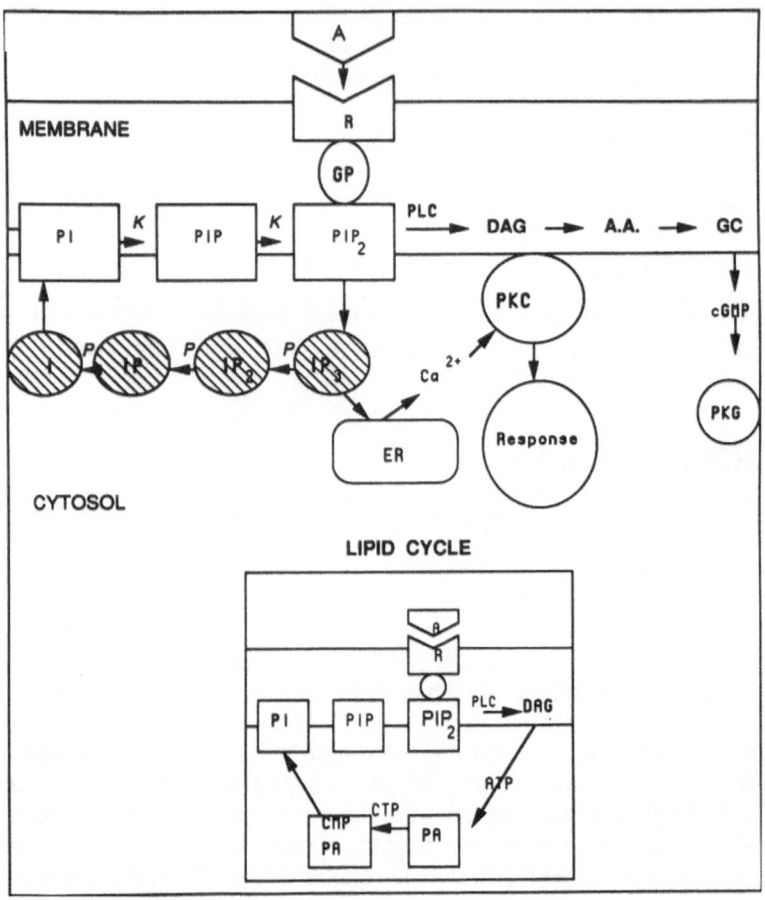

FIGURE 1. Schematic illustration of the phosphatidylinositol pathway and lipid synthesis. The agonist (A) binds to the receptor (R) and stimulates the phosphodiesterase, phospholipase C (PLC) to cause the hydrolysis of phosphatidylinositol biphosphate (PIP_2), and the release of inositol triphosphate (IP_3), though the intervention of the putative GTP-binding protein (G_p). Diacylglycerol (DAG) remains in the membrane. Calcium, released from the endoplasmic reticulum (ER) by IP_3, with DAG, activates protein kinase C (PKC), and a subsequent cell response. IP_3 is converted to inositol biphosphate (IP_2), inositol phosphate (IP), and inositol (I), and to phosphatidylinositol (PI) by a series of phosphatases (P). PI and phosphatidylinositol biphosphate (PIP) are converted to PIP_2 by a series of kinases (K), to complete the cycle. Phospholipids are replenished through the lipid cycle (see inset), when DAG and ATP form phosphatidic acid (PA). PA and cytosine triphosphate (CTP) form PA-CMP, which forms the new PI.

phospholipid pathway; and 4) does an alteration of hydrolysis result in an alteration of cell response.

We have also used the rat basophilic leukemia (RBL-2H3) cell as a model in these studies, since the PI pathway in these cells has been throughly investigated, which allows for the manipulation of the pathway. Also, activation of the pathway leads to an easily measured physiological response (e.g., cell secretion). The stimulation of hydrolysis and a functional response to hydrolysis could thus be measured in one cell type.

MATERIALS AND METHODS

Materials

The solutions used for the isolation and culture of endothelial cells were: 1) a collagenase solution that contained 2 U/ml collagenase type II and 0.5 mg/ml soybean trypsin inhibitor (Cooper Biomedical, Malvern, PA) in medium 199 (M199) [Grand Island Biological Co., (GIBCO), Grand Island, NY]; 2) a supplemented M199 (S-M199), which contained 1.3 mg L-glutamine, 10 ml Eagles basal medium (BME) amino acid solution (100×), 100 ml minimal essential medium (MEM) vitamin solution (100×), 100 ml penicillin-streptomycin, 20% fetal bovine serum (FBS, GIBCO), 5 g bacto-peptone (DIFCO Laboratories, Detroit, MI), and 9 g D-glucose (Sigma Chemical Co., St. Louis, MO) in a volume of 1,000 ml; and 3) 0.25% trypsin (GIBCO). The solution used for the maintenance of RBL-2H3 cells in culture was minimum essential medium (MEM) supplemented with 2 mM L-glucose, 15% FBS, and 10% penicillin-streptomycin. RBL-2H3 cells were subcultured using 0.25% trypsin-EDTA (GIBCO). For experiments with endothelial cells, buffer A was used and contained the following (in mM): NaCl 133, KCl 3.6; MgCl$_2$ 0.4, D-glucose 16, CaCl$_2$ 1.0, LiCl 10, and HEPES 3.0 (GIBCO), pH adjusted to 7.4. For experiments with intact RBL-2H3 cells, buffer B was used, which contained (in mM): NaCl 119, KCl 5, CaCl$_2$ 1, MgSO$_4$ 1, piperazine-N,N'-bis(2-ethane-sulfonic acid) (PIPES) 25, glucose 5.6, NaOH 40, and 0.01% w/v bovine serum albumin, with and without 10 mM LICL, pH 7.2. For experiments with permeabilized RBL-2H3 cells, buffer C was used, with the following contents: 113.7 mM K glutamate, 1 μM CaCl$_2$, 110 mM LICL, 2 mM K PIPES, 1 mM MgEDTA, 5 mM ATP, and 5 mM glucose at pH 6.8.

The following chemicals and drugs were used in the individual experiments: myo-[2-^{3}H]-inositol, 16.9 Ci/mmol; [^{125}I]-angiotensin II (AII), 2,200 Ci/mmole; ^{45}calcium, 2 μCi/mmole; and [^{3}H]-serotonin ([^{3}H]-5HT) from Dupont New England Nuclear, Boston, MA; unlabelled AII acetate salt, lithium chloride, sodium pentobarbital, secobarbital, and phenobarbital from Sigma Chemical Company, St. Louis, MO; bacitracin from Aldrich Chemical Co., Milwaukee, WI; dowex-1 formate resin from Bio-Rad Laboratories, Richman, CA; 0.2% pentex bovine serum albumin, fraction V from Miles Laboratories, Inc., Elkhart, IN; reduced streptolysin O from Welcome Diagnostics, Dartford, IN; guanosine 5'-(3-O-thio)-adenosine triphosphate (GTPγS) from Boehringer Mannheim Biochem., Indianapolis, IN; and 5'-(N-Ethyl) carboxamidoadenosine (NECA) from Sigma Chemical Company.

Dinitrophenol-conjugated bovine serum albumin (DNP$_{24}$-BSA), and rat monoclonal anti-dinitrophenol specific IgE (DNP-specific IgE) were generous gifts of Dr. Henry Metzger, National Institute of Arthritis, Musculoskeletal and Skin Diseases, National Institutes of Health, Bethesda, MD). The optical isomers of pentobarbital (R+ and S−), and the structural analogue of pentobarbital (1'RS,3'SR,3'-hydroxypentobarbital) were gifts of Dr. F. Ivy Carroll, National Institute of Environmental Health Sciences, Research Triangle Park, NC.

Cell Isolation and Culture

Rat thoracic aortic endothelial cells were isolated and cultured exactly as previously described.[14] Briefly, vessels cleaned of fat and incubated in the collagenase solution (20 min, 37°C) were stripped of adventitia and minced into sections of <1 mm^2. Sections were planted into culture flasks and incubated (37°C) for 14 to 25 days with S-M199. Cell outgrowths were detached with trypsin, filtered through a nylon mesh, and seeded into culture flasks (10^6 cells/flask). The medium was changed after 1 hr (to remove contaminating smooth muscle cells) and cultures were grown to confluency. Confluent cultures were subcultured, and the medium changed within 30 min of subculture for 3 successive passages. Cultures were fed every 48-72 hrs. Cultures were characterized as endothelial by light, electron and fluorescent microscopy.[23]

RBL-2H3 cells[24] were a gift of Dr. M.A. Beaven (National Heart Lung and Blood Institute, National Institutes of Health, Bethesda, MD) and were maintained in supplemented MEM.

Cell Permeabilization

Cells were permeabilized as described by Ali et al.[25] Cells labelled with [^{3}H]-inositol were washed with buffer C and incubated with reduced streptolysin O (0.2 U/ml, 37°) for 7 min. The medium was aspirated and buffer containing stimulants and/or anesthetics were added. The degree of cell porosity was assessed by staining with fluorescein diacetate/ethidium bromide.[26] The fraction of cells permeabilized was 85-97%.

Preparation of Anesthetics and Gas Delivery

Stock solutions of barbiturates were prepared as previously described,[14] in warm buffer A or B at concentrations of 5 mM or lower. Stock solutions of lidocaine were prepared in water at 0.1 M, mixed for 5 min (37°) for solubilization of the anesthetic, and the pH adjusted to 7.4 or 7.2, for experiments with endothelial cells or RBL-2H3 cells, respectively, by the slow addition of 0.25 N NaOH. For inositol phospholipid hydrolysis experiments involving one concentration of halothane, halothane was delivered from a calibrated vaporizer to cultured cells in an air-tight container, and then allowed to exit through an outlet port. For experiments involving more than one halothane concentration, halothane was bubbled into the experimental buffer and allowed to equilibrate (10 min, 22°C). The halothane-buffer mixture was added to cells in glass 24-well culture plates, and the plates quickly covered with fitted glass slides. In both types of experiments, concentrations of halothane in the medium was measured before and after experiments, by use of a gas chromatograph.[27] Concentrations ranged from 1-4% before incubation with cells and 0.5 to 2% after a 15 min incubation (37°). The latter values were used as final concentrations (i.e., 1% halothane = 0.27 mM). Stock solutions of stimulants were prepared in warm buffer. Bacitracin (100 μM) was added to all solutions containing AII. The solutions were prepared immediately before each experiment.

Biochemical Assays

Inositol phospholipid hydrolysis experiments were performed exactly as previously described.[14,28] Briefly, cells in 24-well culture plates were incubated with culture medium for 48 hrs and incubated for an additional 16-17 hrs (37°) with 6 μCi/ml [^{3}H]-inositol. The medium was removed and cultures washed with buffer A or B for endothelial cells and RBL-2H3 cells, respectively.

Inositol phospholipid breakdown was stimulated and the reaction halted by the addition of chloroform:methanol (1:2) to each culture well. Cell extracts were separated into 2 phases by the addition of chloroform and water, vortex mixing, and centrifugation. The upper phase ([^{3}H]-inositol phosphates in total) was transferred to dowex-1-formate resin in columns, washed and eluted with Na$^+$ formate. [^{3}H]-inositol lipids in the lower phase were washed with methanol:1 M KCL (1:1), and the solvent evaporated. The ^3H content was determined by liquid scintillation counting.

^{45}Calcium uptake was measured by incubating RBL-2H3 cells overnight in 24-well culture plates, with MEM and antigen-specific IgE, or MEM alone. Cultures were washed with buffer B and incubated for 5 min (37°) with $^{45}Ca^{2+}$, barbiturates, and/or agonists. Cultures were placed on ice and washed with buffer. Cells were lysed with water overnight at 22° C. The $^{45}Ca^{2+}$ content was counted by liquid scintillation.[29]

Intracellular Ca^{2+} ($[Ca^{2+}]_i$) levels were measured when cell suspensions were centrifuged, washed and resuspended to 10^6 cells/ml. Cells were loaded with quin2 AME (50 min, 37°C), harvested, and resuspended in buffer. Fluorescent measurements and calculations of % saturation of quin2 were performed with a Perkin Elmer MPF66 spectrophotometer and computer system (excitation wavelength for quin2 = 339 nm). Calibration for the estimation of $[Ca^{2+}]_i$ levels were performed by lysing cells with triton X-100, and the addition of $MnCl_2$ to quench fluorescence, exactly as described.[30]

Cell secretion of $[^3H]5$-HT was measured when RBL-2H3 cells were incubated overnight in 24-well culture plates, in complete medium containing $[^3H]5$-HT. Cells were washed in buffer B and test reagents or buffer were added at timed intervals. Cultures were placed on ice and ^3H content in the medium determined by liquid scintillation counting.[31]

Measurement of Specific [^{125}I]Angiotensin II Binding

[^{125}I]Angiotensin II (AII) binding was measured in endothelial cells.[14] Cultures in 6-well culture plates (10^6 cells/well) were incubated for 48 hrs in supplemented M199 without FBS, washed in buffer A (with LiCl), and incubated for 10 min at 37°C. The medium was removed and cells with [^{125}I]AII, pentobarbital (0.1 to 1 mM), bacitracin (100 μM) and/or buffer were incubated (22°C) for various timed intervals. The reaction was halted with 4 ml ice-cold buffer. Cells were scraped from wells and collected by filtration on millipore filters (0.4 μ). Culture wells and filters were air-dried. The ^{125}I content of filters and filtrate was determined by gamma counting. Specific binding is reported as the amount of total labelled AII minus the amount of radioactivity bound in the presence of 2 μM unlabelled AII.

Expression of Results

For studies on inositol phospholipid hydrolysis with both intact and permeabilized cells, data were normalized as previously discussed.[14,28] For studies on $^{45}Ca^{2+}$ uptake, $[Ca^{2+}]_i$, inositol, and $[^3H]5$-HT secretion data were normalized and expressed as reported by Ali et al.[29] and Beaven et al.[31]

RESULTS

Effect of Pentobarbital on PI Hydrolysis and Angiotensin II Binding in Endothelial Cells

Analysis of the effect of pentobarbital on AII-induced inositol phospholipid hydrolysis by Lineweaver-Burke plot (figure 2A) showed a mixed type inhibition by pentobarbital ($K_i = 0.45$ mM), with no effect, however, on levels of membrane inositol phospholipid levels, or on spontaneous hydrolysis. These findings were substantiated

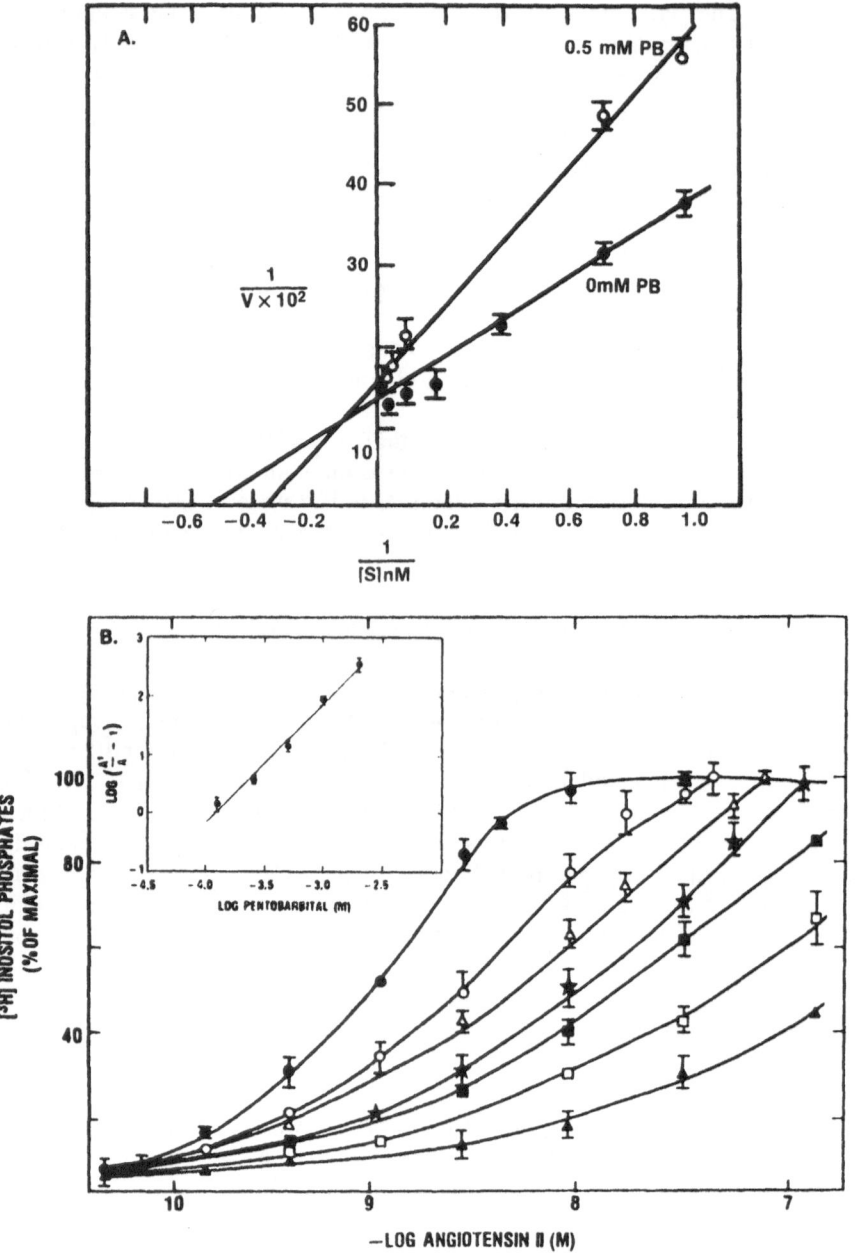

FIGURE 2. Kinetic analysis of the inhibition of inositol phospholipid hydrolysis by pentobarbital. (A) Lineweaver-Burke plot and (B) log-concentration curves of inositol phospholipid hydrolysis induced by angiotensin II (AII) in the presence and absence of pentobarbital. [³H]-labelled endothelial cells were first incubated (37°) with pentobarbital or buffer for 10 min, and incubated a further 5 min with pentobarbital, angiotensin II and bacitracin. In panel A, V represents the release of [³H]-inositol phosphates over 5 min, [S] = the concentration of AII, and PB = pentobarbital. In Panel B, pentobarbital concentrations were: 0.12 mM (o), 0.25 mM (Δ), 0.5 mM (⋆), 0.8 mM (■), 1.0 mM (□), and 2 mM (▲). Solid circles (●) represent controls (no pentobarbital). Percent maximal hydrolysis is the percent of the maximal hydrolysis generated by AII alone. The inset represents a Schild regression of the data for the log concentration curves. Results are plotted as the log of the dose ratio

276

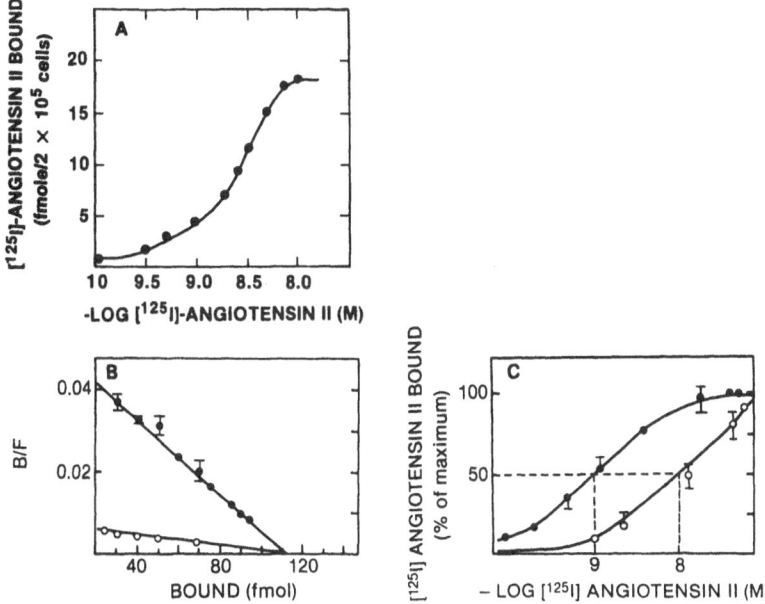

FIGURE 3. [^{125}I]AII binding to cultured endothelial cells: effect of pentobarbital. Endothelial cell cultures (10^6 cells/well) were incubated for 5 min with [^{125}I]AII (45 nCi/ml), diluted with unlabelled AII at 22°C. Non-specific binding was determined in the presence of 2 μM unlabelled AII. For the highest and lowest concentrations of labelled AII used, non-specific binding was 3 and 0.04%, respectively. All experiments included bacitracin. Specific binding (corrected for non-specific binding) was plotted as follows: A) log binding isotherm; B) Scatchard plot of binding in the presence (○) and absence (●) of pentobarbital (0.5 mM); and C) [^{125}I]AII binding in the presence (○) and absence (●) of pentobarbital (0.5 mM). For panel C, the data are expressed as a percentage of the maximal values (i.e., in 5×10^{-8} M AII). Results in panel A are the mean ± SEM of 6 experiments and in panels B and C, 3 experiments. Reprinted from A. J. Robinson-White, S. M. Muldoon and F. C. Robinson, *Eur J Pharmacol* 172:291-303 (1989) with permission.

by experiments performed with AII in the presence of increasing concentrations of pentobarbital (figure 2B). The non-parallel nature of the curves indicate a deviation from a competitive type inhibition. Schild regression analysis (figure 2B, inset) confirmed the non-parallel nature of the curves (slope = 1.74 ± 0.04; n = 4).

[^{125}I]AII binding to endothelial cells was saturable and highly specific (>97% of the total bound; see figure 3A). The Hill plot (data not shown) indicated an average slope (1.51 ± 0.04) that was greater than unity. Pentobarbital unexpectedly inhibited this binding (figures 3B and C). Scatchard analysis of [^{125}I]AII binding indicated a competitive type inhibition by pentobarbital (figure 3B), with a decrease in the K_d for AII from 1.2 nM to 11.2 ± 1.5 nM in the presence of pentobarbital

((A'/A) − 1, where A' = 50% maximal hydrolysis in the presence of AII and pentobarbital, and A = 50% maximal hydrolysis by angiotensin II alone) vs. the negative log of the pentobarbital concentration. The slope of the regression line = 1.54 ± 0.04; r = 0.996. Values in panel A represent mean ± SEM of 5 separate experiments, and in panel B of 3 experiments (three wells / experimental point). Reprinted from A. J. Robinson-White, S. M. Muldoon, F. C. Robinson, *Eur J Pharmacol* 172:291-303 (1989) with permission.

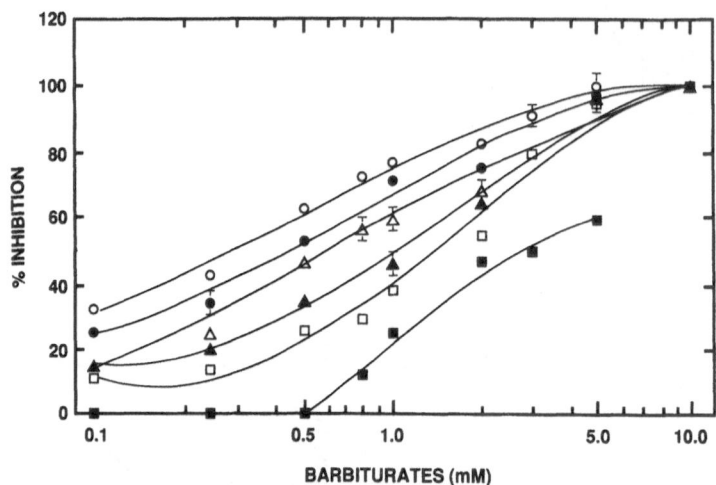

FIGURE 4. Efficacy of different barbiturates in inhibiting antigen-induced [³H]inositol phosphate release in intact rat RBL-2H3 cells. Cells were incubated with the barbiturates for 10 min before addition of DNP_{24}-BSA (10 ng/ml, 15 min), and experiments were performed as stated in "Materials and Methods." Results are expressed as percent inhibition of the release of [³H]inositol phosphates and are the mean ± SEM of six experiments (three wells per experimental point). The individual curves were significantly different from each other ($P < 0.05$). Symbols: ■ = (1'RS,3'SR)-3'hydroxypentobarbital; □ = phenobarbital; ▲ = R(+) optical isomer of pentobarbital; △ = racemic pentobarbital; ● = S(−) optical isomer of pentobarbital; ○ = secobarbital. Reprinted from A. J. Robinson-White, S. M. Muldoon, L. Elson and D. M. Collado-Escobar, *Anesthesiology* 72:996-1004 (1990) with permission.

(0.5 mM). Pentobarbital (0.5 mM) also produced a parallel shift to the right in the [¹²⁵I]AII binding curve (figure 3C).

Effects of Barbiturates on PI Hydrolysis, Ca²⁺ Uptake, [Ca²⁺]ᵢ Levels and 5-HT Secretion in RBL-2H3 Cells

Intact RBL-2H3 cells, when stimulated by the antigen DNP_{24}-BSA (10 ng/ml), released [³H]-labelled inositol phosphates into the cell cytosol.[28] Various barbiturates inhibited this release in an increasing order of 1'RS,3'SR,3'-hydroxypentobarbital < phenobarbital < R(+) isomer of pentobarbital < racemic pentobarbital < S(−) isomer of pentobarbital < secobarbital (figure 4), with a corresponding decrease in IC_{50} values of 2.9 ± 0.3 to 0.28 ± 0.07. Maximal inhibition by all barbiturates, except 1'RS,3'SR,3'-hydroxypentobarbital, occurred between 6 to 8.5 mM. No effect was seen on the levels of membrane inositol lipids with any of the barbiturates.

GTPγS-induced inositol phospholipid hydrolysis in permeabilized RBL-2H3 cells in the presence (figure 5A) and absence (figure 5B) of Ca²⁺. Calcium alone produced little or no stimulation of hydrolysis (1 mM = 2.4 %). Pentobarbital (0.5 mM) inhibited the GTPγS-induced hydrolysis whether in the presence or absence of Ca²⁺, while pentobarbital alone had no effect (data not shown). As in the intact cell, pentobarbital did not alter levels of labelled membrane lipids. However, unlike in the intact cell, spontaneous hydrolysis was inhibited by the most potent barbiturates (*i.e.*, secobarbital, R($-$) and R($+$) pentobarbital, and racemic pentobarbital).

$^{45}Ca^{2+}$ uptake (figures 6A and B) and $[Ca^{2+}]_i$ levels in intact cells (figures 6C and D) were stimulated with DNP_{24}-BSA and the adenosine analogue NECA. Pentobarbital (0.5 mM) inhibited $^{45}Ca^{2+}$ uptake and $[Ca^{2+}]_i$ levels when cells were stimulated by DNP_{24}-BSA (figure 6A and C), but not when cells were stimulated by NECA (figures 6B and D). The inhibition of $^{45}Ca^{2+}$ uptake and $[Ca^{2+}]_i$ levels was maximal (56%) at 10 ng/ml DNP_{24}-BSA and at 60 sec (45%), respectively.

RBL-2H3 cells released $[^3H]5$-HT in response to 10 ng/ml DNP_{24}-BSA in a time-dependent (figure 7) and concentration-dependent manner (data not shown). The inhibition by pentobarbital was time and concentration dependent as well (figure 7), with an IC_{50} of 0.8 mM for pentobarbital. Pentobarbital alone, however, did not alter release.

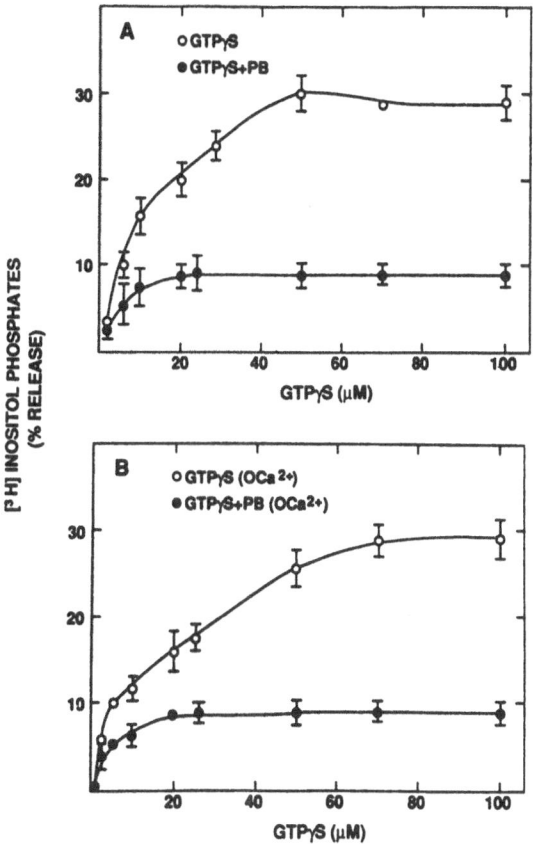

FIGURE 5. Inhibition by pentobarbital of GTPγS-induced $[^3H]$inositol phosphate release in permeabilized RBL-2H3 cells. Permeabilized cells were treated for 10 min with pentobarbital or buffer. Cells were treated for a further 15 min with GTPγS. In panel A, this incubation occurred in the presence of 1 μM Ca^{2+}, with or without pentobarbital (0.5 mM). In panel B, conditions were the same except for the absence of Ca^{2+}. Results are expressed as percent inositol phosphate release, and data represent the mean ± SEM of 3 to 4 experiments. PB = pentobarbital. Reprinted from A. J. Robinson-White, S. M. Muldoon, L. Elson and D. M. Collado-Escobar, *Anesthesiology* 72:996-1004 (1990) with permission.

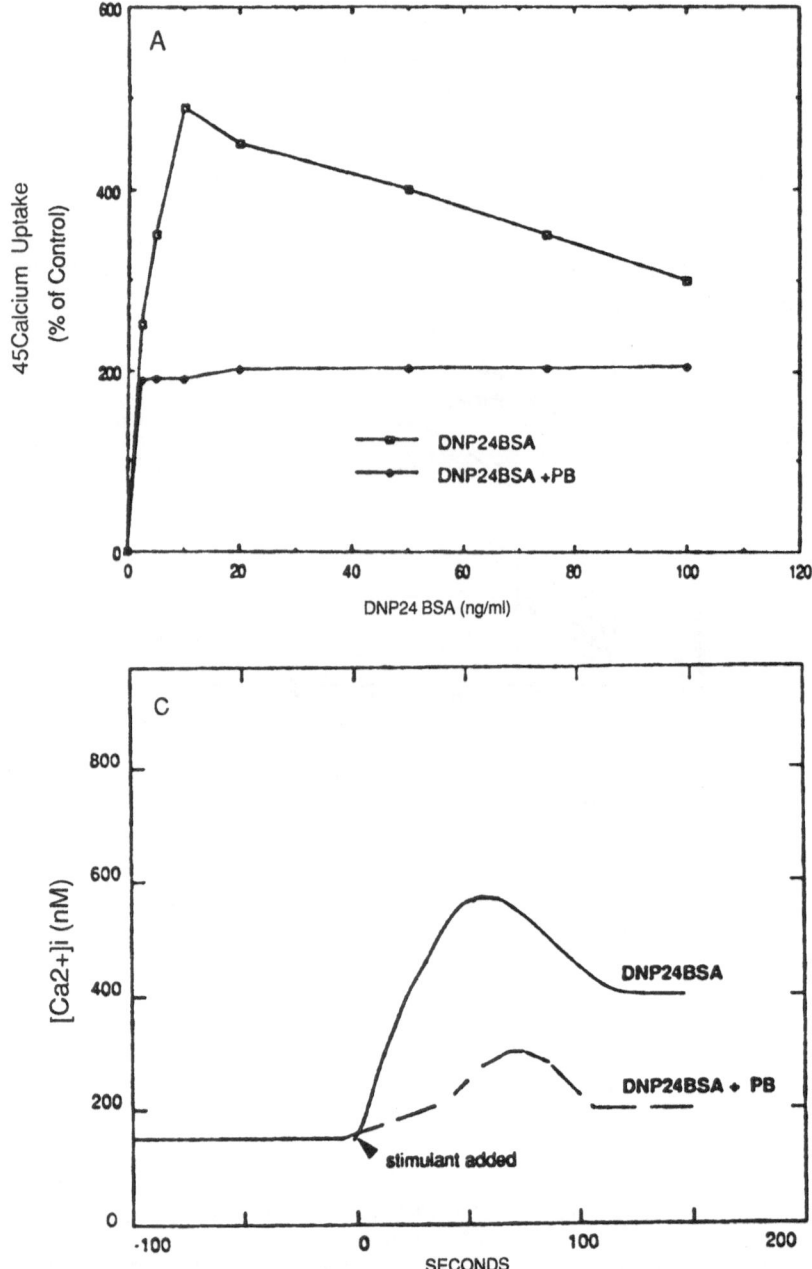

FIGURE 6. Inhibition of $^{45}Ca^{2+}$ uptake and $[Ca^{2+}]_i$ levels by pentobarbital in RBL-2H3 cells. For panels A and B, cells were grown in 24-well culture plates and incubated for 5 min (37°C) with DNP_{24}-BSA (A) or NECA (B), with and without pentobarbital (0.3 mM) and $^{45}Ca^{2+}$. For panels C

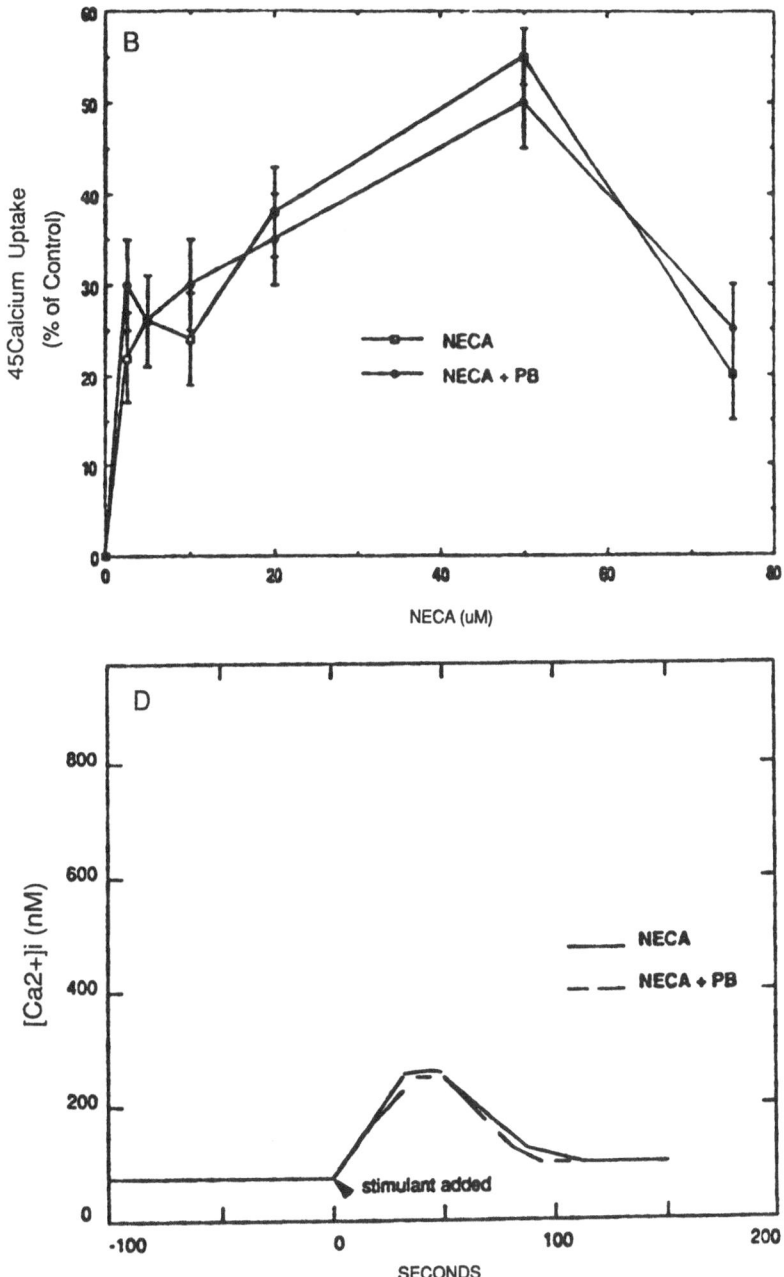

and D, cell suspensions (10^6 cells/ml) were loaded with quin2 for 50 min, and stimulated with DNP_{24}-BSA (C) or NECA (D), in the presence or absence of pentobarbital (0.5 mM). $[Ca^{2+}]_i$ was measured by fluorescent spectroscopy. Results shown in each panel are the mean ± SEM of 3 experiments.

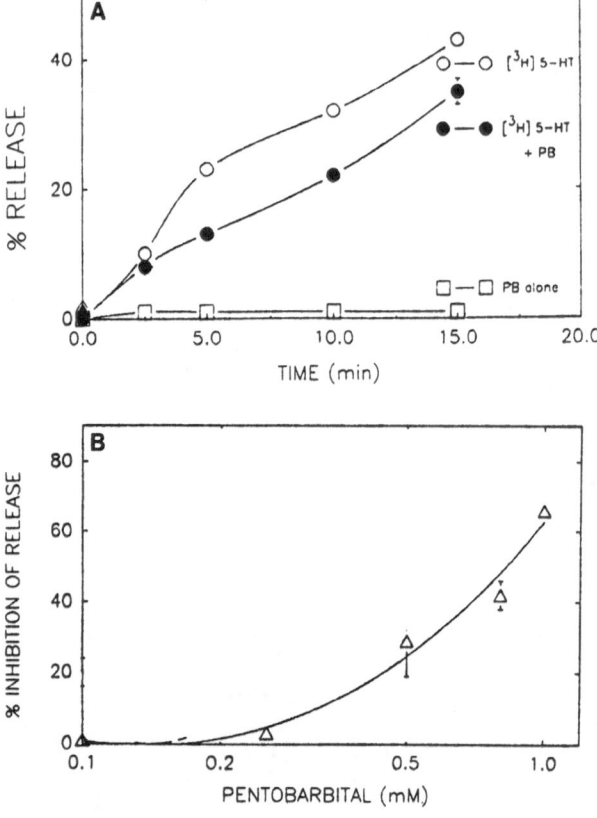

FIGURE 7. Inhibition of [³H]5-HT release in RBL-2H3 cells by pentobarbital. Cells in 24-well culture plates were loaded with [³H]5-HT (1 μCi/ml) for 18 hr (37°C). Panel A shows the time course of inhibition by pentobarbital (0.5 mM). Panel B illustrates the amount of inhibition of (DNP₂₄-BSA)-induced release caused by pentobarbital at 15 min. The concentration of DNP₂₄-BSA was 5 ng/ml. Results are the mean ± SEM of 3 experiments for each panel.

Effects of Local and Volatile Anesthetics on PI Hydrolysis

Lidocaine altered AII-induced inositol phosphate release in a biphasic manner with no effect on spontaneous hydrolysis (figure 8A). Lidocaine, however, increased the amount of labelled inositol phospholipids in the membrane in the presence or absence of the agonist (figure 8B).

When halothane (1%) was delivered to RBL-2H3 cells in the air-tight container (figure 9A and C), no effect on spontaneous hydrolysis was observed. However, inhibition (10-38%) of DNP₂₄-BSA (0.25 to 50 ng/ml)-induced inositol phosphate

FIGURE 8. Effect of lidocaine on inositol phospholipid hydrolysis in endothelial cells. Cells, in 24-well culture plates, were labelled with [³H]inositol (6 μCi), washed and incubated with lidocaine or buffer for 10 min, followed by incubation (5 min) with angiotensin II (AII; 5 ng/ml) alone, AII and pentobarbital (0.5 mM) or pentobarbital alone. Effect of pentobarbital on AII-induced inositol phosphate release is shown in panel A, and on labelled membrane lipids in panel B. Results are the mean ± SEM of 5 experiments in panel A and 2 experiments in panel B.

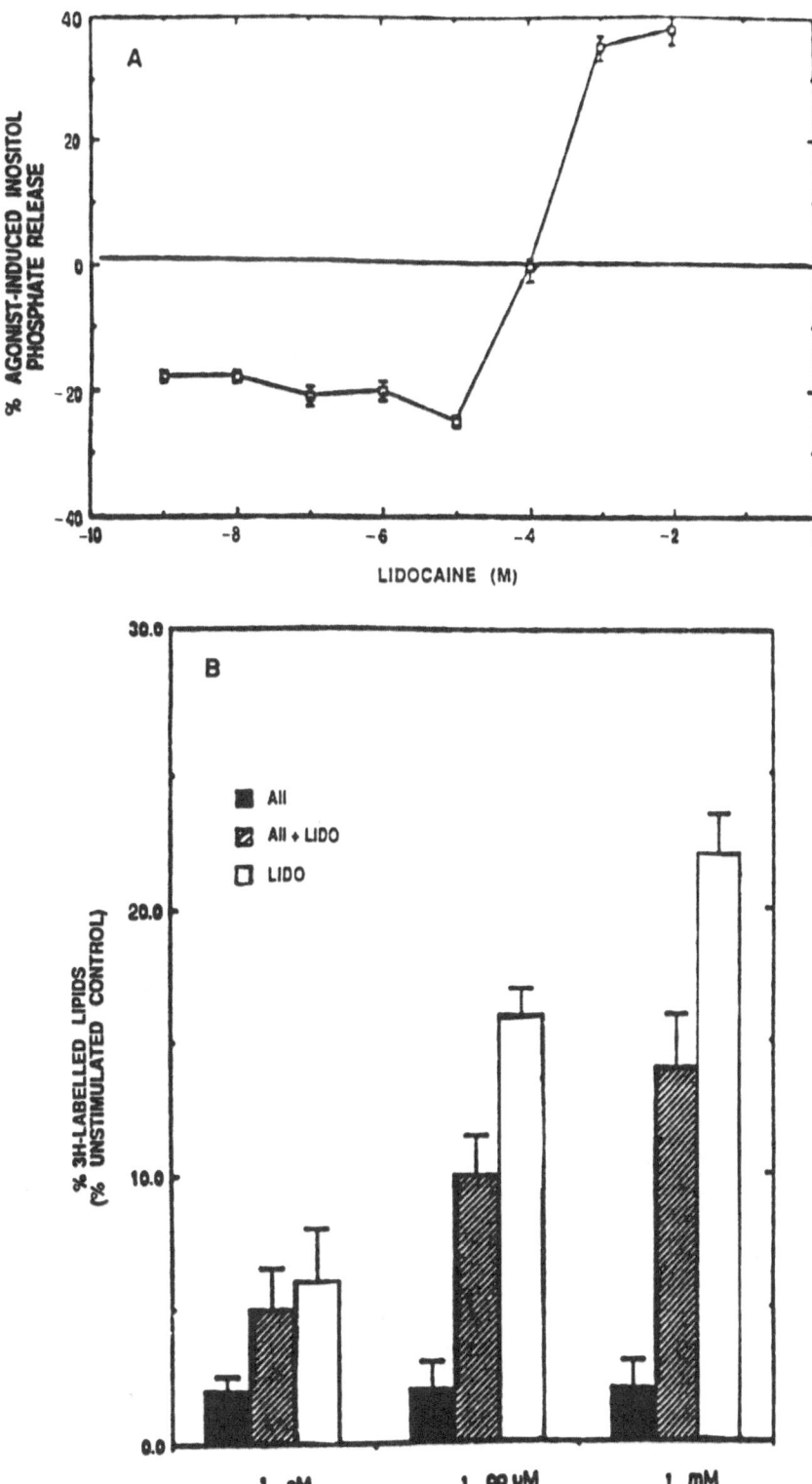

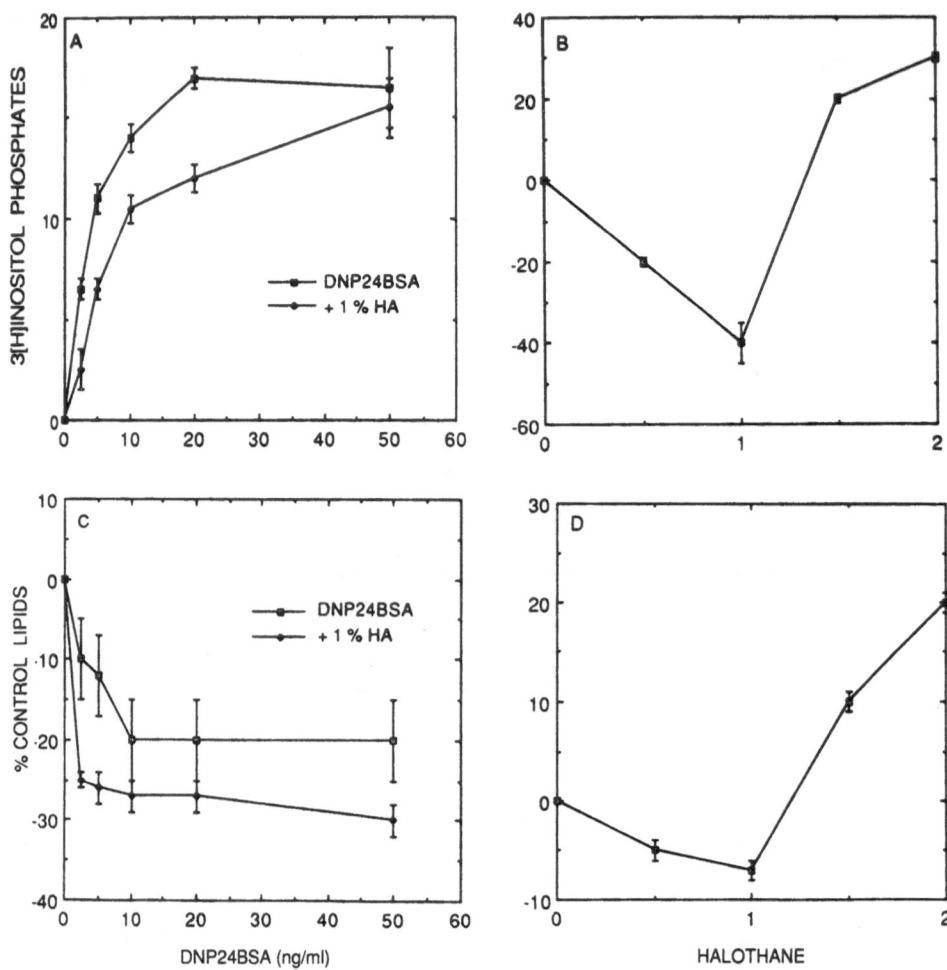

FIGURE 9. Effect of halothane on inositol phospholipid hydrolysis in RBL-2H3 cells. Cells were grown in 24-well plastic culture plates (panels A and C), and 24-well glass plates (panels B and D), and labelled with [³H]inositol for 16-17 hrs. Experiments were performed in an air-tight gas container (panels A and C) and in glass covered glass plates (panels B and D). Panels A and B represent percent of [³H]inositol phosphates released as stimulated by DNP_{24}-BSA in the presence of 1% halothane, or 0.5 to 2%, respectively. Panels C and D represent the percent of control lipids.

release was seen. Halothane, unlike lidocaine, did not alter lipid levels in the absence of the agonist. When halothane, at increasing concentrations, was delivered to cells in 24-well glass culture plates, a biphasic effect on inositol phosphate levels and on membrane lipids occurred (figure 9B and D, respectively). An inhibition was observed at concentrations of 0.5 to 1%, and a stimulation at 1.5 to 2% halothane.

DISCUSSION

The three classes of anesthetics may alter inositol phospholipid hydrolysis by separate mechanisms. Barbiturates inhibit inositol phosphate release in a highly specific manner, and consequently cell secretion is decreased, possibly by an effect on the G-protein, G_p, or on the G-protein-PLC complex. They, however, do not alter membrane inositol phospholipids. Barbiturates may inhibit the G-proteins involved in the stimulation of $^{45}Ca^{2+}$ uptake and increased $[Ca^{2+}]_i$ levels as induced by different agonists, in a discriminate manner. The local anesthetic, lidocaine, alters agonist-induced inositol phosphate release in a biphasic manner, with a corresponding effect on labelled inositol lipids, in the presence and even in the absence of the agonist. Our preliminary studies show that the volatile anesthetic, halothane, alters stimulated hydrolysis in a biphasic manner, similar to the effect of lidocaine. Yet, unlike lidocaine, halothane does not alter inositol lipid levels in the absence of the agonist.

Kinetic analysis of the inhibition of AII-induced hydrolysis (figure 2A) by pentobarbital showed a mixed type inhibition, which was confirmed by Schild regression analysis of AII-induced hydrolysis as inhibited by increasing pentobarbital concentrations (figure 2B). This data indicated that pentobarbital acts in a more complex manner than by simple competitive inhibition. Studies with $[^{125}I]$AII, however, did show competitive inhibition of binding by pentobarbital, as indicated in the Scatchard plot (figure 3B). This finding was unexpected since the structure of AII and pentobarbital are not similar. One possible explanation for the difference in inhibition type could be that barbiturates may alter the activity of the G-protein or G-protein complex that regulates receptor binding and inositol phospholipid hydrolysis. G-proteins can alter the affinity of the receptor for the ligand,[32-33] and when inhibited, give the appearance of competitive inhibition of ligand binding.

Studies with permeabilized RBL-2H3 cells confirmed that pentobarbital alters the activity of G-proteins. Cells made permeable by the bacteriolysin streptolysin O to large molecules such as GTPγS were stimulated by GTPγS to bypass the receptor and directly activate the G-protein, G_p. The presence of extracellular Ca^{2+}, although necessary for receptor-stimulated hydrolysis in intact RBL-2H3 cells, is not necessary for the stimulation by GTPγS in permeabilized cells.[25,34] Our data show that the inhibition of GTPγS-induced simulation by pentobarbital is not related to any effect that the barbiturate may have on Ca^{2+} influx. Further kinetic analysis of the inhibition of GTPγS-induced hydrolysis shows that pentobarbital acts in an uncompetitive manner in permeabilized RBL-2H3 cells.[28] These data point away from a competitive interaction of the barbiturate with GTPγS and instead to an alteration of the function of the G-protein, PLC or the coupling of the two components.

The effect of barbiturates in the intact RBL-2H3 cell was highly specific. Inhibition of hydrolysis by the racemic pentobarbital was intermediate of that of its R(+) and S(−) optical isomers, with little or no inhibition by the structural analogue of pentobarbital (1'RS,3'SR,3'-hydroxypentobarbital). Our data (figure 6) also suggest that pentobarbital can discriminate between different G-proteins in RBL-2H3 cells. In the RBL-2H3 cell, pentobarbital inhibits both antigen-induced $^{45}Ca^{2+}$ uptake and $[Ca^{2+}]_i$ levels, but not responses to the adenosine analogue, NECA. NECA-induced $[Ca^{2+}]_i$ and $^{45}Ca^{2+}$ uptake are altered by pertussis and cholera toxins; however, the toxins have no effect on responses to the antigen. This difference in effect is attributed to the regulation of the two receptor types. IgE and NECA by different G-proteins.[29,34-35]

In the RBL-2H3 cell, pentobarbital inhibited a functional response to hydrolysis, $[^3H]$5-HT secretion (figure 7). This result was expected since cell secretion depends on increases in $[Ca^{2+}]_i$ levels, which were inhibited by pentobarbital in RBL-2H3 cells. The data does point to the possibility that a functional response to hydrolysis (i.e., release of cell factors that alter smooth muscle cell function) in endothelial cells may also be inhibited by pentobarbital.

The effects observed with pentobarbital were not limited to one cell or receptor type. Hydrolysis in both the rat endothelial cell and RBL-2H3 cell was inhibited by pentobarbital. This effect was also seen in various other endothelial cells from other species and vessel types (author's unpublished data) and in smooth muscle cells. Similarly, pentobarbital inhibited the activity of several receptor types (*e.g.*, serotonergic, adrenergic, histaminergic) when cells were stimulated by different agonists (ref. 1 and unpublished data). Interestingly, in the RBL-2H3 cell, pentobarbital also inhibited hydrolysis that was accomplished by a different mechanism of receptor stimulation than that in the endothelial cell. In the RBL-2H3 cell, inositol phospholipid hydrolysis involves the aggregation of receptors for immunoglobulin IgE. The aggregation initiates the activation of PLC, followed by activation of the G-protein.[36]

The data with the local anesthetic, lidocaine, and our preliminary studies with the volatile anesthetic, halothane, suggest that these anesthetics may interfere with hydrolysis by separate mechanisms. Since lidocaine alters lipid levels, even in the absence of the agonists, the data suggests that lidocaine may alter inositol phosphate release by an effect on lipid synthesis through the lipid cycle (figure 1, inset),[15] or on the function of membrane kinases or phosphomonoesterases that allow the interconversion of inositol phospholipids.[15-16] The results with the volatile anesthetic, halothane (figure 9), were similar to those with lidocaine, since halothane exerted a biphasic effect on inositol phosphate release and on labelled membrane lipids. However, no effect was observed with halothane alone on lipid levels. Although the results with halothane are different from those with the barbiturates and the local anesthetic, lidocaine, these results are too preliminary to suggest a mechanism for this effect. Nevertheless, comparing these data to those with the barbiturates, three separate mechanism of action for the three classes of anesthetics on inositol phospholipid hydrolysis appear to exist.

REFERENCES

1. B. M. Altura, B. J. Altura, A. Carella, P. D. M. V. Turlapathy and J. Weinberg, Vascular smooth muscle and general anesthetics, *Fed Proc* 39:1584 (1980).
2. R. D. Miller, "Anesthesia," Churchill Livingstone, New York (1981).
3. N. T. Muth and P. C. Smith, Circulatory effects of modern inhalation anesthetic agents", *in:* "Handbook of Experimental Pharmacology," M. B. Clenoweth, ed., Springer-Verlag, Berlin, (1972).
4. D. E. Longnecker and P. D. Harris, Dilation of small arteries and veins in the bat during halothane anesthetics, *Anesthesiol* 37:423 (1971).
5. P. D. Harris, L. F. Hodoval and D. E. Longnecker. Quantitative analysis of microvascular diameters during pentobarbital and thiopental anesthesia in the bat, *Anesthesiol* 35:337 (1972).
6. J. E. Faber, P. D. Harris and D. L. Wiegman, Anesthetic depression of microcirculation, central hemodynamics, and respiration in decerebrate rats, *Am J Physiol* 243:H837 (1982).
7. B. M. Altura and B. T. Altura, Effects of local anesthetics, antihistamines, and glucocorticoids on peripheral blood flow in vascular smooth muscle, *Anesthesiol* 41:197 (1974).
8. P. Seeman, The membrane actions of anesthetics and tranquilizers, *Pharmacol Rev* 24:583 (1972).
9. F. S. Labella, Is there a general anesthesia receptor? *Can J Physiol Pharmacol* 59: 432 (1981).
10. P. R. Andrews and L. C. Mark, Structural specificity of barbiturates and related drugs, *Anesthesiol* 57:314, (1982).
11. E. M. Vaughn Williams, A classification of anti-arrhythmic actions reassessed after a decade of new drugs, *J Clin Pharmacol* 24:129 (1984).
12. C. F. Stamer, Theoretical characterization of ion channel blockade: Competitive binding to periodically accessible receptors, *Biophys J* 52:405 (1987).
13. D. Kenny, L. R. Pelc, H. L. Brooks, J. P. Kampine, W. T. Schmeling and D. C. Warltier, Calcium channel modulation of α_1 and α_2-adrenergic pressor responses in conscious and anesthetized dogs, *Anesthesiol* 72:874 (1990).

14. A. J. Robinson-White, S. M. Muldoon and F. C. Robinson, Inhibition of inositol phospholipid hydrolysis in endothelial cells by pentobarbital, *Eur J Pharmacol* 172:291 (1989).

15. M. J. Berridge, Inositol triphosphate and diacylglycerol as second messengers, *Biochem J* 220:345 (1984).

16. M. J. Berridge and R. F. Irvine, Inositol triphosphate a novel second messenger in cellular signal transduction, *Nature* 312:315 (1984).

17. S. Cockcroft and B. D. Gomperts, Role of guanine nucleotide binding protein in the activation of phosphoinositide phosphodiesterase, *Nature* 314:534 (1985).

18. C. D. Smith, C. C. Cox and R. Snyderman, Receptor-coupled activation of phosphoinositide-specific phospholipase C by an N protein, *Science* 232:97 (1986).

19. Y. Nishizuka, The role of protein kinase C in cell surface signal transduction and tumor promotion, *Nature* 308:693 (1984).

20. R. F. Furchgott and J. V. Zawadzki, The obligatory role of endothelial cells in the relaxation of arterial smooth muscle by acetylcholine, *Nature* 288:373 (1980).

21. R. F. O'Brien, R. J. Robbins and I. V. McMurty, Endothelial cells in culture produce a vasoconstrictor substance, *J Cell Physiol* 132:263 (1980).

22. M. Yanagsawa, H. Karihara, S. Kimara, Y. Tomobe, M. Kobayashi, M. Mitsui, Y. Goto and T. Maski, A novel potent vasoconstrictor peptide produced by vascular endothelial cells, *Nature* 332:411 (1988).

23. S. M. Schwartz, Selection and characterization of bovine aortic endothelial cells, *In Vitro* 14:966 (1986).

24. E. L. Barsumian, C. Sersky, M. G. Petrino and R. P. Siraganian, IgE-induced histamine release from rat basophilic leukemia lines: Isolation of releasing and non-releasing clones. *Eur J Immunol* 11:317 (1981).

25. H. Ali, J. R. Cunha-Melo and M. A. Beaven, Receptor-mediated release of inositol 1,4-bisphosphate in rat basophilic leukemia RBL-2H3 cells permeabilized with streptolysin O. *Biochem Biophys Acta* 1010:88 (1989).

26. F. J. Leonetti, S. M. Hunt, P. S. Lin, S. R. Kurtz and C. R. Valeri, Preservation of human granulocytes obtained by counterflow centrifugation, *Transfusion* 17:465 (1977).

27. S. M. Muldoon, P. M. Vanhoutte, R. R. Lorenz and R. A. Van Dyke, Venomotor changes are caused by halothane acting on the sympathetic nerves, *Anesthesiol* 43:41 (1975).

28. A. J. Robinson-White, S. M. Muldoon, L. Elson and D. M. Collado-Escobar, Evidence that barbiturates inhibit antigen-induced responses through interactions with a GTP-binding protein in rat basophilic leukemia (RBL-2H3) cells, *Anesthesiol* 72:996 (1990).

29. H. Ali, J. R. Cunha-Melo, W. Saul and M. A. Beaven, Activation of phospholipase C via adenosine receptors provides synergistic signals for secretion in antigen-stimulated RBL-2H3 cells, *J Biol Chem* 265:745 (1990).

30. M. A. Beaven, J. P. Moore, G. A. Smith, T. R. Hasketh and J. C. Metcalfe, The calcium signal and phosphatidylinositol breakdown in 2H3 cells, *J Biol Chem* 259:7137 (1984).

31. H. Ali, D. M. Collado-Escobar and M. A. Beaven, The rise in concentration of free Ca^{2+} and of pH provide sequential synergistic signals for secretion in antigen-stimulated rat basophilic leukemia (RBL-2H3) cells, *J Immunol* 143:2626 (1989).

32. R. J. Lefkowitz, M. G. Caron and G. L. Stetes, Mechanisms of membrane-receptor regulation. Biochemical, physiological, and clinical insights derived from studies of the adrenergic receptor, *New Engl J Med* 310:1570 (1984).

33. A. G. Gilman, G-Proteins: Transducers of receptor-generated signals, *Ann Rev Biochem* 56:60 (1987).

34. V. Narasimihan, D. Holowky, C. Fewtrell and B. Baird, Cholera toxin increases the rate of antigen-stimulated calcium influx in rat basophilic leukemia cells, *J Biol Chem* 263:19626 (1988).

35. H. Ali, D. M. Collado-Escobar, J. R. Cunha-Melo, H. M. S. Gonzago, F. L. Huang, K-P. Haung and M. A. Beaven, Studies of protein kinase C in the rat basophilic leukemia (RBL-2H3) cell reveals that antigen-induced signals are not mimicked by the actions of phorbol myristate acetate Ca^{2+} ionophore, *J Immunol* 143:2617 (1989).

36. K. Macyama, R. J. Hohman, H. Metzger and M. A. Beaven, Quantitative relationships between aggregation of IgE receptors, generation of intracellular signals, and histamine secretion in rat basophilic leukemia (2H3) cells, *J Biol Chem* 261:2583-2592 (1986).

26

Direct Actions of Volatile Anesthetics on the Coronary Vasculature

David R. Larach

INTRODUCTION

Blood vessels determine the distribution of cardiac output to the various body organs, and successful performance of this function necessitates a high degree of regulation of the vascular system. These regulatory processes involve both neurohumoral factors under nervous system control, and local control of vascular resistance within each bed being perfused. Thus, when attempting to understand the mechanisms of vascular action of hemodynamically active drugs such as halothane, it is important to distinguish their direct vascular effects (mediated by direct interaction of the anesthetic with the blood vessel) from their indirect vascular actions.

Indirect vascular actions of anesthetics may be mediated by the nervous system through reflexes (*e.g.*, the baroreflex) or by anesthetic modulation of sympathetic outflow from the neuraxis. Additionally, indirect vascular effects can manifest as actions upon the nonvascular cells of an organ (*e.g.*, negative myocardial inotropy mediated by cardiomyocytes), which in turn change vascular tone through autoregulatory processes. Considerable prior research has explored the indirect actions of halogenated anesthetics on the vasculature (see Seagard *et al.*[1] for review). Yet only recently has comparable study begun to be devoted to the direct vascular effects of these drugs. In the heart, coronary vascular resistance is tightly linked to local myocardial metabolism. This fact can confound efforts to distinguish direct (vascular) from indirect (non-vascular) drug actions affecting vascular tone, and is particularly relevant to drugs possessing negative inotropic properties such as the halogenated anesthetics.

Consider figure 1, in which a hypothetical anesthetic generates multiple opposing vascular responses in an *in vivo* preparation. First, coronary vascular resistance is altered directly (this being the effect of primary interest). Second, the anesthetic causes hypotension that lowers coronary perfusion pressure and induces an indirect autoregulatory vasodilation.[2] Third, decreased afterload, cardiac contractility, and preload all lower myocardial O_2 demand; this leads to indirect coronary vasoconstriction by flow-metabolism coupling.[2] Finally, reduced preload and contractility lessen the systolic compression of intramural vessels, indirectly lowering the apparent vascular resistance. In order to deduce the direct coronary action of the

DAVID R. LARACH, Department of Anesthesia, College of Medicine, The Pennsylvania State University, Hershey, PA 17033.

Mechanisms of Anesthetic Action in Skeletal, Cardiac, and Smooth Muscle
Edited by T.J.J. Blanck and D.M. Wheeler, Plenum Press, New York, 1991

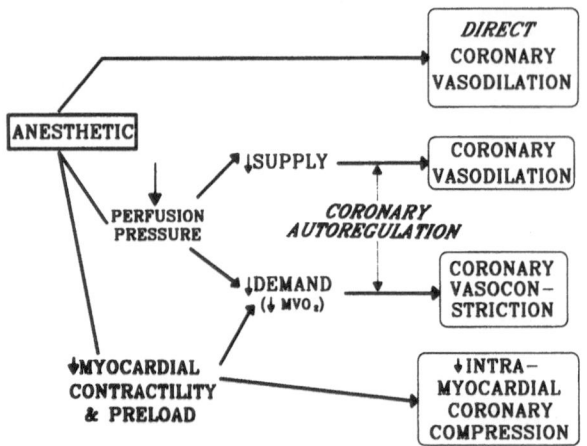

FIGURE 1. Schematic diagram showing the interrelationships between the direct and the indirect factors that influence coronary vascular resistance in the whole animal or human. A hypothetical anesthetic is shown having multiple vascular actions, including: 1) direct coronary vasodilation; 2) indirect vasodilation due to pressure-flow autoregulation; 3) indirect vasoconstriction due to reduced myocardial O_2 consumption (MVO_2) through the process of cardiac flow-metabolism coupling; and 4) indirect vasodilation due to reduced systolic compression of intramyocardial coronary branches. See text for details.

anesthetic, it is necessary to eliminate the indirect vascular resistance effects of the drug.

The blood vessels often are embedded within organs that provide a metabolically active environment (such as the myocardium). The complexity of the vascular wall is much greater than previously thought, and involves mutual interactions among vascular smooth muscle cells, endothelial cells, formed elements of the blood, circulating hormones, hydrostatic and hydrodynamic factors (*e.g.*, vessel stretch and blood flow effects), and neural and connective tissue elements within the adventitia.[3,4,5] Three general types of laboratory studies on whole blood vessels can be performed: *in vivo*, *in vitro*, or *in situ* isolated organ. *In vivo* experiments using intact animals to study vascular actions often are the most physiological; these are preferred for studies in which intact inter-organ regulation is important, and may provide the best potential for extrapolation to the clinical setting. Mechanistic studies also can utilize *in vivo* preparations effectively, however it may be difficult to control for changes in the neurohumoral milieu brought about by reflex actions and by direct anesthetic effects on other organ systems. Recent technical advances permitting sophisticated study of conscious, chronically-instrumented animals has removed many of the problems in prior literature relating to lack of an appropriate anesthetic-free control state.

In vitro preparations utilize isolated blood vessels in a tissue bath or perfusion chamber. The vessel has been removed both from the animal and from its normal perivascular environment. The primary advantages include precise control of external conditions such as the electrolyte and hormonal composition of the perfusate, mechanical stretch, and lack of neuronal activity. Lost, however, is the normal interaction with the tissue within which the vessel ordinarily is embedded. This can be particularly important in organs displaying marked vascular regulation such as the heart, because excised vessels will not possess the normal flow-metabolism coupling normally seen when coronary vessels are in close proximity to cardiomyocytes. Vascular damage during excision, especially to the endothelium, may occur also.

Despite these limitations, *in vitro* vascular studies are of great value for exploring mechanisms of anesthetic action. Using isolated preparations, the investigator has total control of environmental variables, can artificially reconstruct various vessel-wall layers, and can use toxic concentrations of pharmacological probes and antagonists in order to test specific mechanistic hypotheses.

The *in situ* isolated organ type of preparation consists of vascular tissue that lies undisturbed within an isolated organ that undergoes artificial perfusion. Examples include the coronary circulation within the isolated perfused heart (*ex vivo* Langendorff preparation), or the perfused rat hindquarter. *In situ* perfusion provides the advantage of maintenance of a normal peri-vascular cellular environment, with the ability to carefully control the experimental perfusion conditions. In this way, vascular function in the presence of qualitatively preserved flow-metabolism coupling can be examined in a controlled preparation.

A number of studies have investigated the net actions of volatile anesthetics on coronary vascular resistance. However, lack of isolation of the indirect vascular effects often masked the important <u>direct</u> action of these agents on coronary resistance vessels. For example, halothane has been shown to decrease,[6] to increase,[7] or to leave unchanged[8] the coronary vascular resistance in animal studies, depending upon the systemic hemodynamics and the pathophysiology. Even in a study using isolated perfused working hearts under controlled conditions, the indirect vascular actions of volatile anesthetics appeared to alter vascular tone profoundly.[9] Only one prior study has purported to identify a direct coronary resistance-vessel vasodilating action by halothane; however, that older investigation suffers from serious limitations of methodology and reporting.[10]

EXPERIMENTAL STUDIES

My colleagues and I have developed a new experimental technique utilizing the *in situ* coronary resistance vessels within the isolated perfused rat heart for examining the direct coronary resistance vessel actions of drugs possessing negative inotropic activity on the myocardium.[11] The basis for this method is the reduction of myocardial oxygen consumption ($M\dot{V}O_2$) to low basal levels by inducing cardiac arrest, such that the administration of a negative inotropic drug is not capable of causing any further reduction in $M\dot{V}O_2$. Together with control of each of the other indirect factors that affect coronary resistance (see figure 1), this preparation permits the elucidation of the direct vascular actions of anesthetics.

Briefly, isolated rat hearts were subjected to retrograde aortic perfusion at constant pressure, without prior anesthetic exposure. The recirculating perfusate was a modified Krebs-Hensleit bicarbonate solution with glucose, pyruvate, albumin and insulin at pH 7.40 and 37°C. Each heart was arrested and the left ventricular cavity was vented to minimize preload and intramyocardial coronary compression. The influence of halothane or isoflurane at concentrations between 0.5 and 2.0 × MAC (monitored by mass spectrometry) on coronary flow was measured at steady state, and expressed as an anesthetic concentration-response curve. Due to the constant pressure perfusion conditions, the flow values are inversely related to coronary vascular resistance. Statistical analysis included iterative four-parameter logistic fitting of families of curves, with $P < 0.05$ considered significant. Data are mean ± SEM. Please see Larach *et al.*[11] for further details.

Cardiac arrest was induced to lower $M\dot{V}O_2$ to basal values. We avoided the most commonly-used methods of cardioplegia such as hyperkalemia, hypocalcemia, or hypermagnesemia because these techniques would alter vascular reactivity due to the importance of membrane voltage-dependent calcium channels in vascular excitation-contraction coupling. Instead, we arrested the rat hearts with the sodium-channel blocking toxin, tetrodotoxin. The lack of effect of tetrodotoxin on smooth muscle contraction due to direct pharmacological activation has been demonstrated in numerous isolated, buffer-perfused tissues in the literature (cited in reference 11). Preliminary studies in my laboratory appear to confirm this lack of effect of tetrodotoxin in the rat aorta and porcine coronary artery when several different agonists are used to induce vasoconstriction.

Figure 2 demonstrates the dose-dependent direct coronary vasodilation responses that were induced by both halothane and isoflurane in the arrested vented rat heart.[11] The median effective dose (ED_{50} for coronary vasodilation) was 1.31 ± 0.08 for halothane, and 1.53 ± 0.12 for isoflurane; $P = 0.06$, where the units are multiples of MAC (minimum alveolar concentration: [ED_{50}] to achieve anesthesia). The vasodilation was completely reversible (compare zero anesthetic point with the left pair of bars). The administration of the potent coronary vasodilator adenosine after anesthetic recovery (right pair of bars) generated higher coronary flows than did either halothane or isoflurane.

DISCUSSION

These data indicate that halothane and isoflurane are equipotent in their direct coronary resistance vessel vasodilating activity in the isolated perfused arrested rat heart. A number of prior studies have revealed differences in the coronary

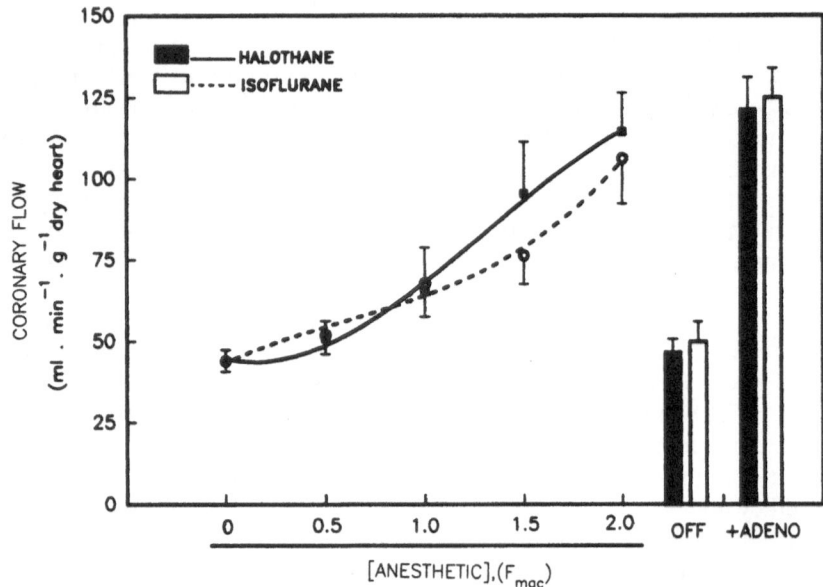

FIGURE 2. Direct coronary flow responses induced by anesthetics in the isolated drained tetrodotoxin-arrested rat heart preparation, during constant-pressure perfusion at 90 mmHg. Anesthetic concentration–response curves are shown for halothane and isoflurane over a range of MAC multiples (F_{mac}) between 0 and 2.0. Note equipotent vasodilation induced by both agents. "OFF" indicates recovery after discontinuing anesthetic. "+ADENO" shows maximal vasodilation by 5×10^{-5} M adenosine, administered following the "OFF" values. F_{mac} data use MAC values of 1.0% (halothane) or 1.4% (isoflurane) for the rat. Redrawn using data from Larach et al.[11]

vasodilation induced by halothane and isoflurane in various working heart preparations. For example, Sahlman et al.[9] demonstrated that halothane reduced coronary flow in the working rat heart, while isoflurane increased coronary flow in the same setting. It is likely that the differences between the coronary direct flow responses induced by halothane and isoflurane in our arrested heart preparation and those responses seen in other experiments in the literature reflect differences in how each of these anesthetics affect the indirect processes that alter coronary vascular tone. Further studies are needed to clarify this point.

The data in figure 2 may appear to indicate that both halothane and isoflurane nearly abolish coronary flow reserve, because adenosine produced only slightly more flow than did either anesthetic. However, conclusions regarding flow reserve cannot be drawn from those data due to a small but significant time-dependent decrease in maximal adenosine-induced flow that prevents direct comparison between the anesthetic and the adenosine data.[11]

Recent experiments in my laboratory have utilized a different protocol with the tetrodotoxin-arrested isolated drained rat heart preparation to measure directly the coronary flow reserve and its modification by halogenated anesthetics. During steady-state anesthetic vasodilation, adenosine 5×10^{-5} and 10^{-4} M are added to the perfusion circuit to induce maximal vasodilation at a constant perfusion pressure. The increment in coronary flow generated by the adenosine is the coronary flow reserve. Preliminary data reveal that both halothane and isoflurane reduce flow reserve in a dose-dependent manner, such that nearly 90% of the reserve is abolished with $3 \times MAC$ concentrations. In contrast, sevoflurane, an investigational halogenated methyl-isopropyl-ether anesthetic,[12] appears to preserve coronary flow reserve significantly more than halothane or isoflurane. This response by sevoflurane appears to be caused by the generation of less direct coronary vasodilation at high MAC fractions than in the presence of halothane or isoflurane, while maximal adenosine vasodilation responses tend to remain unchanged by sevoflurane.

Coronary flow reserve defines the ability of the microcirculation to dilate and maintain organ perfusion in response to a flow-limiting stress, such as stenosis of a conducting coronary artery. Flow reserve is affected by the autoregulatory function of the vascular bed, as well as changes in epicardial coronary resistance, perfusion pressure, inotropic state, heart rate, perfusate viscosity, intramyocardial coronary compression, and the presence of cardiac hypertrophy.[13] Thus, preservation of flow reserve by an anesthetic may afford theoretical advantages in the clinical setting, but this possibility must be subjected to further study. In particular, the lack of significant coronary hemodynamic differences between sevoflurane and isoflurane in the chronically instrumented dog reported by Bernard et al.[14] must be reconciled with our observations in the arrested rat heart. In addition, further research is needed to explore the basic mechanisms underlying the direct and indirect vascular actions of volatile anesthetics on the coronary microcirculation under conditions of health and disease.

ACKNOWLEDGEMENTS

The author is grateful for the valuable contributions of H. Gregg Schuler, Thomas M. Skeehan, M.D., and Christopher J. Peterson, M.D. to this research. The support of NIH grant R29-GM-36593, Maruishi Pharmaceutical Co., Ltd., and the Department of Anesthesia are acknowledged.

REFERENCES

1. J. L. Seagard, Z. J. Bosnjak, F. A. Hopp, Jr., K. J. Kotrly, T. J. Ebert and J. P. Kampine, Cardiovascular effects of general anesthesia, *in:* "Effects of Anesthesia," B. G. Covino, H. A. Fozzard, K. Rehder, G. Strichartz, eds., American Physiological Society, Bethesda, pp. 149-177 (1985).
2. J. I. E. Hoffman and J. A. E. Spaan, Pressure-flow relations in coronary circulation. *Physiol Rev* 70:331-390 (1990).
3. R. F. Furchgott and P. M. Vanhoutte, Endothelium-derived relaxing and contracting factors. *FASEB J* 3:2007-2018 (1989).
4. P. M. Vanhoutte, Endothelium and control of vascular function: State of the art lecture. *Hypertension* 13:658-667 (1989).
5. J. A. Bevan and E. H. Joyce, Flow-induced resistance artery tone: balance between constrictor and dilator mechanisms. *Am J Physiol* 258:H663-H668 (1990).
6. C. W. Buffington, J. L. Romson, A. Levine, N. C. Duttlinger and A. H. Huang, Isoflurane induces coronary steal in a canine model of chronic coronary occlusion. *Anesthesiology* 66:280-292 (1987).
7. R. G. Merin, P. D. Verdouw and J. W. de Jong, Dose-dependent depression of cardiac function and metabolism by halothane in swine (Sus scrofa). *Anesthesiology* 46:417-423 (1977).
8. B. A. Cason, E. D. Verrier, M. J. London, D. T. Mangano and R. F. Hickey, Effects of isoflurane and halothane on coronary vascular resistance and collateral myocardial blood flow: their capacity to induce coronary steal. *Anesthesiology* 67:665-675 (1987).
9. L. Sahlman, B. Å. Henriksson, J. Martner and S. E. Ricksten, Effects of halothane, enflurane, and isoflurane on coronary vascular tone, myocardial performance, and oxygen consumption during controlled changes in aortic and left atrial pressure: studies on isolated working rat hearts in vitro. *Anesthesiology* 69:1-10 (1988).
10. R. J. Domenech, P. Macho, J. Valdes and M. Penna, Coronary vascular resistance during halothane anesthesia. *Anesthesiology* 46:236-240 (1977).
11. D. R. Larach, H. G. Schuler, T. M. Skeehan and C. J. Peterson, Direct effects of myocardial depressant drugs on coronary vascular tone: anesthetic vasodilation by halothane and isoflurane. *J Pharmacol Exp Ther* 254:58-64 (1990).
12. R. F. Wallin, B. M. Regan, M. D. Napoli and I. J. Stern, Sevoflurane: a new inhalational anesthetic agent. *Anesth Analg* 54:758-766 (1975).
13. J. I. E. Hoffman, Maximal coronary flow and the concept of coronary vascular reserve. *Circulation* 70:153-158 (1984).
14. J.-M. Bernard, P. F. Wouters, M.-F. Doursout, B. Florence, J. E. Chelly and R. G. Merin, Effects of sevoflurane and isoflurane on cardiac and coronary dynamics in chronically instrumented dogs. *Anesthesiology* 72:659-662 (1990).

27

Effects of Volatile Anesthetics on the Coronary Circulation in Chronically Instrumented Dogs

Robert G. Merin, Marie-Françoise Doursout, Jacques E. Chelly

INTRODUCTION

Although it had been generally assumed that the potent inhalation anesthetics could directly relax vascular smooth muscle including coronary,[1-3] there had been few published accounts of the direct effects of anesthetics on the coronary vasculature. The initial experiments concerning the effect of volatile anesthetics on the coronary circulation suggested that the metabolic depression produced by the potent inhalation anesthetics overrode the direct vasodilating effects, resulting in a dose-dependent decrease in coronary blood flow without appreciable effect on calculated coronary vascular resistance.[4-8] However, the study in humans by Reiz and colleagues challenged this universal effect of inhalation anesthetics[9] and subsequent studies in animals under controlled conditions have shown that isoflurane is indeed a coronary vasodilator even in the chronically instrumented intact animal.[10-13] Since coronary vasodilation can produce deleterious redistribution of myocardial perfusion especially in collateralized hearts,[14] there has been great interest in this coronary vasodilating effect of isoflurane. Although Buffington demonstrated that isoflurane can produce this maldistribution of myocardial perfusion under carefully defined conditions,[15] others have disputed his contention.[16,17] Part of the controversy has stemmed from a question about the magnitude of the coronary vasodilation produced by isoflurane. With the advent of two new inhalation anesthetics, sevoflurane and the isoflurane analog desflurane, it seemed appropriate and necessary to compare the coronary dynamic effects of the new anesthetics with isoflurane.

METHODS

Our basic approach involved the use of chronically instrumented dogs well trained to allow awake basal measurements. Then the anesthetics being tested were administered, in random order and on different days, to the same chronically

ROBERT G. MERIN, MARIE-FRANÇOISE DOURSOUT, JACQUES E. CHELLY, Depatment of Anesthesiology, University of Texas Medical School at Houston, Houston, Texas 77030.

Mechanisms of Anesthetic Action in Skeletal, Cardiac, and Smooth Muscle
Edited by T.J.J. Blanck and D.M. Wheeler, Plenum Press, New York, 1991

Table 1. Effect of Inhaled Anesthetics on Coronary Blood Flow and Vascular Resistance

	Awake	Anesthetic Concentration (MAC)			
		1.1-1.2	1.4	1.7-1.8	2.0
Coronary Blood Flow (ml/min)					
Isoflurane	41	56*	58*	61*	66*
Sevoflurane	45	56*			52
Desflurane	42	55	57*	58*	58
Coronary Vascular Resistance (mm Hg · min^{-1} · ml^{-1})					
Isoflurane	2.4	1.5*	1.4*	1.3*	1.1*
Sevoflurane	2.1	1.4			1.3*
Desflurane	2.3	1.4*	1.3*	1.2*	1.2*

Asterisk (*) indicates significant difference ($P < 0.05$) compared to awake value.

instrumented animals in order to provide a more powerful statistical approach. The methodology has been previously published.[11,18] Briefly, healthy mongrel dogs were instrumented in order to measure cardiac output (electromagnetic flow probe on the pulmonary artery); aortic and left atrial pressures (silastic catheters in the aorta and left atrium); left ventricular pressure and dP/dt (Konigsberg manometer implanted into the left ventricle); and coronary blood flow (pulsed Doppler flow meters on the circumflex coronary artery). During anesthesia, ventilation and F_iO_2 were adjusted to match awake arterial oxygen and carbon dioxide tensions, and different concentrations of the anesthetics were given in random order. Statistical analysis utilized analysis of variance, the Bonferroni modification of the paired t, and Dunnett's t-test.

RESULTS

All three anesthetics produced significant increases in coronary blood flow when compared to the awake controls (table 1). Inasmuch as the anesthetics also produced a dose-related decrease in mean arterial and hence coronary perfusion pressure, coronary vascular resistance was significantly decreased by all three anesthetics as well. At no dose (MAC multiple) was there a significant difference between the three anesthetics either on coronary blood flow or coronary vascular resistance.

DISCUSSION

In our chronically instrumented animals, the effect of isoflurane, sevoflurane and desflurane on the coronary circulation appeared to be practically identical when the anesthetics were administered to the same animals on different days. Furthermore, when the effect of the three anesthetics on systemic hemodynamics were compared (heart rate, mean arterial pressure, left ventricular dP/dt, cardiac output and stroke volume), there were also no significant differences at any comparable anesthetic dose.[18,19] However, complete characterization of a true coronary vasodilator includes not only producing a significant decrease in coronary vascular resistance, but also uncoupling the relationship between myocardial oxygen demand and coronary blood flow. The most potent control of coronary blood flow in the intact animal is metabolic so that the oxygen demands of the heart are coupled to coronary blood flow.[20] If such coupling is efficient, there will be no change in myocardial oxygen extraction at any dose of a drug. If on the other hand a drug is a coronary vasodilator, coronary blood flow will be greater than necessary to maintain myocardial oxygen delivery and myocardial oxygen extraction will decrease. This has been shown to be true with the prototype vasodilator, adenosine,[15] and with isoflurane.[9,15] Our experimental design did not allow sampling of coronary venous oxygen contents, so we could not ascertain whether sevoflurane and desflurane also uncouple coronary blood flow from myocardial oxygen demand. However, considering the marked similarity between the effect of the three anesthetics on the coronary circulation, it seems likely that such would be the case.

In a similar study as yet unpublished, Pagel and colleagues from the Medical College of Wisconsin have also examined the effect of volatile anesthetics on the coronary circulation in the chronically instrumented dog.[21] Their protocol was practically identical to ours in that they also tested the anesthetics in the same animals on different days using the same type of coronary flow meter. They did not study sevoflurane but did look at the other two available volatile anesthetics, halothane and enflurane. They did not convert the Doppler transduced flow velocity to flow so that their results are expressed in hertz rather than ml/min. However, their results with isoflurane and desflurane were very similar to ours (table 2). They only examined two concentrations, 1.25 and 1.75 × MAC, but showed that both anesthetics produced a significant increase in coronary blood flow at both anesthetic concentrations and, hence, a dose-related decrease in coronary vascular resistance. In their study, however, desflurane produced less decrease in mean arterial pressure (and, hence, coronary perfusion pressure) than isoflurane. Consequently, the numerical values for diastolic coronary vascular resistance were greater at both anesthetic concentrations with

Table 2. Effect of Inhaled Anesthetics on Diastolic Blood Flow Velocity and Vascular Resistance

| | | Anesthetic Concentration (MAC) | |
	Awake	1.25	1.75
Diastolic Coronary Blood Flow (Hz · 10^2)			
Halothane	33	29	29
Enflurane	34	37	34
Isoflurane	33	43*	44*
Desflurane	35	41*	41*
Diastolic Coronary Vascular Resistance (mm Hg · min^{-1} · ml^{-1})			
Halothane	2.61	2.11	1.98
Enflurane	2.61	1.66*	1.57*
Isoflurane	2.64	1.43*	1.04*
Desflurane	2.35	1.70*	1.41*

Data taken from Pagel et al. (ref. 21). Asterisk (*) indicates P < 0.05 for comparison to awake state.

Table 3. Comparison of the Effects of Four Inhaled Anesthetics on Coronary Blood Flow and Coronary Vascular Resistance

	Anesthetic Concentration (MAC)						
	1.2		1.4	1.75		2.8	
(Diastolic Coronary Blood Flow Velocity) or Coronary Blood Flow in % of awake values							
Halothane	(88)		97	(88)		87	
Enflurane	(109)	98		(100)	83		
Isoflurane	(130)	137		(133)	149		
Desflurane	(117)	131		(117)	138		
(Diastolic Coronary Vascular Resistance) or Coronary Vascular Resistance in % of awake values							
Halothane	(81)		79	(76)		59	
Enflurane	(64)	77		(60)	50		
Isoflurane	(54)	64		(39)	54		
Desflurane	(72)	59		(60)	53		

Data in parentheses from Pagel, Schmeling and Warltier (ref. 21).

desflurane than with isoflurane, although the difference between the two anesthetics was not statistically significant. It is of note that although neither halothane nor enflurane significantly changed coronary blood flow under these circumstances, there was a decrease in coronary vascular resistance produced by both anesthetics. The hierarchy resembled that suggested by Hickey's group,[13] namely, that isoflurane was the most potent coronary vasodilator followed by enflurane, with halothane producing the least effect on coronary vasculature. If our previous studies using chronically instrumented animals are included (although the experimental design is not as powerful as that of Pagel et al), we also saw the same sort of hierarchy with the inhalation anesthetics (table 3). Halothane and enflurane produced no significant change in coronary blood flow, but did decrease calculated coronary vascular resistance. In fact, the decrease in coronary vascular resistance produced by enflurane at 1.7 to 1.75 MAC when expressed as % of awake controls was approximately the same as that produced by isoflurane and desflurane. Studies in humans by Reiz' group also suggested that enflurane was a potent coronary vasodilator.[22]

It is of some interest that the effects of isoflurane on the coronary circulation appear to be species related. The coronary vasodilation produced in humans has also been produced by all other investigators looking at the dog.[10,11,12,13,15,18,19,21] However, the effect of isoflurane on the coronary circulation in intact swine (unpublished observations)[16,23] and the rat[24] appears to be different since no coronary vasodilation has been observed in either species. There has not been a dose-dependent study in the intact rat, although Larach and colleagues have shown a direct coronary vasodilating effect by both halothane and isoflurane in the isolated non-working arrested rat heart[25] (see also Chapter 26). Other investigators have also now shown a direct depressant effect of volatile anesthetics on coronary vascular smooth muscle;[26,27] further discussion can be found in Chapters 21 and 24.

In summary, in the chronically instrumented dog, when the effects of volatile anesthetics on coronary dynamics in the well trained, awake animal are compared, isoflurane, sevoflurane and desflurane all produce a significant increase in coronary blood flow and a decrease in coronary vascular resistance. Although neither halothane nor enflurane produce a significant increase in coronary blood flow, they also produce a less marked decrease in coronary vascular resistance. These effects in the intact chronically instrumented animal are undoubtedly related to a combination of the direct coronary vascular smooth muscle relaxation produced by these drugs and the metabolic regulation which is at least partially intact, particularly with halothane and less so with the other four inhalation anesthetics.

REFERENCES

1. M. L. Price, H. L. Price, Effects of general anesthetics on contractile responses of rabbit aortic strips, *Anesthesiology* 23:16-20 (1962).
2. D. H. Sprague, J. C. Yang, S. H. Ngi, Effects of isoflurane and halothane on contractility and cyclic 3-5-adenosine monophosphate system in the rat aorta, *Anesthesiology* 40:162-167 (1974).
3. B. M. Altura, B. T. Altura, A. Carella *et al.*, Vascular smooth muscle and general anesthetics, *Fed Proc* 39:1584-1591 (1980).
4. R. G. Merin, Myocardial hemodynamics and metabolism in the halothane depressed canine heart, *Anesthesiology* 31:20-27 (1969).
5. R. G. Merin, H. H. Borgstedt, Myocardial function and metabolism in the methoxyflurane-depressed canine heart, *Anesthesiology* 34:562-568 (1971).
6. R. G. Merin, T. Kumazawa, N. L. Luka, Myocardial function and metabolism in the conscious dog and during halothane anesthesia, *Anesthesiology* 44:402-416 (1976).
7. R. G. Merin, T. Kumazawa, N. L. Luka, Enflurane depressed myocardial function, perfusion and metabolism in the dog, *Anesthesiology* 45:501-507 (1976).
8. R. G. Merin, S. Basch, Are the myocardial function and metabolic effects of isoflurane really different from those of halothane and enflurane? *Anesthesiology* 55:398-408 (1981).
9. S. Reiz, E. Balfors, M. B. Sorensen *et al.*, Isoflurane—a powerful coronary vasodilator in patients with coronary artery disease, *Anesthesiology* 59:91-97 (1983).
10. S. Gelman, K. D. Fowler, L. R. Smith, Regional blood flow during isoflurane and halothane anesthesia, *Anesth Analg* 63:557-565 (1984).
11. K. Rogers, E. S. Hysing, R. G. Merin *et al.*, Cardiovascular effects of and interaction between calcium blocking drugs and anesthetics in chronically instrumented dogs. II. Verapamil, enflurane and isoflurane, *Anesthesiology* 64:568-575 (1986).
12. J. C. Sill, A. A. Bove, M. Nugent *et al.*, Effects of isoflurane on coronary arteries and coronary arterioles in the intact dog, *Anesthesiology* 66:273-279 (1987).
13. R. F. Hickey, P. E. Sybert, E. D. Verrier *et al.*, Effects of halothane, enflurane and isoflurane on coronary blood flow, autoregulation and coronary vascular reserve in the canine heart, *Anesthesiology* 68:21-30 (1988).
14. W. Schaper, P. Lewi, W. Flameng, L. Gijpen, Myocardial steal produced by coronary vasodilation in chronic coronary artery occlusion, *Basic Res Cardiol* 68:3-20 (1973).
15. C. W. Buffington, J. L. Romson, A. Levine *et al.*, Isoflurane induces coronary steal in a canine model of chronic coronary occlusion, *Anesthesiology* 66:280-292 (1987).
16. M. Gilbert, S. L. Roberts, M. Mori *et al.*, Comparative coronary vascular reactivity and hemodynamics during halothane and isoflurane anesthesia in swine, *Anesthesiology* 68:243-253 (1988).
17. B. A. Cason, E. D. Verrier, M. J. Lunden *et al.*, Effects of isoflurane and halothane on coronary vascular resistance and collateral myocardial blood flow: Their capacity to induce coronary steal, *Anesthesiology* 67:665-675 (1987).
18. J.-M. Bernard, P. F. Wouters, M.-F. Doursout *et al.*, Effects of sevoflurane and isoflurane on cardiac and coronary dynamics in chronically instrumented dogs, *Anesthesiology* 72:659-662 (1990).
19. R. G. Merin, J.-M. Bernard, M.-F. Doursout *et al.*, Comparison of the effects of isoflurane and desflurane on cardiovascular dynamics and regional blood flow in the chronically instrumented dog, *Anesthesiology* (1991) in press.
20. E. O. Feigl, Coronary physiology, *Physiol Rev* 63:1-205 (1983).

21. P. S. Pagel, J. P. Kampine, W. T. Schmeling, D. C. Warltier, Comparison of systemic and coronary hemodynamic actions of desflurane, isoflurane, halothane and enflurane in the chronically instrumented dog, *Anesthesiology* (1991) in press.

22. A. Rydvall, S. Haggmark, S. Mark, H. Myhman *et al.*, Effects of enflurane on coronary hemodynamics in patients with ischemic heart disease, *Acta Anaesthesiol Scand* 28:690-695 (1984).

23. G. Lundeen, M. Manohar, C. Parks, Systemic distribution of blood flow in swine while awake and during 1.0 and 1.5 MAC isoflurane anesthesia with or without 50% nitrous oxide, *Anesth Analg* 62:499-512 (1983).

24. W. C. Seyde, M. E. Durieux, D. E. Longnecker, The hemodynamic response to isoflurane is altered in genetically hypertensive as compared with normotensive rats, *Anesthesiology* 66:798-804 (1987).

25. D. R. Larach, H. G. Schuler, T. M. Skeehan, C. J. Peterson, Direct effects of myocardial depressant drugs on coronary vascular tone: Anesthetic vasodilation by halothane and isoflurane, *J Pharmacol* 254:58-64 (1990).

26. B. A. Bollen, J. H. Tinker, K. Hermsmeyer, Halothane relaxes previously constricted isolated porcine coronary artery segments more than isoflurane, *Anesthesiology* 66:748-752 (1987).

27. G. Blaise, J. C. Sill, M. Nugent *et al.*, Isoflurane causes endothelium-dependent inhibition of contractile responses of canine coronary arteries, *Anesthesiology* 67:513-517 (1987).

Contributors

Jill Beech
Department of Clinical Studies
University of Pennsylvania
School of Veterinary Medicine
New Bolton Center
Kennett Square, Pennsylvania 19348

Michael R. Berman
Department of Anesthesiology and Critical
    Care Medicine
The Johns Hopkins Hospital
Baltimore, Maryland 21205

R. Betto
Istituto di Patalogia generale
Universita' di Padova
    and
NRC Unit for Muscle Biology and
    Physiopathology
Via Trieste 75
Padova, Italy

Gilbert A. Blaise
Department d'Anesthesie
Universite de Montreal
Hopital Notre Dame
Montreal, Quebec H2L 4K8

Thomas J. J. Blanck
Department of Anesthesiology and Critical
    Care Medicine
The Johns Hopkins Hospital
Baltimore, Maryland 21205

Zeljko J. Bosnjak
Departments of Anesthesiology and
    Physiology
Medical College of Wisconsin
Milwaukee, Wisconsin 53226

Nedjilka Buljubasic
Department of Anesthesiology
Medical College of Wisconsin
Milwaukee, Wisconsin 53226

Eugenie S. Casella
Department of Anesthesiology and Critical
    Care Medicine
The Johns Hopkins Hospital
Baltimore, Maryland 21205

S. Ceoldo
Istituto di Patalogia generale
Universita' di Padova
    and
NRC Unit for Muscle Biology and
    Physiopathology
Via Trieste 75
Padova, Italy

Jacques E. Chelly
Department of Anesthesiology
University of Texas Medical School
    at Houston
Houston, Texas 77030

Marie-Françoise Doursout
Department of Anesthesiology
University of Texas Medical School
    at Houston
Houston, Texas 77030

Benjamin Drenger
Department of Anesthesiology
Cardiothoracic Anesthesia Unit
Hadassah Hospital
Jerusalem, Israel

G. Fachechi-Cassano
Istituto di Patalogia generale
Universita' di Padova
    and
NRC Unit for Muscle Biology and
    Physiopathology
Via Trieste 75
Padova, Italy

Jeffrey E. Fletcher
Departments of Anesthesiology
    and Biochemistry
Hahnemann University
Philadelphia, Pennsylvania 19102-1192

Noel Flynn
Department of Anesthesiology
Medical College of Wisconsin
Milwaukee, Wisconsin 53226

William Freas
Department of Anesthesiology
Uniformed Services University
   of the Health Sciences
Bethesda, Maryland 20814

Diane Golightly
Department of Anesthesiology
Uniformed Services University
   of the Health Sciences
Bethesda, Maryland 20814

Qi-Hua Gong
Department of Anatomy
Hahnemann University
Philadelphia, Pennsylvania 19102

Ravi Gutta
Department of Anesthesiology
University of Michigan Medical Center
Ann Arbor, Michigan 48109-0572

Michael F. Haney
Department of Anesthesiology
University of Michigan Medical Center
Ann Arbor, Michigan 48109-0572

Jayne Hart
Department of Anesthesiology
Uniformed Services University
   of the Health Sciences
Bethesda, Maryland 20814

Paul Hoehner
Department of Anesthesiology and Critical
   Care Medicine
The Johns Hopkins Hospital
Baltimore, Maryland 21205

Philippe R. Housmans
Department of Anesthesiology
Mayo Medical School and Mayo Foundation
Rochester, Minnesota 55905

Roger A. Johns
Department of Anesthesiology
University of Virginia Health Sciences Center
Charlottesville, Virginia 22908

John P. Kampine
Department of Anesthesiology
Medical College of Wisconsin
Milwaukee, Wisconsin 53226

Ana Katz
Department of Anesthesiology
Soroka Medical Center
Beer Sheva, Israel

Masayuki Katsuoka
Ihara Chemical Company
Shizuoka, Japan

Paul R. Knight
Department of Anesthesiology
University of Michigan Medical Center
Ann Arbor, Michigan 48109-0572

Hirochika Komai
Department of Anesthesiology
University of Wisconsin
Madison, Wisconsin 53792

David R. Larach
Department of Anesthesia
College of Medicine
The Pennsylvania State University
Hershey, Pennsylvania 17033

Rocio LLave
Department of Anesthesiology
Uniformed Services University
   of the Health Sciences
Bethesda, Maryland 20814

Carl Lynch III
Department of Anesthesiology
University of Virginia Health Sciences Center
Charlottesville, Virginia 22908

Robert G. Merin
Department of Anesthesiology
University of Texas Medical School
   at Houston
Houston, Texas 77030

Sheila M. Muldoon
Department of Anesthesiology
Uniformed Services University
   of the Health Sciences
Bethesda, Maryland 20814

Isabelle Murat
Department of Anesthesia
Hopital Saint-Vincent de Paul
82 av. Denfert Rochereau
F-75674 Paris, France

John Nagel
Department of Anesthesiology
Uniformed Services University
   of the Health Sciences
Bethesda, Maryland 20814

R. Nelson
Department of Anesthesiology
Mayo Clinic and Mayo Foundation
Rochester, Minnesota 55905

Thomas E. Nelson
Department of Anesthesiology
University of Texas Health Science Center
Houston, Texas 77030

S. Tsuyoshi Ohnishi
Philadelphia Biomedical Research Institute
King of Prussia, Pennsylvania 19406

M. Ozhan
Department of Anesthesiology
Mayo Clinic and Mayo Foundation
Rochester, Minnesota 55905

Mary Quigg
Department of Anesthesiology and Critical
    Care Medicine
The Johns Hopkins Hospital
Baltimore, Maryland 21205

Appavoo Rengasamy
Department of Anesthesiology
University of Virginia Health Sciences Center
Charlottesville, Virginia 22908

R. Todd Rice
Department of Anesthesiology and Critical
    Care Medicine
The Johns Hopkins Hospital
Baltimore, Maryland 21205

A. Robinson-White
Department of Anesthesiology
Uniformed Services University
    of the Health Sciences
Bethesda, Maryland 20814

Henry Rosenberg
Department of Anesthesiology
Hahnemann University
Philadelphia, Pennsylvania 19102-1192

Susan Riggs Runge
Department of Anesthesiology and Critical
    Care Medicine
The Johns Hopkins Hospital
Baltimore, Maryland 21205

Ben F. Rusy
Department of Anesthesiology
University of Wisconsin
Madison, Wisconsin 53792

G. Salviati
Istituto di Patalogia generale
Universita' di Padova
    and
NRC Unit for Muscle Biology and
    Physiopathology
Via Trieste 75
Padova, Italy

J. Christopher Sill
Department of Anesthesiology
Mayo Clinic and Mayo Foundation
Rochester, Minnesota 55905

Guilherme Suarez-Kurtz
Departamento de Farmacologia
    Basica e Clinica
Universidade Federal do Rio de Janeiro
Rio de Janeiro, RJ - 21941, Brazil

Roberto Takashi Sudo
Departamento de Farmacologia
    Basica e Clinica
Universidade Federal do Rio de Janeiro
Rio de Janeiro, RJ - 21941, Brazil

C. Uhl
Department of Anesthesiology
Mayo Clinic and Mayo Foundation
Rochester, Minnesota 55905

Renée Ventura-Clapier
Laboratoire de Physiologie Cellulaire
    Cardiaque
INSERM U-241
Universite Paris-Sud
F-91405 Orsay, France

Yvonne Vulliemoz
Departments of Anesthesiology
    and Pharmacology
College of Physicians and Surgeons
Columbia University
New York, New York 10032

David M. Wheeler
Department of Anesthesiology and Critical
    Care Medicine
The Johns Hopkins Hospital
Baltimore, Maryland 21205

Steven J. Wieland
Department of Anatomy
Hahnemann University
Philadelphia, Pennsylvania 19102

Dixon W. Wilde
Department of Anesthesiology
University of Michigan Medical Center
Ann Arbor, Michigan 48109-0572

Index

Anesthetics, intravenous (continued)
  pentobarbital (continued)
    and G proteins (smooth), 285
    and inositol phosphates (smooth), 211,
      278
    and inositol phospholipid hydrolysis
      (smooth), 275-279, 285, 286
    and serotonin secretion (smooth), 278,
      279, 282
  thiopental
    and cooling contractures (cardiac), 119,
      120
    and sarcoplasmic reticulum calcium
      content (cardiac), 122
Anesthetics, local
  bupivacaine
    and slow action potential (cardiac), 164
  lidocaine
    and cooling contractures (skeletal), 48-50
    and inositol phospholipid hydrolysis
      (smooth), 282-284, 286
    and slow action potential (cardiac), 164
  procaine
    and cooling contractures (skeletal), 48-50
    and slow action potential and contraction
      (cardiac), 161-164
    and sarcoplasmic reticulum calcium release
      (cardiac), 93, 164, 165
  tetracaine
    and slow action potential and contraction
      (cardiac), 160-162
Angiography, see Methods and techniques
Angiotensin II, 211, 275-278, 282, 284, 285
Annexin, 6
Aorta, 209-211, 216-220, 224, 229-233, 264,
  273, 291, 292, 296
Arachidonic acid, 11, 14-17, 64, 170, 172, 175,
  215, 219
Arginine, see L-arginine
Arginine analogs, see N-monomethyl
  L-arginine and Nitro-L-arginine
ATP
  and cytosolic calcium concentration
    (smooth), 265, 266
  and stimulation of EDRF production, 216,
    221, 222
Atropine, 171-173

Barbiturates, see Anesthetics, intravenous
Barium, 10, 98, 201
Benzyl alcohol, see Alcohols
Bradykinin, 208-209, 217, 219, 229, 230, 261
4-bromphenacyl bromide, 58, 171-174
Bupivacaine, see Anesthetics, local
Butanol, see Alcohols

Caffeine
  and calcium concentration, intracellular
    (cardiac), 78-81, 128-131
  and calcium transient (cardiac), 78, 79, 81
  and contraction of left ventricle, 80, 82

Caffeine (continued)
  contractures, in malignant-hyperthermia-
    susceptible muscle, 59, 60
  and cooling contractures (skeletal), 45, 46,
    50
  and sarcoplasmic reticulum calcium content
    (cardiac), 79-83, 128-131, 150, 151
  and sarcoplasmic reticulum calcium release
    (skeletal), 36, 37, 48, 76, 77
Calcium ATPase, see Calcium pump
Calcium channel blockers
  D600, 48, 92, 109-113
  diltiazem, 139, 164
  nifedipine, 113, 164, 165, 177
  nitrendipine, 92, 109-113, 134, 135, 139
  verapamil, 112, 148, 149, 164
Calcium channels (sarcolemmal)
  in cardiac cells, 100, 101, 103
  in cerebral artery smooth muscle cells, 241,
    243, 244
  current measurements, 100, 101, 103, 241,
    243, 244
  and G proteins (cardiac), 176, 178
  and radioligand binding (cardiac), 109-113
  and sarcoplasmic reticulum calcium release
    (cardiac), 102
  and slow action potentials (cardiac), 159,
    168
  and spontaneous contractions (cardiac),
    134, 138, 139
  in transverse tubule (skeletal), 5, 6
  and volatile-anesthetic-induced negative
    inotropy, 92
Calcium channels (sarcoplasmic reticulum),
  see Sarcoplasmic reticulum
Calcium content
  cell, 77-79, 81, 84, 94, 105, 144, 187, 201
  sarcoplasmic reticulum, 32, 49, 74, 78-80,
    83, 84, 92, 93, 104, 150-152, 201
  tissue, 80-84
Calcium current, 92, 97-101, 104, 105, 109,
  139, 143, 152, 164, 237, 241, 243
Calcium efflux (from sarcoplasmic reticulum),
  31-37, 39, 40
Calcium indicators
  aequorin, 93, 99-105, 187, 201, 203
  fura-2, 77, 126-129, 132, 135
  indo-1, 153, 210, 259, 265-267
  quin2, 147, 149-152, 275, 280
Calcium paradox, 94, 105
Calcium pump
  sarcolemma, 79, 83, 84, 92, 93
  sarcoplasmic reticulum, 5, 7, 28, 29, 31-33,
    39, 93
Calcium release channel, see Sarcoplasmic
  reticulum
Calcium transient, 91-94, 97, 201-203
  and caffeine, 78, 79, 128-131, 150, 151
  and calcium content, 81
  and enflurane, 80, 101-104, 135, 136, 138
  and halothane, 78-80, 101-104, 134-137

Phospholipase A2, 58-63, 65, 66, 170, 172, 174
Phospholipase C, 170, 174, 210, 211, 258, 261-263, 271, 272
Photodynamic therapy, 247, 252
Potassium channels, 4, 5, 15, 17, 176, 177, 230, 231
Procaine, *see* Anesthetics, local
Propranolol, 169, 171-173
Prostaglandin, 17, 170, 172, 208, 229, 233
Protein kinase C, 17, 210, 211, 258, 262, 263, 268, 271
Pulmonary artery, 207-209, 216, 218, 219, 222, 296
Pyrophosphate, 32-35, 37, 58

Quin2, *see* Calcium indicators

Radioligand binding, *see* Methods and techniques
Reactive oxygen intermediates, *see* Free radical
Rested state contraction, 143, 147, 148, 151, 152, 157-165
Resting membrane potential, *see* membrane potential
Ruthenium red, 24, 29, 33-35, 39
Ryanodine
  binding to isolated sarcoplasmic reticulum and fatty acids, 61, 62, 66
    measurement of, 58
  contraction in cardiac muscle in the presence of, 116-119, 143, 144, 148, 149
  contracture in skeletal muscle induced by, 24, 25
  and malignant-hyperthermia-susceptible skeletal muscle, 24, 25
  and slow action potentials and contractions (cardiac), 162-165
  in tetanic stimulation experiments (cardiac), 193-195

Salicylates, 50, 51
Sarcoplasmic reticulum
  and alcohols (cardiac), 165
  calcium pump, 5, 7, 28, 29, 31-33, 39, 93
  calcium release channel, 5-7, 10, 21-29, 31, 33-40, 49, 50, 53, 57, 58, 73, 74, 77-80, 83, 84, 143, 144, 148, 149, 152, 162, 164, 208
  calcium uptake, 31, 33, 34, 104, 105, 151
  and cooling contractures (cardiac), 119, 120
  and cooling contractures (skeletal), 43, 48-51
  and denervation (skeletal), 53
  and enflurane (cardiac), 119, 120, 129-131, 137-139, 150
  and halothane (cardiac), 79, 80, 83, 117-120, 128-130, 137-139, 148-150

Sarcoplasmic reticulum (coninued)
  and halothane (skeletal), 39, 40, 49, 74, 75, 83
  and halothane (smooth), 266
  and inositol 1,4,5-trisphosphate (cardiac), 174
  and inositol 1,4,5-trisphosphate (smooth), 258
  and isoflurane (cardiac), 117-120, 128-131, 137-139, 150
  isolated, 21, 22, 28, 57, 58, 61, 65, 66
  and local anesthetics, 162, 164, 165
  longitudinal tubules, 32, 39
  and malignant hyperthermia, 6, 9, 10, 47, 57, 74-76
  and skinned fibers (cardiac), 183
  and tension development, time course of (cardiac), 162
  terminal cisternae, 5, 6, 21, 22, 31, 33, 37-40, 53, 58, 62
Serotonin, 208-210, 229, 231, 238-244, 259, 266, 268, 273
Single channel recording, *see* Methods and techniques
SK&F 86002, 171, 172, 178
Skinned fibers, *see* Methods and techniques
Slow action potential, *see* Membrane potential
Sodium channels, 4-6, 9, 10, 15-17, 66, 165
Sodium current, 5, 6, 13-17
Sodium-calcium exchange, 84, 92, 93, 102, 105, 115, 128, 131, 156, 203
Sodium nitroprusside, *see* Nitroprusside
Stearic acid, *see* Fatty acids
Succinylcholine, 53, 57-61, 64, 65

Terminal cisternae, *see* Sarcoplasmic reticulum
Tetanic stimulation, *see* Methods and techniques
Tetracaine, *see* Anesthetics, local
Tetrodotoxin, 13, 16, 211, 292, 293
Thiopental, *see* Anesthetics, intravenous
Tissue culture, *see* Methods and techniques
Transverse tubule, 5, 6, 21, 76, 77
Triglyceride lipase, 58, 59, 61-64, 66
Triglycerides, 59, 61, 64, 66, 67
Tropomyosin, *see* Contractile proteins
Troponin, *see* Contractile proteins

Vasopressin, 210, 265-267
Volatile anesthetics, *see* Anesthetics, inhalational
Voltage clamp, *see* Methods and techniques

WY 50295, 171, 172, 178